IET TRANSPORTATION SERIES 26

Traffic Information and Control

Other related titles

Volume 1	**Clean Mobility and Intelligent Transport Systems** M. Fiorini and J.-C. Lin (Editors)
Volume 2	**Energy Systems for Electric and Hybrid Vehicles** K.T. Chau (Editor)
Volume 5	**Sliding Mode Control of Vehicle Dynamics** A. Ferrara (Editor)
Volume 6	**Low Carbon Mobility for Future Cities: Principles and applications** H. Dia (Editor)
Volume 7	**Evaluation of Intelligent Road Transportation Systems: Methods and results** M. Lu (Editor)
Volume 8	**Road Pricing: Technologies, economics and acceptability** J. Walker (Editor)
Volume 9	**Autonomous Decentralized Systems and Their Applications in Transport and Infrastructure** K. Mori (Editor)
Volume 11	**Navigation and Control of Autonomous Marine Vehicles** S. Sharma and B. Subudhi (Editors)
Volume 12	**EMC and Functional Safety of Automotive Electronics** K. Borgeest
Volume 16	**ICT for Electric Vehicle Integration with the Smart Grid** N. Kishor and J. Fraile-Ardanuy (Editors)
Volume 17	**Smart Sensing for Traffic Monitoring** N. Ozaki (Editor)
Volume 18	**Collection and Delivery of Traffic and Travel Information** P. Burton and A. Stevens (Editors)
Volume 25	**Cooperative Intelligent Transport Systems: Towards high-level automated driving** M. Lu (Editor)
Volume 38	**The Electric Car** M.H. Westbrook
Volume 45	**Propulsion Systems for Hybrid Vehicles** J. Miller
Volume 79	**Vehicle-to-Grid: Linking electric vehicles to the smart grid** J. Lu and J. Hossain (Editors)

Traffic Information and Control

Edited by
Ruimin Li and Zhengbing He

The Institution of Engineering and Technology

Published by The Institution of Engineering and Technology, London, United Kingdom

The Institution of Engineering and Technology is registered as a Charity in England & Wales (no. 211014) and Scotland (no. SC038698).

© The Institution of Engineering and Technology 2021

First published 2020

This publication is copyright under the Berne Convention and the Universal Copyright Convention. All rights reserved. Apart from any fair dealing for the purposes of research or private study, or criticism or review, as permitted under the Copyright, Designs and Patents Act 1988, this publication may be reproduced, stored or transmitted, in any form or by any means, only with the prior permission in writing of the publishers, or in the case of reprographic reproduction in accordance with the terms of licences issued by the Copyright Licensing Agency. Enquiries concerning reproduction outside those terms should be sent to the publisher at the undermentioned address:

The Institution of Engineering and Technology
Michael Faraday House
Six Hills Way, Stevenage
Herts, SG1 2AY, United Kingdom

www.theiet.org

While the authors and publisher believe that the information and guidance given in this work are correct, all parties must rely upon their own skill and judgement when making use of them. Neither the authors nor publisher assumes any liability to anyone for any loss or damage caused by any error or omission in the work, whether such an error or omission is the result of negligence or any other cause. Any and all such liability is disclaimed.

The moral rights of the authors to be identified as authors of this work have been asserted by them in accordance with the Copyright, Designs and Patents Act 1988.

British Library Cataloguing in Publication Data
A catalogue record for this product is available from the British Library

ISBN 978-1-83953-025-8 (hardback)
ISBN 978-1-83953-026-5 (PDF)

Typeset in India by MPS Limited

Contents

About the editors   xi

1 Introduction   1
Ruimin Li and Zhengbing He

    1.1 Motivation   1
    1.2 Purpose   1
    1.3 Scope   2
    1.4 Book structure   2

Part I: Modern traffic information technology   3

2 Traffic analytics with online web data   5
Yisheng Lv, Yuanyuan Chen, Xueliang Zhao and Hao Lu

    2.1 Introduction   5
    2.2 Literature review   6
    2.3 Methodology   8
       2.3.1 System overview   8
       2.3.2 Main algorithms and models   10
    2.4 Some results   17
       2.4.1 Traffic sentiment analysis and monitoring system   18
       2.4.2 Traffic event detection   18
       2.4.3 Traffic status prediction   24
       2.4.4 Semantic reasoning for traffic congestion   24
    2.5 Conclusion   27
    References   27

3 Macroscopic traffic performance indicators based on floating car data: formation, pattern analysis, and deduction   31
Lu Ma and Jiaqi He

    3.1 Introduction   31
    3.2 A macroscopic traffic performance indicator: network-level trip speed   33
       3.2.1 The mathematical form of the NLT speed   33
       3.2.2 The empirical data for analyses   34
       3.2.3 Descriptive analyses of influential factors   37

		3.2.4	Correlative relationships between variables	38
	3.3	Methods of time series analysis		39
		3.3.1	The concept and basic features of the time series	39
		3.3.2	The exponential smoothing method	40
		3.3.3	The ARIMA method	41
		3.3.4	The support vector machine (SVM) method	43
	3.4	Analyses of the NLT speed time series		43
		3.4.1	Evaluation criteria of the modeling performance	43
		3.4.2	The decomposition of the NLT speed time series	44
		3.4.3	The analysis based on exponential smoothing methods	44
		3.4.4	The analysis based on ARIMA models	46
		3.4.5	The analysis based on a hybrid ARIMA–SVM Model	50
	3.5	Conclusions		53
	References			55

4 Short-term travel-time prediction by deep learning: a comparison of different LSTM-DNN models   **59**
Yangdong Liu, Lan Wu, Jia Hu, Xiaoguang Yang and Kunpeng Zhang

4.1	Introduction		59
4.2	Traffic time series estimation with deep learning		60
	4.2.1	Recurrent neural network	61
	4.2.2	Convolutional neural networks	63
	4.2.3	Generative adversarial networks	65
4.3	The LSTM-DNN models		66
4.4	Experiments		68
	4.4.1	Datasets	68
	4.4.2	Evaluation metrics	68
	4.4.3	Hyperparameter settings for LSTM-DNN models	69
	4.4.4	Comparison between LSTM-DNN models and benchmarks	70
4.5	Conclusion and future work		72
References			75

5 Short-term traffic prediction under disruptions using deep learning   **79**
Yanjie Dong, Fangce Guo, Aruna Sivakumar and John Polak

5.1	Introduction		79
5.2	Literature review		81
	5.2.1	Traffic prediction under normal conditions	81
	5.2.2	Traffic prediction under disrupted conditions	82
	5.2.3	Review of traffic prediction using deep learning techniques	86
	5.2.4	Summary	89
5.3	Methodology		89
	5.3.1	Traffic network representation on a graph	89
	5.3.2	Problem formulation	90
	5.3.3	Model structure	90

			Contents vii

		5.3.4 Quantification of prediction accuracy	95
	5.4	Short-term traffic data prediction using real-world data in London	96
		5.4.1 Traffic speed data	96
		5.4.2 Preparation for the prediction model	96
		5.4.3 Short-term traffic speed prediction under non-incident conditions	102
		5.4.4 Short-term traffic data prediction under incidents	105
	5.5	Conclusions and future research	108
	References		109
6	**Real-time demand-based traffic diversion**		**115**
	Pengpeng Jiao		
	6.1	Model of path choice behavior of driver under guidance information	116
		6.1.1 Discrete probability selection model	117
		6.1.2 Prospect theory model	118
		6.1.3 Fuzzy logic model	119
		6.1.4 Other models	120
	6.2	Optimization of traffic diversion strategy	120
		6.2.1 Responsive strategy	121
		6.2.2 Iterative strategy	122
	6.3	Research on dynamic O–D estimation	123
		6.3.1 Intersection model	124
		6.3.2 Expressway model	124
		6.3.3 Network model	125
	6.4	Dynamic traffic diversion model based on dynamic traffic demand estimation and prediction	126
		6.4.1 DODE model of urban expressway	127
		6.4.2 Traffic diversion model of urban expressway	129
		6.4.3 Dynamic traffic diversion model based on DODE	131
		6.4.4 Model solution	133
		6.4.5 Case study	134
	6.5	Conclusion	141
	References		141
7	**Game theoretic lane change strategy for cooperative vehicles under perfect information**		**147**
	Andres Ladino and Meng Wang		
	7.1	Introduction	147
	7.2	Problem formulation	150
	7.3	Highway traffic system dynamics	150
		7.3.1 Longitudinal dynamics	151
		7.3.2 Lateral dynamics	152

	7.3.3		Lane change and dynamic communication topology	152
	7.3.4		Closed-loop dynamics	152
7.4		Game theoretic formulation of the lane change decision problem	153
	7.4.1		Dynamic lane change game formulation	153
	7.4.2		Existence of equilibrium	155
	7.4.3		Properties of the lane change dynamic game	156
7.5		Numerical examples		158
	7.5.1		Experimental setting	158
	7.5.2		Scenario 1: delayed merge	158
	7.5.3		Scenario 2: courtesy lane change	159
7.6		Conclusion		161
References			161

8 Cooperative driving and a lane change-free road transportation system 163
Zhengbing He

8.1		Introduction	163
8.2		Cooperative driving strategies at intersections	166
	8.2.1	Safety driving pattern-based strategy	166
	8.2.2	Reservation-based strategy	167
	8.2.3	Trajectory optimization-based strategy	168
8.3		Cooperative driving strategies at on-ramps	170
	8.3.1	Virtual vehicle mapping strategy	170
	8.3.2	Slot-based strategy	171
8.4		Lane change-free road transportation system	172
	8.4.1	Lane change-free road transportation system: an illustration	173
	8.4.2	System design	174
	8.4.3	Simulation test	179
8.5		Conclusion and future direction	179
Acknowledgements		181
References		181

Part II: Modern traffic signal control 185

9 Urban traffic control systems: architecture, methods and development 187
Fusheng Zhang and Lu Wei

9.1		Introduction	187
	9.1.1	Brief description	187
	9.1.2	Classification	190
	9.1.3	Level of traffic control system	191
9.2		SCOOT	191
	9.2.1	Overview	191

	9.2.2	Basic principles	192
	9.2.3	System architecture	193
	9.2.4	Optimization process	194
	9.2.5	Additional features	197
9.3	SCATS		198
	9.3.1	Overview	198
	9.3.2	Basic principles	198
	9.3.3	System architecture	199
	9.3.4	Optimization process	200
9.4	Summaries and limitation analysis		202
9.5	Future analysis of urban traffic control system		202
	9.5.1	Changes in system environments	203
	9.5.2	Standardization	204
	9.5.3	Summary	205
References			205

10 Algorithms and models for signal coordination   207
Hao Wang and Changze Li

10.1	Introduction	207
10.2	Basic MAXBAND approach	211
10.3	Extended MAXBAND approach	214
	10.3.1 Variable bandwidth method	214
	10.3.2 Multimode band method	216
	10.3.3 Path-based method	219
10.4	MAXBAND for network system	222
10.5	Discussion and open issues	227
10.6	Conclusion	229
References		229

11 Emerging technologies to enhance traffic signal coordination practices   235
Zong Tian and Aobo Wang

11.1	Coordination timing development and optimization	236
	11.1.1 Developing cycle length and splits using controller event data	237
	11.1.2 Optimizing offsets and phasing sequences based on travel-run trajectories	239
11.2	Field implementation and timing diagnosis	241
11.3	Performance measures for assessing the quality of signal coordination	243
11.4	Signal timing documentation	248
11.5	Summary	249
References		250

x   Traffic information and control

12 Traffic signal control for short-distance intersections with dynamic reversible lanes   253
Haipeng Shao, Siyuan Song, Juan Yin and Hui Jin

12.1	Introduction	253
12.2	Application of dynamic reversible lane	254
12.3	Model of signal timing	255
	12.3.1 Signal phase and sequence	255
	12.3.2 Signal timing model	257
12.4	Calibration and validation	264
	12.4.1 Simulation scenarios	264
	12.4.2 Validation of the proposed plan	265
12.5	Adaptability analysis	265
	12.5.1 Road conditions	265
	12.5.2 Left-turning traffic proportion	273
12.6	Conclusion	273
References		275

13 Multiday evaluation of adaptive traffic signal system based on license plate recognition detector data   279
Ruimin Li, Fanhang Yang and Shichao Lin

13.1	Introduction	279
13.2	Methodology	282
	13.2.1 Travel time delay	283
	13.2.2 Travel time-based measurements	283
	13.2.3 PCD and related indexes	284
	13.2.4 Travel time reliability indexes	285
13.3	Case description and dataset	286
	13.3.1 Case description	286
	13.3.2 Dataset	286
13.4	Results	288
	13.4.1 Evaluation of travel time delay	288
	13.4.2 Cumulative frequency diagram of travel time	291
13.5	Evaluation of the PCD	293
	13.5.1 Travel time reliability evaluation	293
13.6	Conclusion	296
Acknowledgments		299
References		299

14 Conclusion   301
Zhengbing He and Ruimin Li

Index   305

About the editors

Ruimin Li is a tenured associate professor in the Department of Civil Engineering at Tsinghua University, China. He is a vice chairman and secretary general of the transportation modelling and simulation commission of the China Simulation Federation, and an editorial advisory board member of several international journals. His research interests include intelligent transportation systems, urban transportation planning, and traffic control, safety, and simulation.

Zhengbing He is a professor at the Beijing Key Laboratory of Traffic Engineering, Beijing University of Technology, China. He is an IEEE senior member, an editorial advisory board member of Transportation Research Part C, and an associate editor of IET Intelligent Transport Systems. He has published more than 80 academic papers in the fields of traffic flow theory and intelligent transportation systems.

Chapter 1
Introduction
Ruimin Li[1] and Zhengbing He[2]

1.1 Motivation

Data is the foundation of various transportation studies. Traffic signal control is the core means of urban traffic operation. Over the past decade, the rapid development of Information and Communications Technologies provided the new opportunities for transportation research and practice. On one hand, a variety of newly emerged data such as crowdsourcing data, license plate recognition data and floating car data provided rich and solid foundation for traffic monitor, prediction and control. On the other, artificial intelligence technique provided new and rich tools for transportation community, greatly promoting the efficiency of transportation research and practice.

This book, *Traffic Information and Control*, mainly focuses on the field of traffic information processing and signal control with the support of diverse traffic data.

1.2 Purpose

The aim of this book is to introduce and propose the typical and advanced methods to estimate and predict traffic flow state at different levels (macroscopic, mesoscopic and microscopic), and to optimize traffic signal control for intersections, arterials and areas based on various detection data, so as to provide reference for researchers and engineers.

Potential applications of this book are explained next.

1. To introduce advanced big data-based applications, including online web data-based traffic analytics systems, deep-learning-based traffic flow predictions, etc.;
2. To propose novel traffic management and systems based on advanced information technologies, including real-time demand-based traffic diversion strategy, lane change-free future road transportation systems;

[1]Department of Civil Engineering, Tsinghua University, Beijing, China
[2]Beijing Key Laboratory of Traffic Engineering, Beijing University of Technology, Beijing, China

3. To provide a review of the development of traffic coordination control and the emerging technologies to enhance the traffic coordination control; and
4. To introduce the innovative short-distance intersection coordination control and evaluation of traffic control based on detailed detection data.

Potential audiences of this book include academic staff and students, practicing engineers, in traffic information processing study and traffic signal control.

1.3 Scope

The scope of this book mainly focuses on two aspects: one is the estimation of traffic flow state. Based on emerging more detailed data sources, the traffic flow state can be estimated and predicted more accurately, which provides a solid basis for traffic control and operation. The other is traffic signal control optimization. This part focuses on how to use better data and advanced tools for better signal control, so as to improve the road traffic flow condition.

Supported by various emerging data, there have been many new proposed methods and field practices in the field of traffic state estimation and traffic signal control in recent years. This book is written to introduce some of the latest development in this field and provide reference for researchers and engineers.

1.4 Book structure

This book includes 14 chapters totally. The present chapter highlights the motivation, purpose, scope and book structure. Then this book is mainly divided into two parts, the first is about traffic information analysis with some emerging new data sources and methods. The second part is mainly focused on traffic signal control, especially related to coordinated control for arterials and road networks. Chapter 14, conclusion, summarizes the content of previous chapters and provides future trend of traffic information and control.

Part I

Modern traffic information technology

Chapter 2

Traffic analytics with online web data

Yisheng Lv[1], Yuanyuan Chen[1], Xueliang Zhao[2] and Hao Lu[1]

Social media and other online websites have rich traffic information. How to extract and mine useful traffic information from online web data to address transportation problems has become a valuable and interesting research topic in current data-explosive era. In this chapter, we introduce a traffic analytic system with online web data. The proposed system can collect online data, use machine learning and natural language processing methods to extract traffic events, analyze traffic sentiment, and reason traffic scenarios. We also present some results based on the proposed system and techniques in practice.

2.1 Introduction

Recent advances in technologies like smart mobile devices, mobile internet, and location-based services have made exponential growth of user-generated data [1,2]. In cities, large amounts of smart phone users, social media users, some companies users, etc. are generating very big and diverse data such as travel time, speed, traffic events in texts, location information, and trajectories [3]. To differentiate from traditional traffic data collected from dedicated sensors like loop detectors, we call user-generated traffic-related data as social transportation data that actually embed social signals as in the field of social computing [4–6]. Collecting and mining social transportation data can help one in monitoring, predicting, and understanding city transportation systems [7–9].

Social transportation data are usually easily accessed and are now becoming important traffic information sources. Social transportation data are complementary to traditional traffic data collected in the physical transportation world using physical sensors. Typical social transportation data sources include social media platforms like Twitter,[*] Weibo,[†] and Facebook,[‡] user-generated map services, and also location-based services.

---

[1]The State Key Laboratory for Management and Control of Complex Systems, Institute of Automation, Chinese Academy of Sciences, Beijing, China
[2]Institute of Automation, Chinese Academy of Sciences, Beijing, China
[*]https://mashable.com/category/twitter
[†]https://weibo.com
[‡]https://www.facebook.com

Actually, besides the abovementioned social transportation data sources, other online websites like government official sites, online map service providers (Bing Map,[§] Google Map,[¶] Baidu Map,[‖] Amap,[**] etc.), events publisher websites (sports events, music events, etc.), and weather forecasting websites have rich traffic or traffic-related open-source information. Typical traffic/traffic-related information obtained from online web are traffic events, traffic accidents, traffic congestion, vehicle/bike/pedestrian trajectories, event/activity information, land use information, weather information, holidays, and traffic sentiment information.

Compared with traditional physical traffic sensors like loop detectors and radar, online web data have the following advantages: (1) traffic contents within online web are usually easy to be accessed. For example, Twitter and Weibo provide Application Programming Interface (API) services for third parties to extract microblogs; online map services like Google Map, Bing Map, and Amap provide APIs to extract road traffic conditions and geographical information. (2) There is almost no installment and maintenance cost compared with physical sensors like cameras and loop detectors. (3) It has a bigger coverage of road networks due to the widespread use of smart phones and social media platforms. (4) Traffic data from online web are multimodal, including texts, images, videos, and structured data, which enrich the data formats and can provide more views of descriptions on traffic dynamics.

How to extract and mine useful traffic information from online web data to solve transportation problems has become a valuable and interesting research topic in current data-explosive era. More and more researchers are now attracted to this direction. Online web data have not only been applied into the traditional traffic research problems like traffic event detection, mobility pattern discovery, and traffic prediction, but also into emerging directions like traffic sentiment analysis.

In this chapter, we propose a system to use online web data for traffic analytics and introduce corresponding techniques, including traffic data extraction, traffic event detection, traffic prediction, traffic sentiment analysis, and traffic reasoning.

The rest of this chapter is organized as follows. Section 2.2 reviews the recent advances literature on using online web data for transportation research. Section 2.3 introduces our proposed system and main techniques. Section 2.4 presents some results and applications. Finally section 2.5 concludes this chapter.

2.2 Literature review

Internet is now one of the most important channels for people to share, publish, and access information, which leads to the exponential growth of data explosion. Online web data, especially data of social media, location-based services, and check-in web services, have been widely used in many fields like security, health, and e-commerce.

[§] https://cn.bing.com/maps
[¶] https://www.google.com/maps
[‖] https://map.baidu.com
[**] https://www.amap.com

By conducting a keyword search of multiple electronic database (including IEEE Xplore digital library, ACM digital library, Elsevier Science Direct, Springer, Web of science Core Collection: Citation Indexes, Engineering Village, Wanfang Database, and Google Scholar), we can find that the potential opportunities of using online web data have attracted more and more researchers' attentions (see Figure 2.1). Traditional traffic sensing methods like floating cars, cameras, and loop detectors are apt for collecting traffic variables, i.e., traffic flow, speed, occupancy, and travel time. However, traditional physical traffic detectors are of limited coverage due to the cost. Online web data, especially social media data, have a great size of enormous users and can complement traditional traffic data. The big data nature of online web data requires us to preprocess, analyze, and mine the knowledge contained in the data.

Collecting data from online websites is the first step needed for using online web data for transportation domains. In the beginning, tweets were widely used as traffic data sources. Gradually, besides tweets, researchers also use data from Facebook, transportation agency official websites, online map service providers, check-in websites, etc. There are two commonly used methods in collecting data: API-based and crawler-based methods.

Online web data are typically used for traffic events detection, traffic prediction, traffic sentiment analysis, and mobility behavior identification. Commonly used methods include machine learning methods and natural language processing methods, for example, Latent Dirichlet Allocation (LDA), deep learning, and word embedding [2].

Detecting traffic events from social media is one early and still hot research direction. Gutierrez *et al.* and Tejaswin *et al.* extracted traffic events from Twitter [10,11]. Cui *et al.* developed a traffic event detection system using Sina Weibo data [12]. Wang *et al.* proposed to use LDA methods to extract traffic accidents in Tweets [13,14].

Understanding individual travel behavior and activities can help one in making policies, thus improving transportation systems efficiency. Maghrebi *et al.* used tweets

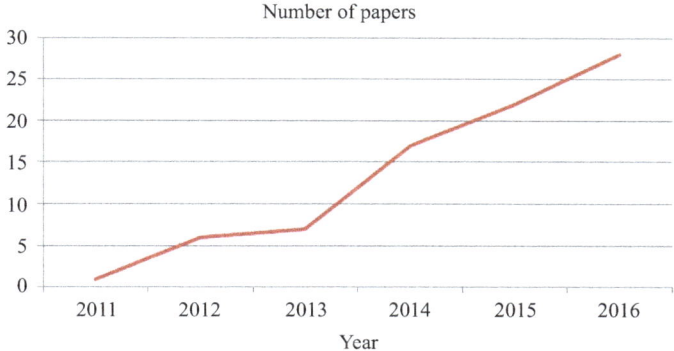

Figure 2.1 The number of papers on social media in transportation over time

in Melbourne metropolitan areas to analyze travel mode choice information [15]. They also used tweets in south and west part of the Sydney metropolitan area to detect travel mode choices [16]. Itoh *et al.* used tweets and subway smart card data to analyze changes of passenger flow behavior [17].

Social media and forums have been the places for people to share their opinions. Extracting such information can help one to improve traffic services. Cao *et al.* proposed the framework for traffic sentiment analysis and demonstrated its effectiveness. Georgiou *et al.* estimated traffic congestion based on the volume of messages and complaints in social media [18]. Giancristofaro and Panangadan used image data and texts from Instagram to evaluate transportation sentiment [19]. Zeng *et al.* analyzed topics related to transportation issues concerning Golden Week in China [20].

Recently, researchers have shown that using online web data can help one to improve traffic prediction. He *et al.* analyzed the correlation between tweet counts and traffic flow, and predicted longer-term traffic flow with tweets [21]. Ni *et al.* studied the short-term traffic prediction problem under special events while using Twitter data [22]. Grosenick extracted traffic events from tweets, and used such information to predict traffic speed [23]. Zhou *et al.* combined weather information, traffic events, and real traffic data to predict road traffic conditions, where traffic events and weather information were extracted from online webs [24]. Rodrigues *et al.* used smart card data and planned events information to predict public transport usage [25]. In addition, some integrated platform, apps, and prototype systems of using online web data for transportation domains have been developed [26–29].

2.3 Methodology

In this section, we present the system overview of traffic analytics with online web data and its corresponding main algorithm models.

2.3.1 System overview

The system architecture of traffic analytics with online web data is shown in Figure 2.2. It mainly has four components: data collection module, data preprocessing module, data modeling/mining module, and applications.

2.3.1.1 Data collection

The data collection module is to collect data from online websites. If websites provide APIs to access its webpage contents, we can use the provided APIs to collect the data. Of course, we can also use crawler-based methods. However, if there is no APIs provided by the service providers, we need to use crawler-based methods to collect data. For example, for Sina Weibo and check-in websites, we can use both the API-based and crawler-based methods; for Amap, an online map service, we can use the API-based methods; and for transportation administration agency websites, we can use the crawler-based methods.

Traffic analytics with online web data 9

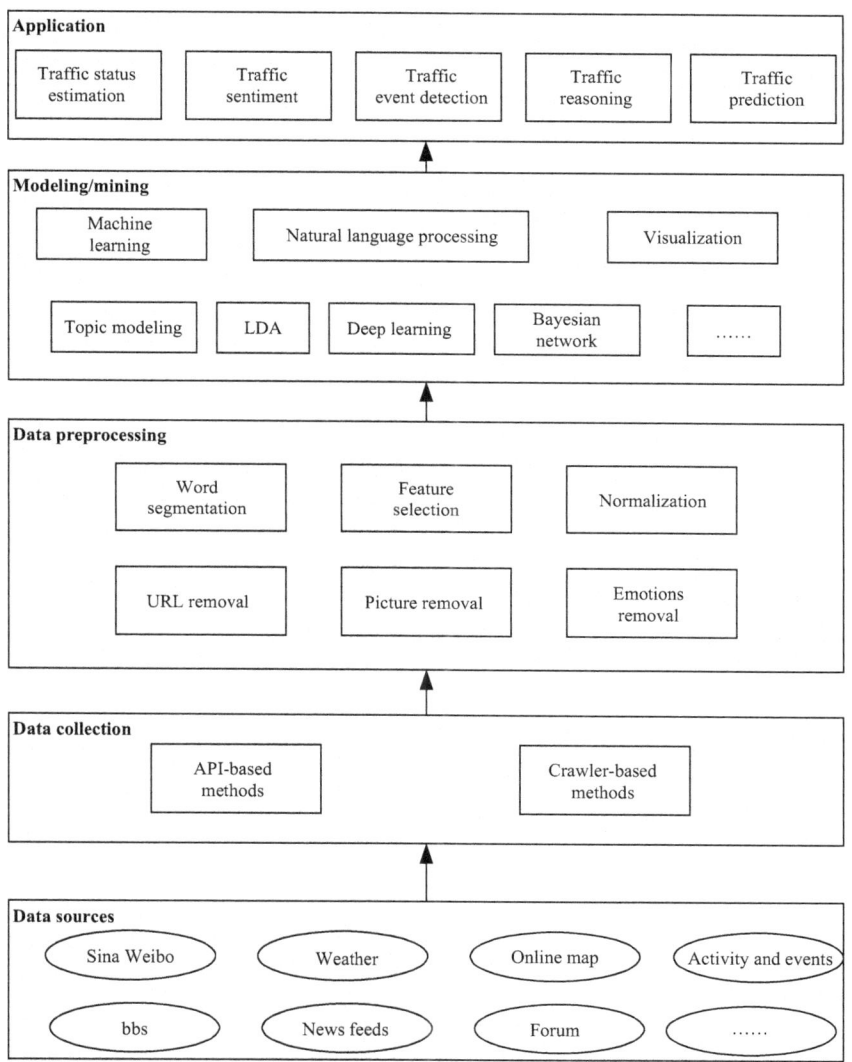

Figure 2.2 System overview of traffic analytics with online web data

2.3.1.2 Data preprocessing

The raw data collected from online websites are usually with noise and unstructured. We need to preprocess the data before deeply analyzing it. The format of collected data is most of text. Here, we describe text preprocessing means as follows: (1) duplicate text removal—this is to remove duplicated and retweeted texts collecting from online websites like Weibo. (2) Word segmentation—this is the first step for natural language processing. The ICTCLAS, a commonly used Chinese word segmentation system, is used to segmenting words. (3) Removal of stop words, URL, emojis, special characters, images. (4) Removal of Weibos with less than ten words. (5) Feature selection. This is to select feature words related to transportation topics.

2.3.1.3 Modeling and mining

After preprocessing data, we can extract traffic events from texts, analyze traffic sentiment, fuse multisource data for traffic prediction, and reason traffic scenarios. Accordingly, a series of techniques are as follows:

1. Topic modeling: Topic modeling is widely used for document classification, text retrieval, and document semantic analysis. The LDA model is widely used for topic modeling. We used LDA models to analyze transportation topics. Specifically, we developed news-LDA models and Weibo-LDA models for longer texts (e.g., news) and short texts (e.g., Weibos), respectively.
2. Word embedding: Traditionally, words are represented as one-hot vectors that cannot capture word semantics. Word embedding, a recently proposed technique, treats words in a continuous vector space that can semantically represent words. We collected 3 billion Weibos and used the Continuous Bag-of-Word (CBOW) model to train word embedding representations.
3. Information fusion: Different data sources have different views of describing the same topic. We cross-validate transportation topics using different online web data sources and off-line data sources. Also, we combine one topic's different views from different data sources to foster the whole view of that topic.
4. Traffic prediction: We use Bayesian networks and deep learning methods (e.g., Long Short-Term Memory neural network, LSTM and Convolutional Neural Network, CNN) for predicting traffic flow, speed, traffic status, and travel time.

2.3.1.4 Applications

Typical applications of online web analytics are traffic event detection, traffic prediction, traffic scenario reasoning, and traffic sentiment analysis. Traffic event detection with online web data is complementary to detect traffic incidents with physical sensor data. Also, online web data, in the form of texts, can be used to reason traffic scenarios. For example, if we find a road is congested, then we can retrieve topics related to the road to determine why it is congested. Incorporating online web data with off-line traffic data can get more insights on traffic prediction, leading to the improved performance of traffic prediction models. Mining public opinions and advices on transportation systems can help one to diagnose planning or operation issues of transportation systems.

2.3.2 Main algorithms and models

2.3.2.1 Latent Dirichlet allocation

LDA and its variants are generative models and widely used for topic modeling and event detection. The LDA model assumes that a document is generated as follows:

1. Choose $N{:}Poisson(\alpha)$.
2. Choose $\theta{:}Dir(\alpha)$, where $Dir(\alpha)$ is a Dirichlet distribution with a parameter α.
3. For each of the N words w_n choose a topic $z_n{:}Multinomial(\theta)$, choose a word w_n from a multinomial probability conditioned on the topic z_n. Due to the different lengths of news and Weibos, we need to establish LDA models for news and Weibos, respectively [30].

The standard LDA model is used for detecting traffic events from news. For detecting traffic events from Weibo, a Weibo-LDA model is developed based on the USER scheme that has good performance in classifying tweets [31]. The generation process for the Weibo-LDA model is listed in Algorithm 1.

Algorithm 1 The w-LDA algorithm for generation process

Input: K, U, T, α, β, where K is the number of topics, U is the number of user profiles, T is the number of Weibo posts, α and β are the latent variables, respectively.

Output: *userProfileSet*, φ_z, ϑ_p

1. For each user $u=1\ldots U$
    For each Weibo post $t=1\ldots T$
        Aggregating the Weibo posts generated by the user into the user profile *userProfileSet(p)*

2. For each topic $z=1\ldots K$
    Draw $\varphi_z \sim$ Dirichlet (β)

3. For each user profile $p=1\ldots P$
    Draw $\vartheta_p \sim$ Dirichlet(α)
    For each word in generated user profile p, $i=1\ldots N_p$
        Draw $z_i \sim$ Multinomial (φ_p)
        Draw $w_i \sim$ Multinomial (ϑ_{z_i})
Return *userProfileSet*, φ_z, and ϑ_p

The collapsed Gibbs sampling technique is used to approximate the distribution of $P(z_i = j | \mathbf{Z}_{-i}, w_i, p_i)$, which is

$$P(z_i = j|\mathbf{Z}_{-i}, w_i, p_i) \propto \frac{n_{-i,j}^{(w_i)} + \beta}{\sum_{i=1}^{V} n_{i,j}^{(w_i)} + V\beta} \times \frac{n_{-i,j}^{(p_i)} + \alpha}{\sum_{j=1}^{K} n_{i,j}^{(p_i)} + K\alpha} \tag{2.1}$$

where $P(z_i = j|\mathbf{Z}_{-i}, w_i, p_i)$ is the probability of the word w_i assigned to the topic j, w_i is the ith word in the vocabulary, p_i is the user profile containing the word w_i, $\mathbf{Z}_{-i}$ denotes the topic j assigned to all other words, $n_{-i,j}^{(w_i)}$ is the number of occurrences of words assigned to the topic j excluding the current word w_i, $n_{-i,j}^{(p_i)}$ is the number of occurrences the topic j assigned to words in the user profiles excluding the current user profile p_i.

For any obtained samples, $\varphi_z^{(j)}$ and $\vartheta_p^{(j)}$ can be estimated as follows:

$$\varphi_z^{(j)} = \frac{n_{i,j}^{(w_i)} + \beta}{\sum_{i=1}^{V} n_{i,j}^{(w_i)} + V\beta} \tag{2.2}$$

$$\vartheta_p^{(j)} = \frac{n_{i,j}^{(p_i)} + \alpha}{\sum_{j=1}^{K} n_{i,j}^{(p_i)} + K\alpha} \tag{2.3}$$

where $n_{i,j}^{(w_i)}$ is the number of occurrences the word w_i assigned to the topic j, $n_{i,j}^{(p_i)}$ is the number of occurrences the topic j assigned to words in the user profile p_i.

2.3.2.2 Word embedding

Word embedding represents a word as a vector of real numbers in a continuous space and has less dimensions compared with the vocabulary size. The CBOW and the Skip-Gram models are two ways of learning word embeddings [32]. The CBOW model tries to learn word vector representations by predicting the current word from a window of surrounding context words, while the Skip-Gram model tries to learn word vector representations by maximizing the probability of predicting surrounding words based on the current word, as shown in Figure 2.3. We chose the CBOW model to train word representations in a dataset containing 3 billion Weibos.

Hierarchical softmax or negative sampling techniques can be used in training the CBOW model. Hierarchical softmax uses a Huffman tree to represent all the words in the vocabulary. When using the hierarchical softmax in the CBOW model, the hidden layer is designed to average the input word vectors, and the output of the hidden layer is

$$h = \frac{1}{C} \sum_{u \in context(w)} v(u) \tag{2.4}$$

where $v(u)$ represents the vector of the word u, $context(w)$ is the set of contextual words of the word w, and C is the cardinality of the set $context(w)$. Given the context of the word w, the conditional probability of the word w is

$$p(w|context(w)) = \prod_{j=1}^{L(w)-1} \sigma(h^T v'_{n(w,j)})^{1-d_{j+1}^w} [1 - \sigma(h^T v'_{n(w,j)})]^{d_{j+1}^w} \tag{2.5}$$

where $n_{w,j}$ is the jth inner point from the root to word w in the Huffman tree, $v'(n)$ is the vector of inner point n, $L(w)$ is the length of the path in Huffman tree for word w, and d_{j+1}^w is the jth bit of Huffman code for word w. Take the logarithm of the

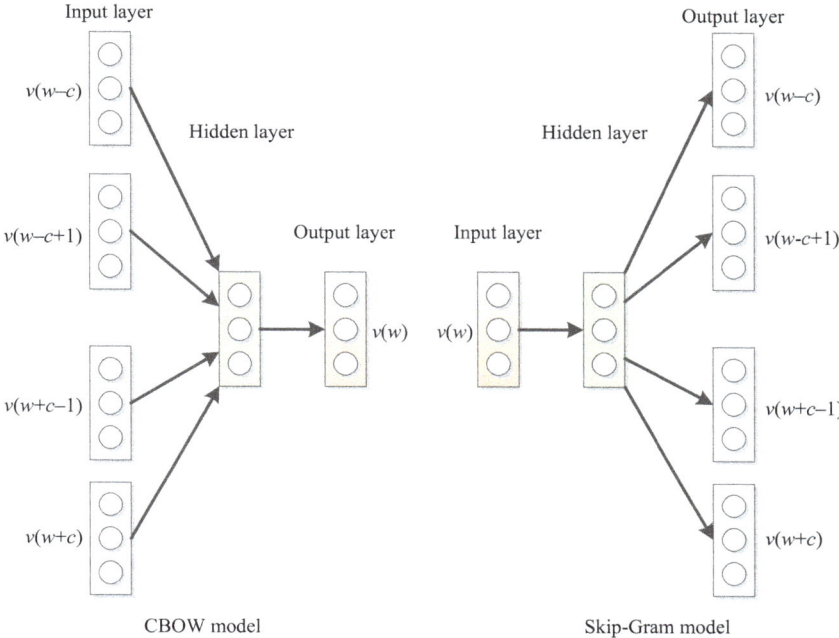

Figure 2.3   The CBOW and the Skip-Gram model architectures

conditional probability and get the loss function:

$$L = \log p(w|context(w))$$
$$= \sum_{j=1}^{L(w)-1} \left\{ \left(1 - d_{j+1}^w\right) \log\left[\sigma\left(h^T v'_{n(w,j)}\right)\right] + d_{j+1}^w \log\left[\sigma\left(h^T v'_{n(w,j)}\right)\right] \right\} \quad (2.6)$$

2.3.2.3   Bayesian network

Transportation systems are complex systems. Many factors could impact traffic prediction. Bayesian network can be used to take into account these factors for traffic prediction. When using Bayesian networks for traffic prediction, traffic variables and corresponding impact factors can be represented as nodes, while edges represent the causal relationships. The model parameters can be estimated via the Expectation Maximization algorithm. Figure 2.4 shows an example Bayesian network structure for traffic status prediction. With the rapid growth of traffic data, traffic prediction needs more and more storage and computing resources. Thus, we can use MapReduce to predict traffic in a distributed computing way.

14 Traffic information and control

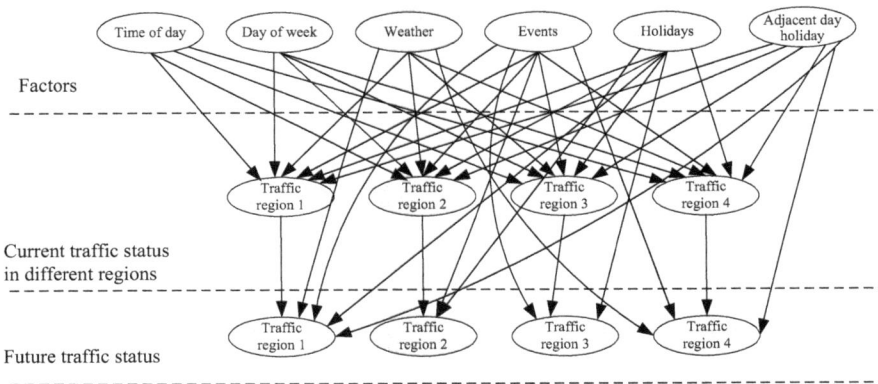

Figure 2.4 An example Bayesian network structure for traffic status prediction

2.3.2.4 Deep learning

Deep learning has shown great potentials for traffic analytics. For example, we can use deep learning models for traffic prediction and traffic event detection from social media. Typical deep learning methods used in traffic analytics are stacked auto-encoders, LSTM, CNN, and hybrid structures like LSTM-CNN.

Long short-term memory neural network (LSTM)
The LSTM model was proposed by Hochreiter and Schmidhuber [33]. Actually, it is a particular type of RNN (Recurrent Neural Network) [34].

RNN is a type of neural network to process sequence data by recursively applying a transition function to its internal hidden state vector h_t of the input sequence. In the traditional RNN architecture, each RNN block consists of neurons, which have both feed-forward and feedback links. The hidden state vector $ht - 1$ in the last time step and the input x are forwarded to the nonlinear activation function; thus, we can get the hidden state vector h_t, as shown in Figure 2.5.

The RNN transition equations can be expressed as follows:

$$h_t = \emptyset(w_h h_{t-1} + w_t x_t) \qquad (2.7)$$

$$y = \sigma(w h_t + b) \qquad (2.8)$$

where x_t is the input at the time step t, $\emptyset$ represents the internal activation function, h_{t-1} is the hidden state vector at the time step $t-1$ and h_t is the hidden state vector at the time step t, w_h, w_t, and w are the weight matrices, σ denotes the logistic sigmoid function, and b is the bias vector.

Unfortunately, the RNN model cannot learn long-term dependencies of sequence data. The LSTM model can address the previous issue. As a specific type of RNN, the LSTM neural network can learn the long-term dependencies of time series data. The LSTM cell has three gates: forget gate, input gate, and output gate. As shown in Figure 2.6, the three gates control the information flow through the neural network. The three gates' states are calculated as follows:

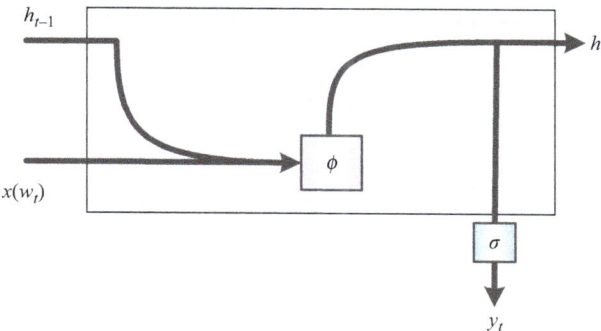

Figure 2.5 The architecture of RNN block

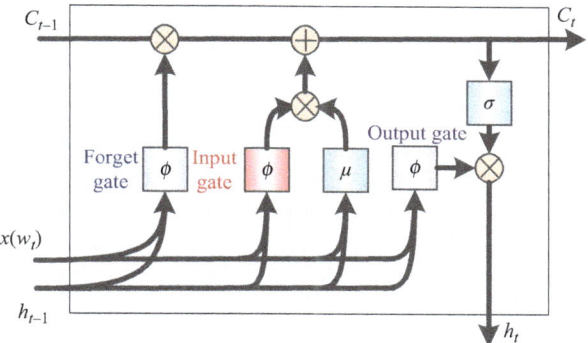

Figure 2.6 Architecture of LSTM block

The forget gate:

$$f_t = \varnothing\left(W_f[h_{t-1}, x(w_t)] + b_f\right) \tag{2.9}$$

The input gate:

$$i_t = \varnothing(W_i[h_{t-1}, x(w_t)] + b_i) \tag{2.10}$$

And the output gate:

$$o_t = \varnothing(W_o[h_{t-1}, x(w_t)] + b_o) \tag{2.11}$$

where f_t, i_t, and o_t represent the output of the forget gate, input gate, and output gate, respectively, W_f, W_i, and W_o are the weight matrices of the three gates, $\varnothing$ represents the internal activation function, and b_f, b_i, b_o mean the bias vectors of the three gates, respectively.

After getting the output of the three gates, the cell state can be calculated as

$$\tilde{C}_t = \mu(W_c[h_{t-1}, x(w_t)] + b_c) \tag{2.12}$$

$$C_t = f_t \otimes C_{t-1} + i_t \otimes \tilde{C}_t \tag{2.13}$$

where W_c is the weight matrix of cell state updating, and b_c is the bias, μ is the activation function, $\otimes$ represents point-wise multiplication. Then we can compute the hidden output:

$$h_t = o_t \otimes \sigma(C_t) \tag{2.14}$$

Convolutional neural network
As early as around 1980, a hierarchical multilayer artificial neural network, named "neocognitron," was designed by imitating the visual cortex of the biological visual system [35]. "Neocognitron" was used to recognize handwritten characters and handle other pattern recognition tasks. The hidden layers of this network consist of simple and complex layers alternately. The simple layers extract image features in the receptive field, and the complex ones respond to the same feature returned by different receptive fields. Simple and complex layers are similar to the convolution and the pooling ones of current CNNs.

In 1988, LeCun *et al.* proposed a more complete CNN LeNet-5, as shown in Figure 2.7 [36], which achieved great success in the recognition of handwritten digits. The success of LeNet-5 attracted attention to the application of CNN. Other applications based on CNN were also developed, such as face and gesture recognitions.

CNN has become a hot research topic in the field of artificial intelligence. It can even be said that CNN is the dominant research technology in fields such as computer vision and natural language processing. Many companies and universities are developing CNN techniques.

In general, CNN is a kind of hierarchical models, which extract high-level semantic information layer-by-layer through a series of operations such as convolution operation, pooling operation, and nonlinear activation function mapping.

CNN generally includes one input layer, several hidden layers, and one output layer. The input layer is responsible for inputting data, such as two-dimensional

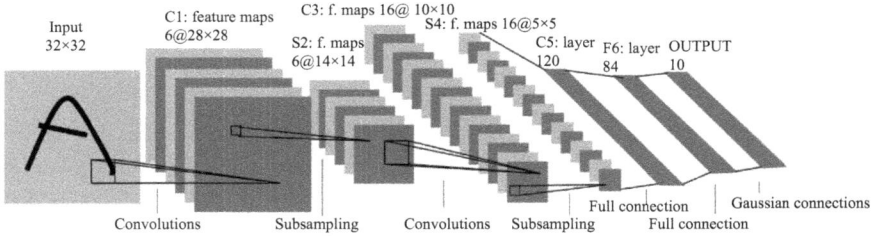

Figure 2.7 Architecture of LeNet-5

pixel data in a flat image and RGB channel information, into the network. Hidden layers usually include three categories: convolutional, pooling, and fully connected. In some new algorithms, some new modules, such as the inception module and the residual block, are also added to the hidden layers.

1. Convolutional layers: The convolution layer is a basic component of a CNN. Its function is to extract features from the input data. It contains multiple convolution kernels and activation functions. Each element that makes up the convolution kernel has a weight and a bias vector. The activation function used by CNN is generally the Rectified Linear Unit or its variants. It is characterized by fast convergence and simple gradient calculation.
2. Pooling layers: After feature extracted from the convolutional layer, the output feature map is passed to the pooling layer for feature selection and information filtering. Commonly used pooling operations are average-pooling and max-pooling.
3. Fully connected layers: In fully connected layers, all neurons between the two layers are connected by the weight vector, which is the same as the connection method of traditional neural networks' neurons. The fully connected layer is located at the last part of the hidden layers of the CNN and transmits signals to other fully connected layers.

The upstream of the output layer in CNN is usually a fully connected layer, so its structure and working principle are the same as the output layer in a traditional feedforward neural network. For image classification problems, the output layer uses a logistic or a softmax function to output classification labels. In the problem of object detection, the output layer can be designed to output the center coordinates, size, and classification labels of the object.

Long short-term memory and convolutional neural network
The LSTM neural network has good performance in processing sequence data and has the potential to learn the long-term correlations among the sequence data. CNN performs well in abstracting deep features of input data. Therefore, it would be a promising method to process sequence data by combining LSTM and CNN. Just like shown in Figure 2.8, we can combine the two kinds of neural network together. The input vectors are fed into the LSTM layer, after getting the useful information, then fed into another LSTM layer. The number of LSTM layers can be designed according to actual requirements. The output of LSTM layers will be the input of CNN, then through the convolution and pooling layer pairs to abstract the deep features, and also there can be more convolution and pooling layer pairs if needed. Finally, the output layer of the LSTM-CNN network model is a fully connected neural network layer.

2.4 Some results

In this section, we present some results on our proposed methods and developed system.

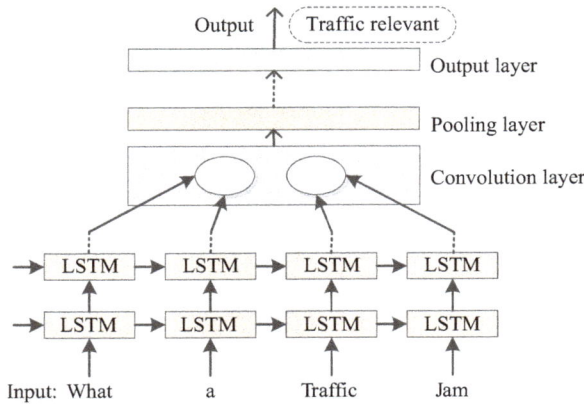

Figure 2.8 Architecture of an LSTM-CNN neural network

2.4.1 Traffic sentiment analysis and monitoring system

Based on the proposed system architecture and main techniques in Section 2.3, we developed a traffic sentiment analysis and monitoring system deployed in Qingdao and several other cities in China. The system has the functions of traffic event detection, traffic sentiment analysis, and traffic scenario analysis and profiling. Figure 2.9 shows the system interface of Qingdao traffic sentiment analysis and monitoring system.

Qingdao traffic sentiment analysis and monitoring system is deployed in Qingdao transportation administrative agencies. Using this system, we can extract traffic events and citizens' opinions on Qingdao transportation systems from Sina Weibo, Qingdao local online forums, newspapers, transportation agency websites, bbs, and so on. The system can collect online web data with the speed of 4 million per hour (Figure 2.10).

2.4.2 Traffic event detection

For traffic event detection from Weibo posts, we first search Weibo posts with manually selected key words, then we classify the searched posts with the deep learning models into traffic-relevant ones and traffic-irrelevant ones. The key words used for searching Sina Weibo are as follows: congestion, traffic accident, sideswipe, accident, detour, car crash, traffic jam, traffic status, traffic regulation, and so on.

We established a dataset with the size of 11,000 samples to develop the LSTM-CNN model for detecting traffic events [37]. In our experiments, the maximum length of word sequences is set as 156 and zero-paddings are used for filling the sequence. The dimension of a word embedding is set 200. To get the word embeddings, we collected 3 billion Weibo posts between 2009 and 2011 covering 1 million and 70,000 user accounts and used the CBOW model to obtain the word embeddings that capture word semantics.

Traffic analytics with online web data 19

Figure 2.9   Qingdao traffic sentiment analysis and monitoring system

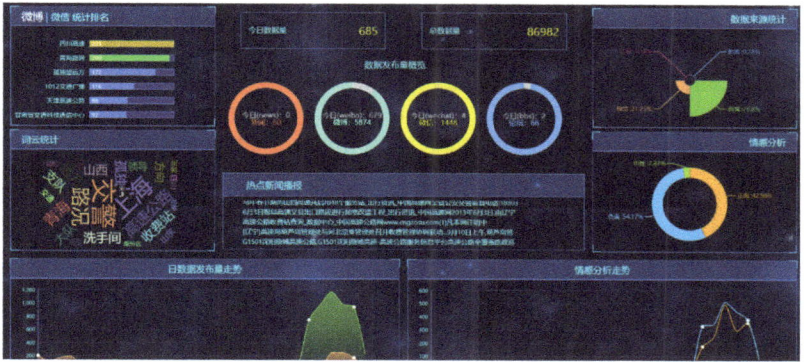

Figure 2.10   The snapshot of traffic sentiment analysis and monitoring system

Performance indexes: To evaluate the performance of the models, we take three measures that are precision, recall, and *F*-measure.

$$\text{Precision} = \frac{TP}{TP + FP}$$

recall:

$$\text{Recall} = \frac{TP}{TP + FN}$$

F-measure:

$$F_\beta = (1+\beta^2) \frac{\text{Precision} \times \text{Recall}}{\beta^2 \times \text{Precision} + \text{Recall}}$$

where *TP*, *FP*, and *FN* are true positive, false positive, and false negative, respectively, and β is a nonnegative real coefficient. And *F*-measure is called F_1 measure when β equals to 1. If we define *m* as the certain class label, true positive represents the number of samples with label *m* correctly assigned with label *m*, false positive represents the number of samples with any other labels except *m* incorrectly assigned with the label *m*, and false negative represents the number of samples with label *m* incorrectly assigned with any other labels except *m*. *F*-measure is the weighted harmonic mean of precision and recall by which we can get a trade-off between precision and recall.

Determination of the LSTM model: When building the LSTM model, the RMSprop optimizer was used to minimize the binary cross-entropy loss. A series of experiments were conducted to obtain better parameters of the model. We tested the learning rate from 0.01 to 0.05 with a stride 0.01, as shown in Figure 2.11. We can see that lower learning rate gives better performances especially for recall and F_1 score as the training going on.

We next tested the effect of the LSTM hidden state size on the model performance. We changed the size of the LSTM hidden states from {5, 10, 20, 50}. The test results are shown in Figure 2.12. We can see that 5 and 10 LSTM hidden state units have better performance and achieve nearly the same performances after epoch 30.

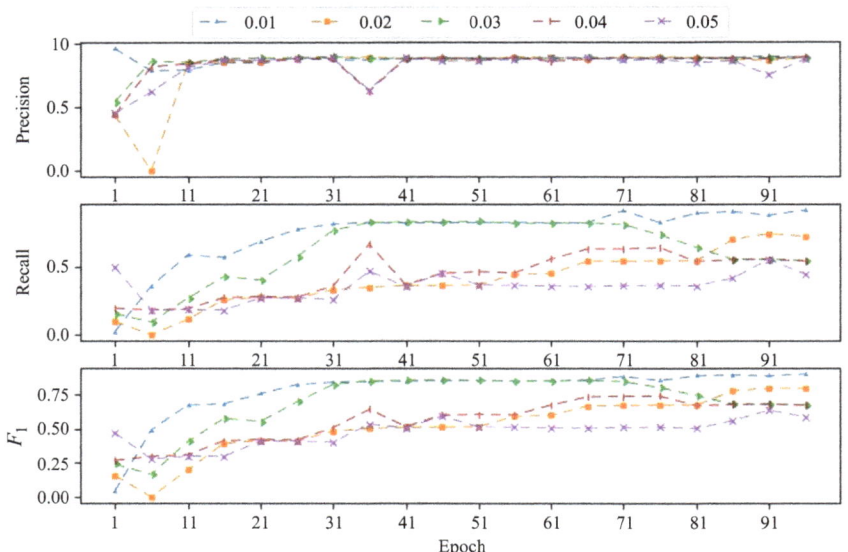

Figure 2.11 Effect of learning rate on precision, recall, and F_1 score

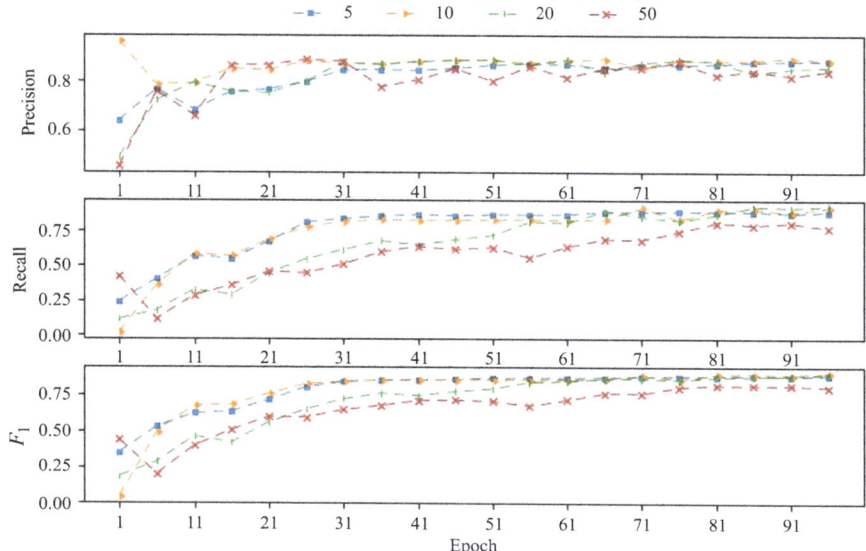

Figure 2.12   Effect of the number of LSTM hidden state on precision, recall, and F_1 score

To get a better LSTM model for detecting traffic events from Weibo posts, we applied the grid search method for the size of LSTM hidden states from {5, 10, 20, 50} and the learning rate from 0.01 to 0.03 with a stride 0.01. We also explored whether to stack two LSTM layers would improve the performances. Experiments show that the model with two stacked LSTM layers that have 10 output units and 0.01 learning rate obtain the best F_1 score.

Determination of the CNN model: For the CNN model, the Adam optimizer was used to minimize the binary cross-entropy loss. To explore the effect of the filter number, we select the filter number from {10, 20, 50, 100, 150}. The test results are shown in Figure 2.13. We can see that increasing the number of filters can yield better performance. It takes longer time to train the model when the number of filters is increased. Also, it seems that 100 filters and 150 filters achieve similar performances.

We next investigate the effect of the filter length when using only one filter length for all filters. We choose the filter length from {2, 3, 4, 5, 6, 10}. The test results are shown in Figure 2.14. We can see that changing the filter length does not improve performances much for our dataset.

To get the best performance of the CNN model for classification, we run the grid search technique over the depth of convolutional layers, pooling strategies, filter number, and filter length of convolutional layers. The best architecture of the CNN model is illustrated in Table 2.1.

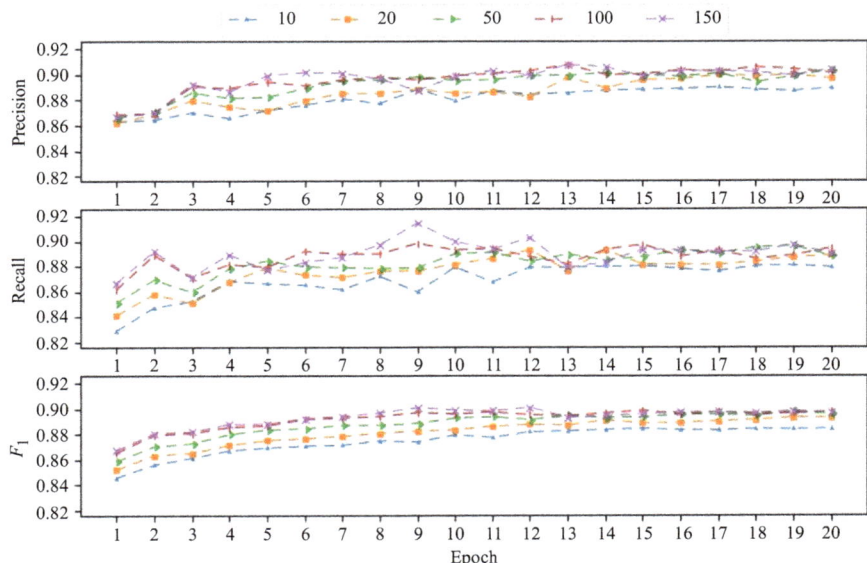

Figure 2.13 Effect of the number of filters on precision, recall, and F_1 score

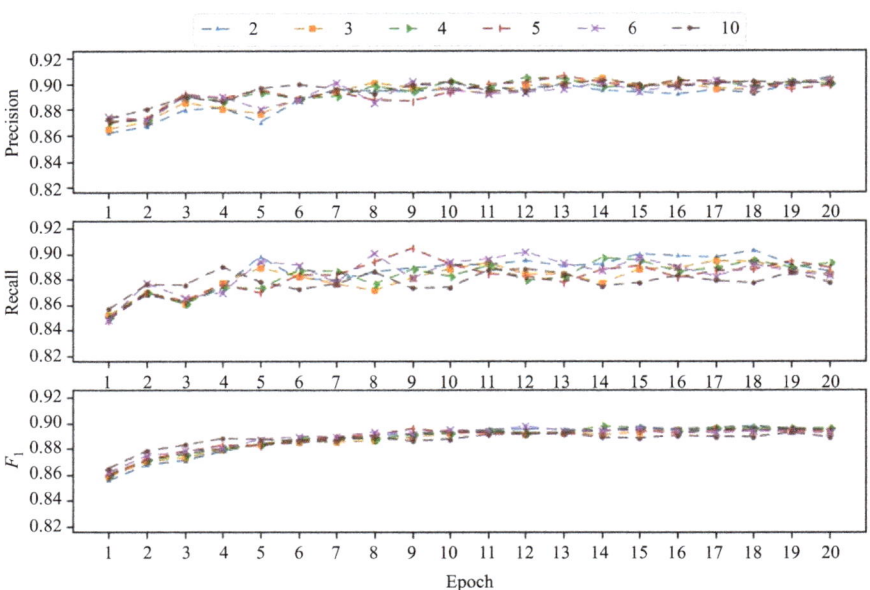

Figure 2.14 Effect of filter length on precision, recall, and F_1 score

Table 2.1 Parameter settings of the best architecture of the CNN model

Layer	Type	Filter	Stride	Output size
0	Input	–	–	(156, 200)
1	Convolution	(4, 200, 100)[a]	1	(153, 100)
2	Max-pooling	(1, 153)[b]	1	(1, 100)
3	Flatten	–	–	(100)
4	Fully connected	–	–	(50)
5	Fully connected	–	–	(1)

[a]The length, width, and the number of the filter are 4, 200, and 100, respectively.
[b]The horizontal size and the vertical size of the pool window are 1 and 153, respectively.

Table 2.2 The performance of LSTM, CNN, and LSTM-CNN in classifying Weibo posts into traffic-related and traffic-unrelated ones

	Traffic-related			Traffic-unrelated		
	Precision	Recall	F_1	Precision	Recall	F_1
CNN	0.8961	0.9006	0.8983	0.9168	0.9130	0.9149
• LSTM	0.8873	0.9182	0.9025	0.9298	0.9028	0.9161
• LSTM-CNN	0.8928	0.9160	**0.9042**	0.9285	0.9083	**0.9183**

Note: Bold values mean the best performance.

Determination of the LSTM-CNN model: We also used the grid search method to determine the best LSTM-CNN model. We chose the size of LSTM hidden state from {5, 10, 20, 50}, selected the filter number of convolutional layer from {10, 20, 50, 100}, set the filter length from 2 to 6 with a stride 1, and chose the learning rate from 0.01 to 0.03 with a stride 0.01. By comparing the F_1 score, we obtain the best parameter settings of the LSTM-CNN model.

Table 2.2 shows the performance of LSTM, CNN, and LSTM-CNN in classifying Weibo posts into traffic-related and traffic-unrelated ones. We can see that all the three models perform well in this task, and the LSTM-CNN model achieves the highest F_1 score. The LSTM model has a higher recall value in traffic relevant class but has the lowest precision value, which means the model has less exactness. The CNN model has the highest recall value in traffic irrelevant class; however, there is poor performance on the other two indexes. Experiments showed that these deep learning methods we used performed well in traffic event detection based on Weibo data, and if we compare the F_1 score, we can find that the LSTM-CNN method outperforms the other two methods.

2.4.3 Traffic status prediction

Nowadays, map applications are widely used by travelers with the increasing use of smart phones. Accurate traffic status prediction can help travelers' route choice. The LSTM model was developed for traffic status prediction [38].

The traffic status data were collected from AMAP between December 28, 2014 and February 3, 2015. The data was collected every 5 min covering 1,649 segments of arterial roads in Beijing, China. The traffic statuses are categorized as three classes, i.e., the unimpeded condition, the slow condition, and the impeded condition. To evaluate the performance of the LSTM model, we used three performance indexes, which are precision, recall, and F_1.

The traffic conditions in the last 30 min were chosen as the input to predict 30-min traffic condition. Given traffic condition sequence:

$$(x_{t-5}, x_{t-4}, x_{t-3}, x_{t-2}, x_{t-1}, x_t)$$

the task is to predict the traffic condition x_{t+5} at time step $t+5$.

We performed the grid search technique to find the best architecture of the LSTM model. The number of output units in an LSTM layer was chosen from {3, 6, 9, 12, 36}. The learning rate was chosen from 0.1 to 1.2 with a step 0.1. We also tested whether to stack three LSTM layers on top of each other or not can help improve the performance. Experiments show that the stacked LSTM with output units' size of each layer is 6, 6, and 6, respectively, is the best architecture for the prediction in the previous dataset.

We compared the performance of the proposed stacked LSTM model with the Multilayer Perceptron (MLP) model, the decision tree model, and the Support Vector Machine (SVM) model. And all baseline methods were well tuned. The prediction results of unimpeded condition, slow condition, and impeded condition are given in Table 2.3, from which we can find that the precision, recall, and F_1 measures of MLP, SVM, and decision tree are close, while the stacked LSTM has the highest score at precision, recall, and F_1 measures higher than the other three methods. It is worth to mention that the sample data are dominated by the instances of unimpeded conditions, because unimpeded traffic exists in most of the time in a normal day. The F_1 measure for slow condition and impeded condition by the stacked LSTM is improved up to 40% and 23%. The stacked LSTM outperformed the other three methods clearly in the previous dataset.

2.4.4 Semantic reasoning for traffic congestion

We used news-LDA and Weibo-LDA to describe the same traffic event from different perspectives. Usually, traffic events are posted on Sina Weibo faster while having less detailed information; traffic events reported in news are slower while having detailed information. Combining news and Weibos to describe the same traffic event can form a complete picture with multi-aspects.

Let us take inferring why a road was congested as an example [30]. We chose the traffic situation in Qingdao on August 4, 2017 as one case. We extracted and fused cross-platform transportation event on that day, manually tagged and mapped

Table 2.3 Performance comparison of MLP, decision tree, SVM, and stacked LST LSTM

Task	Unimpeded condition			Slow condition			Impeded condition		
	Precision	Recall	F_1	Precision	Recall	F_1	Precision	Recall	F_1
MLP	0.9418	0.9721	0.9568	0.5930	0.4503	0.5119	0.6517	0.5913	0.6200
• Decision tree	0.9751	0.9403	0.9574	0.4447	0.6110	0.5147	0.5929	0.6703	0.6292
• SVM	0.9755	0.9397	0.9573	0.4454	0.6084	0.5143	0.5853	0.6801	0.6291
• Stacked LSTM	0.9754	0.9864	0.9808	0.7418	0.6972	0.7188	0.8263	0.7270	0.7734

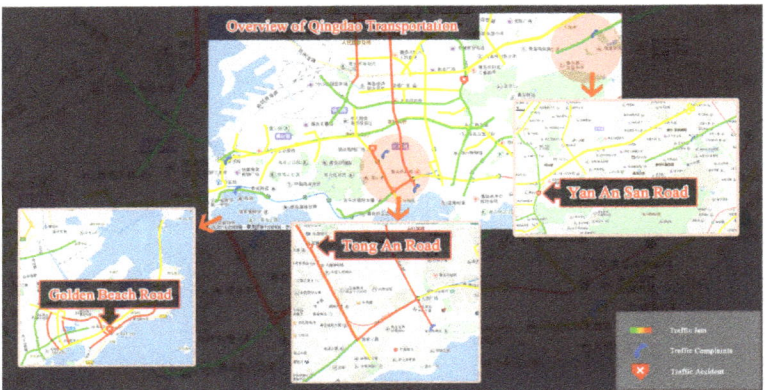

Figure 2.15   Overview of Qingdao transportation on August 4, 2017

all the events to the city roads, which is shown in Figure 2.15, so we can get an intuitive display of overall urban transportation situation. We investigated event details and corresponding causal factors on traffic congestion.

For the Gold Beach Road: we can get (1) traffic-related words like parking place, traffic jam, slow down, and traffic broadcast, (2) activities like beer festive and opening ceremony, and famous stars words like Xiaoming Huang. After reading the Weibo posts and news articles, we found that the annually Qingdao International Beer Festival was opening on the square of Qingdao Beer City where the film star "Xiaoming Huang" and other famous stars presented and had a show for the celebration. In addition, the opening ceremony began at 7.00 PM that was also the evening rush hour. The festival and the rush hour explained why heavy traffic jam happened on the Gold Beach Road.

For the Tong An Road: we can get location word "GuoXin stadium" and singer group words "Mayday," and "concert" using the system. We may speculate that the singing group called Mayday held a concert on the same day. However, it is strange that we found no words about traffic jam or accident but only traffic words like "detour," "traffic regulation," and so on. We further checked the collected Weibo posts and news articles and found that the concert would be held on the next day and the traffic agency early alerted the traffic situation on the road, released the traffic regulation. A lot of fans and audiences forwarded and disseminated these posts in the Weibo platform.

For the Yan An San Road: using the system, we observed the words "traffic jam," "evacuation," and words about fire emergency like "fires status," "fireman," and "fire alert." Meanwhile, we also observed the location word "Petroleum Building." Therefore, we inferred that there was a traffic jam on "Yan An San Road" which was caused by a fire emergency. We double-checked the news article and announcements of administrative agencies and found that the elevator in the Petroleum Building caught fire at 7.00 PM. Lots of Weibo users posted the texts, photos, and videos about the fire emergency. The official agencies also reported the event through news and social media.

2.5 Conclusion

With the rapid development of mobile internet, social signals in online web data will grow dramatically and can complement traditional physical traffic sensor data greatly. In this chapter, we analyze the advantages of online web data for traffic analytics. We propose a traffic analytic system with online web data. The proposed system has four components: data collection module, data preprocessing module, data modeling/mining module, and applications. We also present some main techniques used in the proposed system, including the LDA model, word embedding, Bayesian networks, and deep learning. We give some results on using the developed traffic analytic system with online web data in practice.

References

[1] Z. Zhang, "Fusing Social Media and Traditional Traffic Data for Advanced Traveler Information and Travel Behavior Analysis," Ph.D., the University at Buffalo, the State University of New York, 2017.

[2] Y. Lv, Y. Chen, X. Zhang, Y. Duan, and N. L. Li, "Social Media Based Transportation Research: the State of the Work and the Networking," IEEE/CAA Journal of Automatica Sinica, vol. 4, no. 1, pp. 19–26, 2017.

[3] E. Morgul, H. Yang, A. Kurkcu, et al., "Virtual Sensors: Web-Based Real-Time Data Collection Methodology for Transportation Operation Performance Analysis," Transportation Research Record: Journal of the Transportation Research Board, vol. 2442, pp. 106–116, 2014.

[4] F. Wang, "Scanning the Issue and Beyond: Real-Time Social Transportation With Online Social Signals," Intelligent Transportation Systems, IEEE Transactions on, vol. 15, no. 3, pp. 909–914, 2014.

[5] X. Zheng, W. Chen, P. Wang, et al., "Big Data for Social Transportation," IEEE Transactions on Intelligent Transportation Systems, vol. 17, no. 3, pp. 620–630, 2016.

[6] F.-Y. Wang, "Artificial Intelligence and Intelligent Transportation: Driving Into the 3rd Axial Age With ITS," IEEE Intelligent Transportation Systems Magazine, vol. 9, no. 4, pp. 6–9, 2017.

[7] P. Anantharam and B. Srivastava, "City Notifications as a Data Source for Traffic Management," presented at the 20th ITS World Congress, Tokyo, Japan, 2013.

[8] F.-Y. Wang, "Parallel Control and Management for Intelligent Transportation Systems: Concepts, Architectures, and Applications," IEEE Transactions on Intelligent Transportation Systems, vol. 11, no. 3, pp. 630–638, 2010.

[9] N. Zhang, F.-Y. Wang, F. Zhu, D. Zhao, and S. Tang, "DynaCAS: Computational Experiments and Decision Support for ITS," IEEE Intelligent Systems, vol. 23, no. 6, pp. 19–23, 2008.

[10] C. Gutierrez, P. Figuerias, P. Oliveira, R. Costa, and R. Jardim-Goncalves, "Twitter Mining for Traffic Events Detection," Science and Information Conference (SAI), vol. 2015, pp. 371–378, 2015.

[11] P. Tejaswin, R. Kumar, and S. Gupta, "Tweeting Traffic: Analyzing Twitter for Generating Real-Time City Traffic Insights and Predictions," presented at the Proceedings of the 2nd IKDD Conference on Data Sciences, Bangalore, India, 2015.

[12] J. Cui, R. Fu, C. H. Dong, and Z. Zhang, "Extraction of Traffic Information From Social Media Interactions: Methods and Experiments," 2014 IEEE 17th International Conference on Intelligent Transportation Systems (ITSC), 2014, pp. 1549–1554.

[13] D. Wang, A. Al-Rubaie, J. Davies, and S. S. Clarke, "Real Time Road Traffic Monitoring Alert Based on Incremental Learning From Tweets," 2014 IEEE Symposium on Evolving and Autonomous Learning Systems (EALS), 2014, pp. 50–57.

[14] D. Wang, A. Al-Rubaie, S. Stin, Clarke, and J. Davies, "Real-Time Traffic Event Detection from Social Media," ACM Transactions on Internet Technology, vol. 18, no. 1, pp. 1–23, 2017.

[15] M. Maghrebi, A. Abbasi, and S. T. Waller, "Transportation Application of Social Media: Travel Mode Extraction," 2016 IEEE 19th International Conference on Intelligent Transportation Systems (ITSC), 2016, pp. 1648–1653.

[16] M. Maghrebi, A. Abbasi, T. H. Rashidi, and S. T. Waller, "Complementing Travel Diary Surveys With Twitter Data: Application of Text Mining Techniques on Activity Location, Type and Time," 2015 IEEE 18th International Conference on Intelligent Transportation Systems, 2015, pp. 208–213.

[17] M. Itoh, D. Yokoyama, M. Toyoda, Y. Tomita, S. Kawamura, and M. Kitsuregawa, "Visual Exploration of Changes in Passenger Flows and Tweets on Mega-City Metro Network," IEEE Transactions on Big Data, vol. 2, no. 1, pp. 85–99, 2016.

[18] T. Georgiou, A. E. Abbadi, X. Yan, and J. George, "Mining Complaints for Traffic-Jam Estimation: A Social Sensor Application," 2015 IEEE/ACM International Conference on Advances in Social Networks Analysis and Mining (ASONAM), 2015, pp. 330–335.

[19] G. T. Giancristofaro and A. Panangadan, "Predicting Sentiment toward Transportation in Social Media Using Visual and Textual Features," 2016 IEEE 19th International Conference on Intelligent Transportation Systems (ITSC), 2016, pp. 2113–2118.

[20] K. Zeng, W. Liu, X. Wang, and S. Chen, "Traffic Congestion and Social Media in China," *IEEE Intelligent Systems*, vol. 28, no. 1, pp. 72–77, 2013.

[21] J. He, W. Shen, P. Divakaruni, L. Wynter, and R. Lawrence, "Improving Traffic Prediction With Tweet Semantics," presented at the Proceedings of the Twenty-Third International Joint Conference on Artificial Intelligence, Beijing, China, 2013.

[22] M. Ni, Q. He, and J. Gao, "Using Social Media to Predict Traffic Flow under Special Event Conditions," The 93rd Annual Meeting of Transportation Research Board, 2014.

[23] S. Grosenick, "Real-Time Traffic Prediction Improvement through Semantic Mining of Social Networks," Master, University of Washington, 2012.

[24] T. Zhou, L. Gao, and D. Ni, "Road Traffic Prediction by Incorporating Online Information," presented at the Proceedings of the 23rd International Conference on World Wide Web, Seoul, Korea, 2014.

[25] F. Rodrigues, S. S. Borysov, B. Ribeiro, and F. C. Pereira, "A Bayesian Additive Model for Understanding Public Transport Usage in Special Events," IEEE Transactions on Pattern Analysis and Machine Intelligence, vol. 39, no. 11, pp. 2113–2126, 2017.

[26] J. Cui, R. Fu, C. Dong, and Z. Zhang, "Extraction of Traffic Information From Social Media Interactions: Methods and Experiments," 17th International IEEE Conference on Intelligent Transportation Systems (ITSC), 2014, pp. 1549–1554.

[27] F. Lécué, S. Tallevi-Diotallevi, J. Hayes, *et al.*, "STAR-CITY: Semantic Traffic Analytics and Reasoning for CITY," presented at the Proceedings of the 19th international conference on Intelligent User Interfaces, Haifa, Israel, 2014.

[28] D. Semwal, S. Patil, S. Galhotra, A. Arora, and N. Unny, "STAR: Real-Time Spatio-Temporal Analysis and Prediction of Traffic Insights Using Social Media," presented at the Proceedings of the 2nd IKDD Conference on Data Sciences, Bangalore, India, 2015.

[29] B. Singh, "Real Time Prediction of Road Traffic Condition in London via Twitter and Related Sources," Master, Middlesex University, Middlesex University, 2012.

[30] H. Lu, K. Shi, Y. Zhu, Y. Lv, and Z. Niu, "Sensing Urban Transportation Events From Multi-Channel Social Signals With the Word2vec Fusion Model," Sensors, vol. 18, no. 12, p. 4093, 2018.

[31] L. Hong and B. D. Davison, "Empirical Study of Topic Modeling in Twitter," Proceedings of the SIGKDD Workshop on Social Media Analytics, Washington, DC, USA, 25 June 2010.

[32] T. Mikolov, K. Chen, G. Corrado, and J. Dean, "Efficient Estimation of Word Representations in Vector Space," Computer Science, 2013.

[33] S. Hochreiter and J. Schmidhuber, "Long Short-Term Memory," Neural Computation, vol. 9, no. 8, pp. 1735–1780, 1997.

[34] J. L. Elman, "Finding Structure in Time," Cognitive Science, vol. 14, no. 2, pp. 179–211, 1990.

[35] K. Fukushima, "Neocognitron: A Self-Organizing Neural Network Model for a Mechanism of Pattern Recognition Unaffected by Shift in Position," Biological Cybernetics, vol. 36, no. 4, pp. 193–202, 1980.

[36] Y. LeCun, L. Bottou, Y. Bengio, and P. Haffner, "Gradient-Based Learning Applied to Document Recognition," Proceedings of the IEEE, vol. 86, no. 11, pp. 2278–2324, 1998.

[37] Y. Chen, Y. Lv, X. Wang, L. Li, and F. Wang, "Detecting Traffic Information From Social Media Texts With Deep Learning Approaches," IEEE Transactions on Intelligent Transportation Systems, vol. 20, no. 8, pp. 3049–3058, 2019.

[38] Y. Chen, Y. Lv, Z. Li, and F. Wang, "Long Short-Term Memory Model for Traffic Congestion Prediction With Online Open Data," 2016 IEEE 19th International Conference on Intelligent Transportation Systems (ITSC), Rio de Janeiro, 2016, pp. 132–137.

Chapter 3
Macroscopic traffic performance indicators based on floating car data: formation, pattern analysis, and deduction

Lu Ma[1] and Jiaqi He[1]

3.1 Introduction

Urban traffic is an important part of urban activities. With the rapid development of the economy and the advancement of urbanization in many cities, urban road systems experienced serious traffic congestions, which increases the traffic delay and fuel consumption, aggravates the vehicle exhaust and noise, and seriously damages the urban environment. In order to evaluate the congestion conditions of cities, macroscopic measurements are required to provide quantified indications on evaluating the traffic performance of cities.

The macroscopic traffic performance is referring to the concept of evaluating the overall extent of the traffic congestion for an entire city or a specific region. The idea is similar to the microscopic traffic performance for particular transportation facilities, e.g., an urban road segment, a fraction of freeway or an intersection, but some aspects are substantially different between the two types of traffic performance, especially in terms of the way of measuring, influential factors, and the application scenarios. Such a discrepancy demands an exclusive knowledge framework for macroscopic traffic performance. To some extent, the macroscopic traffic performance quantity is the aggregation on the microscopic performance with a special spatiotemporal scale. Therefore, there could exist various ways to design such aggregation and, in turn, yield many macroscopic performance indicators to assess the level of urban or regional status of road traffic. Different indicators could provide consistent results but different in the emphasized facets of road traffic conditions, e.g., speed or delay.

In contrast to the traffic performance evaluation at the microscopic level, the macroscopic evaluation is more complicated because a network contains many road segments and intersections. For the evaluation at the microscopic level, the congestion level of a road segment/intersection could be readily and naturally measured by traffic flow variables, e.g., speed, volume, density, delay, or their varieties [1–7]. For the evaluation at the macroscopic level, the traffic performance of a region or city is

---

[1]School of Traffic and Transportation, Beijing Jiaotong University, Beijing, China

actually a single quantity aggregating the congestion level for all the segments and intersection of the corresponding network. Through the heavy aggregation, certain microscopic-level factors could be weakened or eliminated. For example, a traffic accident could happen on a certain road segment, which would largely affect the traffic conditions of the road segment, and the impact will last for a period of time. However, the traffic accident would have a negligible impact on city-level traffic performance. Especially, when the time span of evaluation is large, e.g., an entire day or week, the impact of all the occurred accidents tends to be a constant assuming that the risk of traffic accidents does not depend on a particular day. From the other perspective, macroscopic factors would exert substantial impacts on the macroscopic traffic performance, including weather conditions, classification of dates, traffic management policies, socioeconomic characteristics, development of new transportation infrastructures, and even the emerging of new technologies. For example, dates could be classified into weekday, weekend, long vacation, travel restriction on the day, special festival, etc., and it is commonly recognized that traffic conditions on different types of dates could be different. Usually, the overall traffic conditions on weekends are better than those on weekdays due to the existence of commuter trips on weekdays. In Beijing, China, many people are willing to travel outside the city during a long vacation, making a better road traffic condition inside the city than other days. On the days with the strict travel restriction, only about half of the vehicles are allowed to travel. Therefore, these macroscopic factors could largely affect traffic performance for a macroscopic scale.

The analyses on a macroscopic scale have been conducted in a wide range of fields [8–12], and many methods were also developed to aggregate the traffic performance for a macroscopic scale. The prevailing ones include the Travel Time Index [13], TomTom Traffic Index [14], INRIX Congestion Index, Beijing's Urban Road Traffic Performance Index [15,16], etc. The calculation of many of the current indices requires the speed measures of each road segment as input information. The Floating Car Data (FCD) are collections of GPS trajectories for a particular sample of all vehicles traveled within the spatiotemporal boundaries of experiments. The attraction of such data is continuously increasing partly because of the increase in the sample size and availability of many new data resources, including navigation service GPS data and ride-sourcing-trip big data.

In the light of these discussions, this chapter introduces a way of measuring the macroscopic traffic performance based on the ride-sourcing FCD, resulting in an indicator named as the Network-Level Trip (NLT) speed. This indicator actually measures the average speed for each complete trip, which is simple and direct to reflect the network speed. Because this indicator dodges the involvement of transportation facilities, map matching, and the related procedures are not required, which is one of the advantages of this method in contrast to methods that need map-related procedures. For the purpose of illustration, the FCD of a sample of ride-sourcing trips in Beijing, China between 2015 and 2016 were collected and adopted to formulate the NLT speed daily. Besides, several influential factors, including weekend, long vacation, temperature, precipitation, Air Quality Index (AQI), and vehicle ownership, were also adopted serving as covariate variables in the following analyses. Several classic methods of time series analysis, including the exponential smoothing and ARIMA (Autoregressive Integrated Moving Average) model, were applied to the daily NLT speed time series illustrating the statistical patterns in different facets. The results indicate that the ARIMA model with one-time regular differencing and one-time seasonal differencing is superior to other models in terms of Akaike Information Criterion (AIC) and Akaike Information Criterion corrected (AICc). However, for the deduction of future values, those models are deficient due to its incapability or weakness of involving influential factors. To this end, a hybrid model combing the ARIMA model and the Support Vector Machine (SVM) model was applied to overcome such issues. In the hybrid model, the ARIMA model predicts the amount of the NLT speed that can be explained by its historical observations, while the SVM model worked on the residual part with the incorporation of influential factors. The results indicate that the hybrid model works well to make deductions of the NLT speed under giving policy scenarios.

The rest of this chapter is organized as follows. Section 3.2 introduces the proposed NLT speed measurement, including its mathematical form and the correlative relationships with other influential factors. In Section 3.3, several methods for time series analysis are introduced. Section 3.4 presents the empirical application of the time series models on the NLT speed series and reports relevant results. Section 3.5 summarizes the key findings of this study and discusses its limitations and directions for future research.

3.2 A macroscopic traffic performance indicator: network-level trip speed

3.2.1 The mathematical form of the NLT speed

The emerging e-hailing transportation companies recorded the trajectory of the entire itinerary for each ride-sourcing trip, which enables the measurement of the trip speed on the big data basis. This chapter introduces a simple method of evaluating the macroscopic traffic performance for a city or certain regions by taking advantage of the ride-sourcing travel data. In this method, a macroscopic traffic performance indicator called "NLT speed" is developed as in (3.1), where V is the NLT speed for a road network corresponding to a particular area, V_i is the travel speed of an individual trip indexed by i, and n is the total number of trips within the spatiotemporal boundaries. For each trip, its speed is just calculated as the overall path distance divided by the travel time. In fact, the ride-sourcing travel data have the advantage of measuring an accurate travel speed for a trip because the recorded trajectory is complete for each trip. In contrast, the GPS data from the navigation services are usually a fraction of the trip. For the data recorded by taxi vehicles, it is difficult to distinguish and extract the trajectory of each trip because we do not know when the taxi is carrying passengers. Based on the ride-sourcing travel data, the proposed NLT speed directly reflects the averaged speed experienced by all drivers, which also to some extent relates to the delay:

$$V = \frac{\left(\sum_{i=1}^{n} V_i\right)}{n} \tag{3.1}$$

3.2.2 The empirical data for analyses

The ride-sourcing FCD data collected by a transportation network company from January 1, 2015, to December 31, 2016, for trips that occurred within the sixth ring road in Beijing, China, were adopted for the following analysis. According to (3.1), the NLT speed was calculated for each day in the study period. As a result, there are in total 731 measures constituting a time series of the macroscopic traffic performance shown in Figure 3.1. Through visual observations, there are some evident peaks from time to time beyond the background fluctuation of the NLT speed series exhibiting significantly larger values in contrast to other values. In fact, those values correspond to particular dates with special events or vacation arrangements. Table 3.1 reports the events and corresponding dates on which the macroscopic traffic performance might behave differently than on a regular day. Note that there are several peaks caused by the travel restriction policies under which only a half of the vehicles are allowed to travel on each day and the allowed vehicles were chosen according to a rotation based on the last digit of their plate number [17,18]. Specifically, all the vehicles with an odd plate number were allowed to travel on the same day, whereas the other vehicles with an even plate number cannot travel on that day, and such a mechanism continues the next day with the switch of the vehicles. In this chapter, such a policy is called the Odd-Even-Plate-Number (OEPN) travel restriction. Such a strong restriction was rarely implemented unless for the days with important activities or extremely poor air conditions. In order to deal with the continuously increasing car ownership levels of the city, other than those days under the OEPN travel restriction policy, about 20% of vehicles are not allowed to travel on weekdays and no restrictions on weekends or long-vacation days.

The impact of urban traffic policies on traffic performance is particularly obvious. It is self-evident that policies such as travel restriction management substantially affect urban traffic conditions. In this chapter, the influential factors of the NLT speed were considered for the four major aspects, i.e., travel restriction,

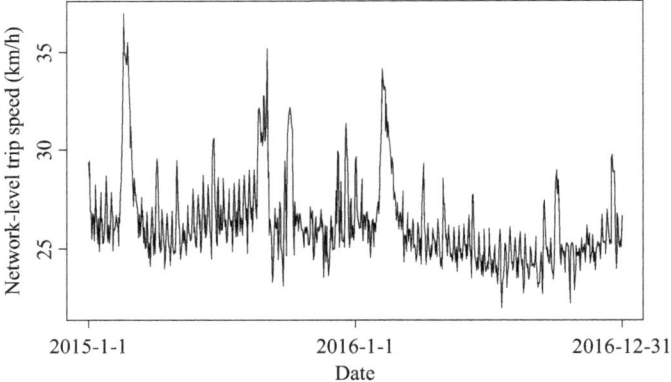

Figure 3.1 The time series of the NLT speed between 2015 and 2016 in Beijing, China

Table 3.1 Details of events occurring on particular dates

No.	Year	Date	Events
1	2015	Jan 1–Jan 3	New Year Holiday
2	2015	Feb 18–Feb 24	Chinese Lunar New Year
3	2015	Apr 04–Apr 06	Qingming Festival
4	2015	May 01–May 03	International Labor Day
5	2015	Jun 20–Jun 22	Dragon Boat Festival
6	2015	Aug 20–Sep 05	OEPN travel restrictions due to military parade
7	2015	Oct 01–Oct 07	National Day
8	2015	Dec 08–Dec 10	OEPN travel restrictions due to serious air conditions
9	2015	Dec 19–Dec 22	OEPN travel restrictions due to serious air conditions
10	2016	Jan 01–Jan 03	New Year Holiday
11	2016	Feb 07–Feb 13	Chinese Lunar New Year
12	2016	Apr 02–Apr 04	Qingming Festival
13	2016	Apr 30–May 02	International Labor Day
14	2016	Jun 09–Jun 11	Dragon Boat Festival
15	2016	July 19–July 21	Rainstorm
16	2016	Sep 15–Sep 17	Mid-Autumn Festival
17	2016	Oct 01–Oct 07	National Day
18	2016	Dec 16–Dec 22	OEPN travel restrictions due to serious air conditions

special date, traffic management, and environment. The date when making trips consider the difference between weekend, long vacation day, and regular weekday. The travel environment mainly includes temperature, precipitation, and AQI. The evolving pattern of the NLT speed is going to be analyzed with respect to those factors. Among these influential factors, travel restriction and special date are categorical variables, and the others are continuous variables.

It is clear that the traffic performance would have evident differences between weekday, weekend, and long vacation day. The traffic performance usually gets better on weekends and long vacation days in Beijing, China. According to the data of 2018, commute and school trips constituted about 51% of all trips in Beijing, China. On weekends, commute trips and the number of vehicles on urban roads have been greatly reduced, and, in turn, increased the traffic performance. However, there was usually no implementation of travel restrictions on weekends leading to the additional possibility of traveling. With the combination of the two reasons, the improvement of traffic performance at weekends is limited. Similar trends have also existed on long vacation days.

The traffic environment also has an important impact on macroscopic traffic performance. Transportation infrastructures, traffic operation, vehicle ownership, and natural environment have a direct or indirect influence on urban traffic performance. By the end of 2018, the number of motor vehicles in Beijing, China reached 6.08 million, an increase of 3.0% over the previous year. The number of private motor vehicles reached 4.89 million, an increase of 2.9% over the previous year. Regarding the new-energy vehicles, there were more than 224 thousand ones, an increase of 35.6% over the previous year. There are more than 17 thousand new-energy trucks, an increase of 49.0% over the previous year. For the composition structure of motor vehicles, small buses account for 85.8%, large and medium buses account for 2.3%, small trucks account for 5.1%, and large and medium trucks account for 1.6%. The urban vehicle ownership directly impacts on the macroscopic traffic performance. With the rapid increase in the total number of vehicles in the city, the number of vehicles on the road will inevitably increase, and, in turn, traffic performance is expected to get worse.

Under the trend of global warming, the time with extremely high temperatures in many areas around the world becomes longer and more frequent. The high temperature will affect transportation equipment and ground facilities. In addition, high temperatures will also bring great changes to the choice of daily travel modes. The impact of low temperature on urban traffic is mainly coming from the increase of rainfall and snowfall, and the frost of the urban road surface. Heavy snow, road icing, and freezing rain will greatly reduce the friction between vehicles and road surface, which will bring safety hazards as well as reducing its efficiency to urban traffic.

Precipitation has a significant impact on macroscopic traffic performance. Heavy rainfall or snowfall could exert on wide areas of the city with a long duration. On rainy days, the pavement surface is wet and slippery, and the road visibility is low, which reduces the travel speed of vehicles. Usually, the traffic performance is lower during rainy or snowy days. However, on days with extremely heavy rain, the travel demand is small probably resulting in a slight improvement in traffic performance.

During wintertime, smoggy weather could frequently happen in Beijing, China usually from the middle of November to the beginning of March. There will be a particularly obvious fluctuation of AQI. The impact of smog on urban traffic performance was mainly reflected in the fact that the visibility of vehicles is compromised and the driver would accordingly adjust the stop-sight distance and running speed.

The choices of travel modes will also affect the macroscopic traffic performance. When more residents choose cars or taxis, the number of vehicles on urban roads will raise, thus reducing the traffic performance, whereas when more residents choose rail transit or bus, the number will drop. However, from a long-term perspective, the travel mode that residents choose might not change significantly because everyone has their preference of travel behaviors. When the change of passenger volume and sharing rate of different modes is very small, the impact on the macroscopic traffic performance might be negligible.

According to the annual report of Beijing traffic in 2018, 32.7% of the total trips occurred in the morning peak (7:00–9:00) and 20.8% in the evening peak (17:00–19:00). During the morning peak period, rail transit trips account for 42.6% of the daily rail transit trips and followed by passenger car trips, which account for 39.1% of the daily passenger car trips. In addition, shuttle bus trips account for 37.9% of the daily trips during the morning peak period. During the evening peak period, shuttle bus trips account for 45.5% of the daily trips, followed by the number of rail transit trips, which account for 39.1% of the total daily rail transit trips.

3.2.3 Descriptive analyses of influential factors

This section illustrates the descriptive statistics of the NLT speed and influential factors as shown in Table 3.2. The average value of the NLT speed is 26.29 km/h. Note that the average value contains the contribution of trips for an entire day. Its minimum value is 22.04 km/h and the maximum value is 36.98 km/h reflecting a relatively large variation. The average temperature is 13.66°C, the average precipitation is 1.73 ml, the average AQI is 117.16, and the average vehicle ownership is 563.67 million. For the categorical variables, 209 days of weekends accounted for 28.6% of all days, 60 days of long vacation days accounted for 8.2% of all days, and 28 days with the OEPN travel restriction accounted for 3.8% of all days.

According to the data of the NLT speed and the corresponding influential factors, the violin plots of those factors are presented in Figure 3.2. The violin plot is similar to the well-known box plot. In Figure 3.2, the white dot is the median, the bar ranges from the lower quartile to the upper quartile, the black line represents the whisker, and the external shape represents the kernel density estimation of the data.

It can be seen that the distribution of the NLT speed is mainly concentrated in the range of 23–28 km/h, and the chance of lower or higher values is relatively small. Figure 3.2 also illustrates the NLT speed on weekends, long vacation days, and the days with the OEPN travel restriction, respectively. The NLT speed was significantly higher on those dates in contrast to on a regular weekday. The median of the NLT speed is 26.68, 29.39, and 30.31 km/h under weekends, long vacation days, and the days with the OEPN travel restriction.

Among the three dichotomous categorical variables, the influence of the long vacation days and the days with the OEPN travel restriction is greater than that of the weekends. On long vacation days, not only most people do not need to go to work or school, but also many people liked to travel outside the city. In sum, all the three variables are important influential factors to the NLT speed.

Table 3.2 Descriptive statistics of variables

Continuous variable	Mean	S. D.	Min.	Max.
Network-level trip speed (km)	26.29	2.17	22.04	36.98
Temperature (°C)	13.66	10.88	−12.5	32.5
Precipitation (ml)	1.73	10.72	0	210.7
AQI	117.16	74.78	23	485
Car ownership (million)	563.67	3.79	559.1	571.7
Categorical variable	**1 (yes)**		**0 (no)**	
Weekend	209 (28.6%)		522 (71.4%)	
Long vacation day	60 (8.2%)		671 (91.8%)	
OEPN travel restriction	28 (3.8%)		703 (96.2%)	

38 Traffic information and control

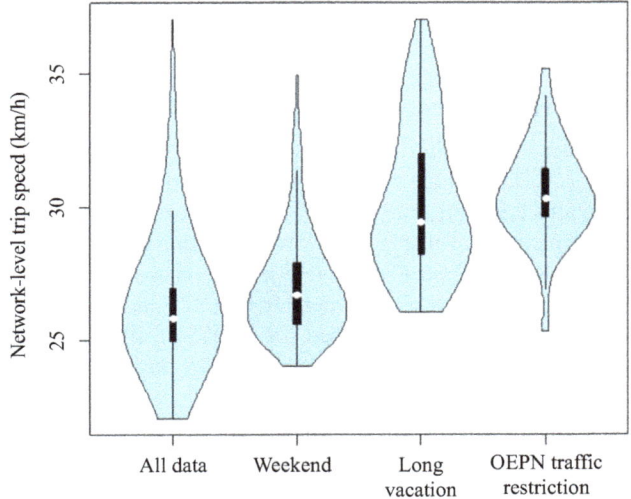

Figure 3.2 The violin plot of the NLT speed data under different types of dates

3.2.4 Correlative relationships between variables

It is interesting and valuable to investigate the correlation between the NLT speed and influential factors, which describes the closeness between two variables. Continuous variables, i.e., temperature, precipitation, AQI, and vehicle ownership, are represented by their nature magnitudes, while categorical variables are represented by zero or one. The value "one" means that the particular date is a weekend, long vacation day, or day with the OEPN travel restriction. In this chapter, the Pearson correlation coefficient was adopted to measure the correlation quantity between variables whose mathematical form is shown in (3.2). Here, $\text{Cov}(X, Y)$ is the covariance of X and Y, and $\text{Var}(\cdot)$ is the variance:

$$r(X, Y) = \frac{\text{Cov}(X, Y)}{\sqrt{\text{Var}(X)\text{Var}(Y)}} \tag{3.2}$$

In order to illustrate the correlation and its significance between the NLT speed and the influential factors, Figure 3.3 plots the distribution of those continuous variables as well as the joint distribution between variables. It also displays the value of the corresponding Pearson correlation coefficient in its upper triangle, and the pentagram is used to indicate the strength of the significance. The scatter plots between variables are drawn in the lower triangle. In fact, the influence on the NLT speed is too complex to be determined by a single factor. Among those correlation coefficients, the one between the NLT speed and long vacation day is the largest, i.e., 0.53. The ones against weekend and day with OEPN travel restriction are relatively larger, i.e., 0.21 and 0.38, respectively. It shows that on weekends, long

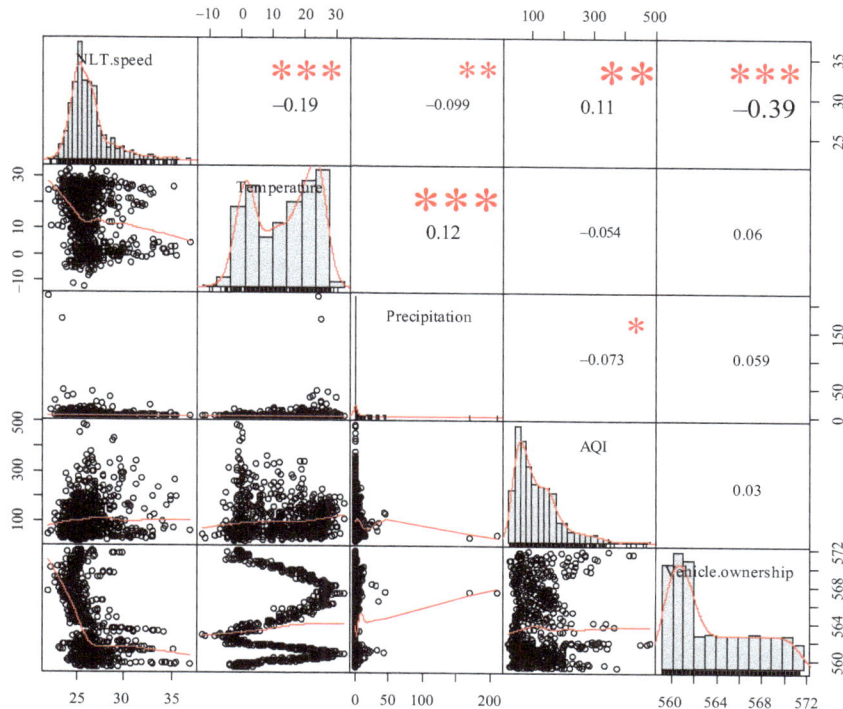

Figure 3.3 The joint distribution and correlative relationship between continuous variables

vacation days, or days with the OEPN travel restriction, the macroscopic traffic performance is better in contrast to a regular weekday. In addition, the correlation between the NLT speed and temperature is −0.19 indicating that the increase of temperature will reduce the NLT speed, probably because cars or taxis are more comfortable modes of transportation on days with a high temperature. In addition, the NLT speed has little correlation against precipitation or AQI.

3.3 Methods of time series analysis

3.3.1 The concept and basic features of the time series

Time series refers to the sequence of the values of statistical indicators following the chronological order [19–21], which is usually the result of measuring an object according to a fixed sampling rate. The time series studied in this chapter is used to arrange the daily NLT speed data into the corresponding sequence according to the chronological order of dates. Time series analysis methods can be used to predict future development according to the past change trend. It usually assumes that some of the evolving characteristics of the research object will continue to the future.

There are regular and irregular features in time series data. The magnitude of each observation in a time series is the result of the interaction of various influential factors at the same time. In terms of the influential direction and magnitudes of these factors, the fluctuation caused by these factors can be classified into three types, i.e., trend, seasonality, and randomness. The trend means that a certain variable moves forward with time or changes in independent variables showing a relatively slow and long-term tendency of continuous rise, fall, and stay of the same nature, but the range of change may not be equal. Seasonality means that a certain factor appears peak and low with the alternation of natural seasons due to external influences.

For a particular time series, two important aspects are reflecting its basic characteristics, i.e., autocorrelation and stationary. The autocorrelation coefficients can be used to measure the correlation between observations for every k unit of time (y_t and y_{t-k}) as defined in (3.3). Here, t is the length of the time series and the collection of all r_k constitutes the Autocorrelation Function (ACF).

$$r_k = \frac{\sum_{t=k+1}^{T}(y_t - \bar{y})(y_{t-k} - \bar{y})}{\sum_{t=1}^{T}(y_t - \bar{y})^2} \tag{3.3}$$

If the expectation and variance of a time series according to a stochastic process are constant in any time, and the covariance between any two observations only depends on the time distance between the two observations and does not depend on the actual time of the two observations, the time series is said to be stationary.

Time series always show a variety of complex features. Sometimes, it is difficult to analyze directly. One method is to decompose a time series into multiple subseries with respect to different patterns, e.g., the Seasonal and Trend decision using Loess (STL) decomposition method [22–24]. Traditional decomposition methods considered long-term trend (T), seasonal variation (S), cyclic variation (C), and irregular variation (I). The long-term trend refers to a trend or state of continuous development and change of the object in a long period; the seasonal variation refers to the regular change of the research object along the time in the development process caused by a seasonal factor; the cyclic variation refers to the periodic continuous change of the research object that usually takes the time greater than a year as the period of the cycle; the irregular variation refers to the irregular change of time series caused by other unknown factors.

3.3.2 The exponential smoothing method

The exponential smoothing method was invented in the 1950s [25–27]. It assumes that the trend of a time series is stable in a long period and the change had certain rules, so the time series can be recursive according to the stability and rules. On this basis, many time series prediction methods have been developed. In order to evaluate different exponential smoothing models, through the study of different examples, it is found that AIC is the most preferable one [28]. Koehler *et al.* [29] proposed an innovative state-space model that can be used to study the influence of outliers on the prediction results of the model and the estimation of model parameters.

The exponential smoothing method uses a weighted average of past observations. It strengthens the effect of the recent observations in the study period on the predicted values and gives different weights to the observations at different times, so as to increase the weights of the recent observations. The predicted values can be quickly adapted to the changes of the recent observations. In addition, the exponential smoothing method has high flexibility in the setting of weights given to historical observations. The exponential smoothing method is commonly used in prediction, which has a wide range of applications.

$$\widehat{y}_{t+1|t} = \alpha y_t + \alpha(1-\alpha)y_{t-1} + \alpha(1-\alpha)^2 y_{t-2} + \cdots + 0 \cdots - e \quad (3.4)$$

The simple exponential smoothing (SES) method is the most basic one where the prediction is obtained by simply calculating a weighted average value of historical observations. The closer the observed value is to the present, the greater the weight is given to it. In (3.4), $\widehat{y}_{t+1|t}$ is the predicted value at the $t+1$ time given all the values prior to it, α is the smoothing parameter. The predicted value is just the weighted average of all historical observations. The rate of weight reduction is controlled by the parameter α. For any α between 0 and 1, the weight of the observed value will decrease exponentially with time, so this method is called the "exponential smoothing method." In this chapter, several more advanced models were also developed, including Holt-Winters' Additive (HWA) model, Holt-Winters' Multiplicative (HWM) model, Holt-Winters' Damped Additive (HWDA) model, Holt-Winters' Damped Multiplicative (HWDM) method, and state-space ETS (Error, trend, seasonality) model for the following analyses [30–32].

3.3.3 The ARIMA method

The ARIMA model is known as the Autoregressive Integrated Moving Average model that is widely used in various fields for time series analyses. For example, Williams [33] applied the ARIMA model to predict the highway traffic flow in France. The prediction results show that the ARIMA model has higher goodness of fit and better precision than the single-variable prediction model. Lippi et al. [34] used the seasonal ARIMA model to predict the traffic flow data for a short period, and the accuracy of the prediction results is also high. Holens [35] compares the ARIMA model and a neural network model and uses two datasets from futures exchange. The results show that the performance of the two models is roughly the same, and they can predict turning points with approximately the same accuracy.

A variety of time series models were widely spread, including the Autoregressive (AR) and Moving Average (MA) models. By combining the AR and MA models, the Autoregressive Moving Average (ARMA) model is obtained, and by further adding differencing steps, the ARIMA model is obtained. The basic idea of the ARIMA model is to treat the data as a random sequence over time and use certain mathematical models to approximate the sequence. Once identified, the model can predict future values from the magnitude of historical observations.

The concept of stationary is important in describing a time series. It refers to the series whose statistical properties will not change with respect to the time. When a time series has a trend or seasonality, it is not stationary. In addition, there are also special cases. When a time series is cyclic without trend and seasonality, the time series can also be considered stationary. In order to transfer a nonstationary time series into a stationary one, differencing steps could be made. Through differencing, the level change in time series can be eliminated, and then the trend and seasonality of time series can be weakened. If it is still a nonstationary series after the first-order differencing, the second-order differencing can be further carried out. The idea of AR comes from linear regression. As its name shows, linear regression analysis predicts the dependent variable through a number of independent variables, while the AR model predicts the variable based on its values from the past times. If there are p past values in the right-hand side of the regression model, such a model is called p-order AR model as shown in (3.5). Here ε_t is the white noise. The formula is similar to the multiple linear regression model but the AR model takes the lag values of y_t as the covariates. The model is also called AR (p) model.

$$y_t = c + \phi_1 y_{t-1} + \phi_2 y_{t-2} + \cdots + \phi_p y_{t-p} + \varepsilon_t \tag{3.5}$$

The AR model requires a few data as the input information. This model has some disadvantages. The variable of the target object must be correlated to itself according to its observations at different times. The autocorrelation coefficient is important in this model. If the autocorrelation coefficient is less than 0.5, the use of the AR may be inappropriate, and the predicted results could be inaccurate. The AR model can be used to predict the economic phenomenon related to its early stages, that is, the economic phenomenon is greatly affected by its historical statuses, such as the mining amount of ore, the output of various natural resources, etc. If the economic phenomenon is greatly affected by other factors, such as social factors, the AR model would be insufficient, and some other models should be selected under this circumstance.

The MA and AR models are both basic models in time series analysis. The AR model studies the impact of the historical data of an object on its future values, while the MA model uses the random interference or the linear combination of prediction errors of the object in the past times to predict the future values. The mathematical formula of the MA model is given in (3.6).

Here ε_t is white noise and q represents the moving order. The model is called MA (q) model. In general, the value of ε_t is not established, so it is not a genuine linear regression in the usual sense. The MA model uses the weighted MA of the past several prediction errors to predict future values, while the exponential smoothing method uses the MA smoothing to estimate the trend period of the past values. By changing the parameter $\theta_1 \cdots \theta_q$, it will get different specifications of models.

$$y_t = c + \varepsilon_t + \theta_1 \varepsilon_{t-1} + \theta_2 \varepsilon_{t-2} + \cdots + \theta_q \varepsilon_{t-q} \tag{3.6}$$

With the combination of the AR and MA models, and using the differenced data, it gives the nonseasonal ARIMA model. Equation (3.7) provides the mathematical expression of the model. Here, y'_t is the series after differencing. The right side of the equation contains the random errors of the lag values and the lag values themselves. It was named ARIMA (p, d, q) model, and d is the number of differencing steps have been applied to the original data.

$$y'_t = c + \phi_1 y'_{t-1} + \cdots + \phi_p y'_{t-p} + \theta_1 \varepsilon_{t-1} + \cdots + \theta_q \varepsilon_{t-q} + \varepsilon_t \tag{3.7}$$

In the formula of the ARIMA model, the constant c and the differencing order d have an important influence on the long-term prediction. In specific, d has a great influence on the prediction interval. When d is larger, the prediction interval grows faster. When d is zero, the standard deviation of the prediction interval is the same as that of the historical data, and the prediction interval will be similar to the historical value.

3.3.4 The support vector machine (SVM) method

The SVM performs as a generalized linear classifier separating the data according to the supervised learning method [36–38]. It determines the decision boundary by calculating the maximum margin hyperplane of the training sample. In the SVM model, when the data are linearly separable, it can solve a maximum interval through it and when the data are nonlinear separable, it can map the data of the original training set to a high-dimensional space through nonlinear mapping. In this new space, it can find an optimal separation hyperplane to classify different samples.

When using the SVM method to make a prediction, the feature vector that the prediction process is based on should be determined in the beginning. The feature vector is usually composed of influential factors. In the following analyses, the SVM method will be applied in combination with the ARIMA model. In specific, the residual from the ARIMA model will serve as the response variable in the SVM method. Several influential factors, including weekend, long vacation, OEPN travel restriction, temperature, precipitation, AQI, and vehicle ownership, will be incorporated in the SVM method.

3.4 Analyses of the NLT speed time series

3.4.1 Evaluation criteria of the modeling performance

In order to compare the quality of various models, several criteria, i.e., AIC, AICc, Bayesian Information Criterion (BIC), and Root Mean Square Error (RMSE), were adopted, among which the first three were used to evaluate the fitting quality of the model, and RMSE was used to compare the prediction accuracy of models. Equations (3.8) and (3.9) provide the mathematical forms of AIC and AICc, respectively. AIC considers the trade-off between the goodness of data fitting and the complexity of the model, where k is the number of free parameters and L is the log-likelihood value at convergence. AIC actually penalizes the log-likelihood value to some extent so that it can overcome the biased judgment from overfitting. AICc is similar to AIC but provides an additional penalty with respect to the k and the sample size n in order to consider the cases with small sample size. When n is far greater than k^2, AICc is close to AIC.

$$\text{AIC} = 2k - 2\ln(L) \tag{3.8}$$

$$\text{AICc} = \text{AIC} + \frac{2k(k+1)}{n-k-1} \tag{3.9}$$

Another adopted criterion is the BIC and its form is shown in (3.10). In conditions with incomplete information, it estimates the unknown part with the Bayesian formula modifying the occurrence probability of the event and finally uses the expected value and the modified probability to make the optimal decision for estimation. Similar to AICc, BIC considers the effect of the number of samples on model fitting. For AIC, AICc, and BIC, a smaller value indicates a better model fitting. Note that the comparing among AIC, AICc, and BIC is meaningless, as they measure the model fitting from different aspects.

$$\text{BIC} = \ln(n)k - 2\ln(L) \tag{3.10}$$

RMSE is the square root of the deviation between the predicted value and the observed value, i.e., the ratio of the square of the residual to the number of observations as shown in (3.11). Here, $\tilde{y}_i$ is the predicted value and y_i is the observed value. According to the definition, a smaller RMSE would indicate a better model fitting.

$$\text{RMSE} = \sqrt{\frac{1}{n}\sum_{i=1}^{n}(\tilde{y}_i - y_i)} \tag{3.11}$$

3.4.2 The decomposition of the NLT speed time series

In Figure 3.4, the NLT speed series is decomposed into trend component, seasonal component, and remainder component using the STL method. Under the influence of various factors such as travel restriction, travel time, and travel environment, the NLT speed series fluctuates. According to the seasonal-component chart, it can be found that the change of the NLT speed follows a strong seasonality with a cycle of seven indicating that the periodic fluctuate was repeated weekly. In more detail, Friday is the day with the smallest NLT speed, and at weekends the NLT speed is generally higher than the one on weekdays.

3.4.3 The analysis based on exponential smoothing methods

This chapter adopted the FCD of ride-sourcing trips from January 1, 2015 to December 31, 2016. Those FCD were transferred to the NLT speed data for the 731 days which were divided into the training set and validation set with the ratio of 4 to 1, namely, 146 observations in the validation set and 585 in the training set. Several classic methods, including the SES, HWA, HWM, HWDA, HWDM, and state-space ETS models, were applied to predict the NLT speed for the validation samples shown in Figure 3.5.

Table 3.3 illustrates the modeling performance of those exponential smoothing methods. The AIC, AICc, and BIC of the state-space ETS model are all the smallest, indicating that it has the best modeling performance, whereas the SES method has

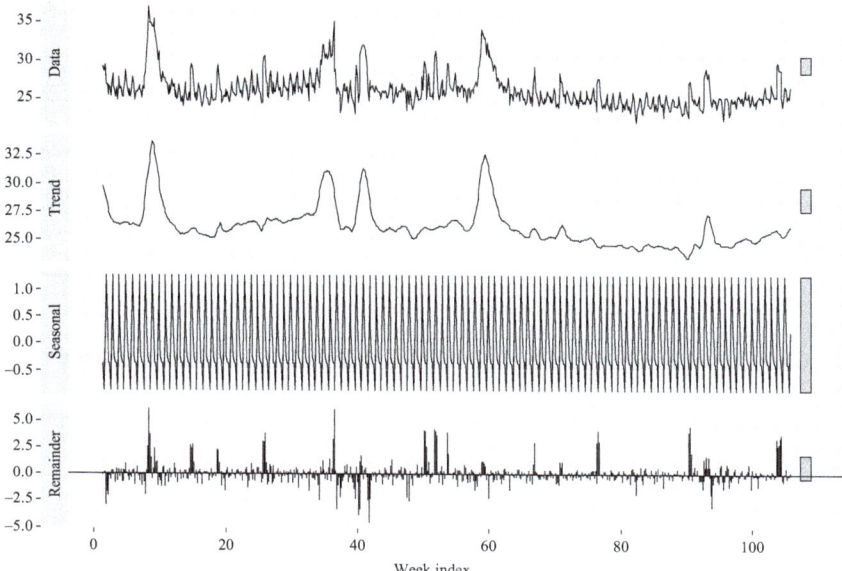

Figure 3.4 The STL decomposition of the NLT speed series

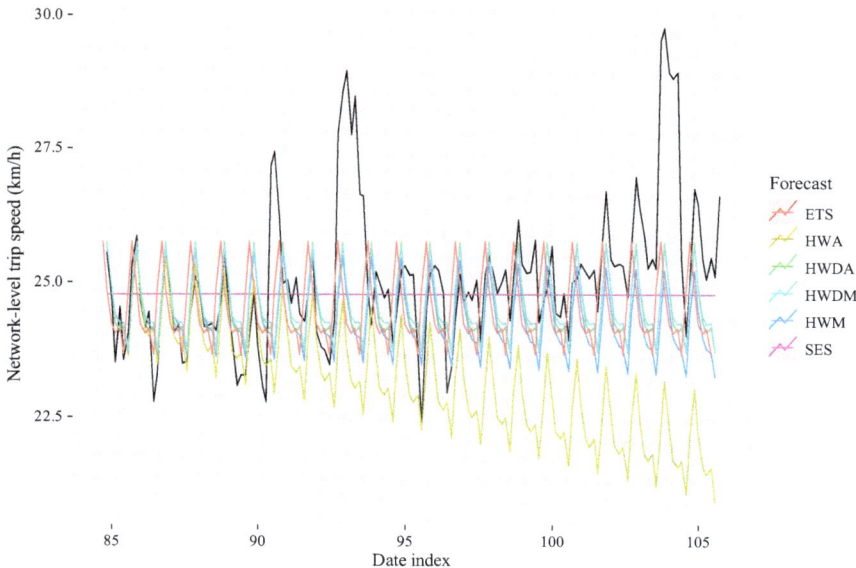

Figure 3.5 The prediction results of the exponential smoothing methods

the worst performance. In addition, the multiplicative models have a slightly better

Table 3.3 The modeling performance of the exponential smoothing methods

Methods	AIC	AICc	BIC	RMSE
Simple exponential smoothing method	4,007.897	4,007.938	4,021.007	1.2729
Holt-Winters' additive method	3,820.208	3,820.755	3,872.647	1.0674
Holt-Winters' multiplicative method	3,792.048	3,792.594	3,844.487	1.075491
Holt-Winters' damped additive method	3,806.717	3,807.356	3,863.526	1.053339
Holt-Winters' damped multiplicative method	3,773.853	3,774.491	3,830.661	1.058246
State-space ETS model	3,759.625	3,760.009	3,803.324	1.052021

performance than the corresponding additive models do, indicating that the multiplicative methods are more suitable for fitting the NLT speed data. In terms of the prediction accuracy, the state-space ETS model, and the HWDA method exhibit the best performance.

3.4.4 The analysis based on ARIMA models

With the aid of the function "auto.arima" in R software [39], it provides an automatic optimization of the model specification. Eventually, the output result is ARIMA (0,1,0) (2,0,0)$_7$ with the AIC, AICc, and BIC of 1,801.64, 1,801.71, and 1,919.11, respectively. According to the Ljung–Box test, the p-value of the model is 0.04235 indicating that there is additional information not mined by the model. So, better models could be found by adjusting the specifications of the model. In addition, a test on its residuals was conducted as shown in Figure 3.6. According to the ACF plot, when the lag order is 3, 9, 14, 27, or 28, the ACF value is beyond the significance level indicating that seasonal effects exist.

In order to find a better ARIMA model, it is necessary to understand the statistical characteristics of the training set of the undifferenced NLT speed series. Figure 3.7 illustrates the series and its ACF and PACF plots, respectively.

The ACF measures the similarity between two observations of a time series but under certain time differences. The partial autocorrelation is the summary of the relationship between the time series after removing the interference factors. Through the ACF and PACF plots of the time series, it is found that the trend in the ACF plot is gradually gentle and most of them are within the confidence region. Except for the first few orders, all of them are in the confidence region and gradually converge to zero with the increase of the order, showing the sign of tailing. However, there are a lot of data in this study, involving a large number of orders; so it is difficult to directly determine the order specifications of the model only through the ACF and PACF plots.

The stationary of the undifferenced series was tested to determine whether it needs differencing. In this chapter, the KPSS unit root test was chosen. The KPSS test is a unilateral test on the right side, with the null hypothesis that the sequence is stationary. If the calculated test statistic is within the critical region, it means that the assumption is correct, i.e., the original sequence is stationary. If it is outside of

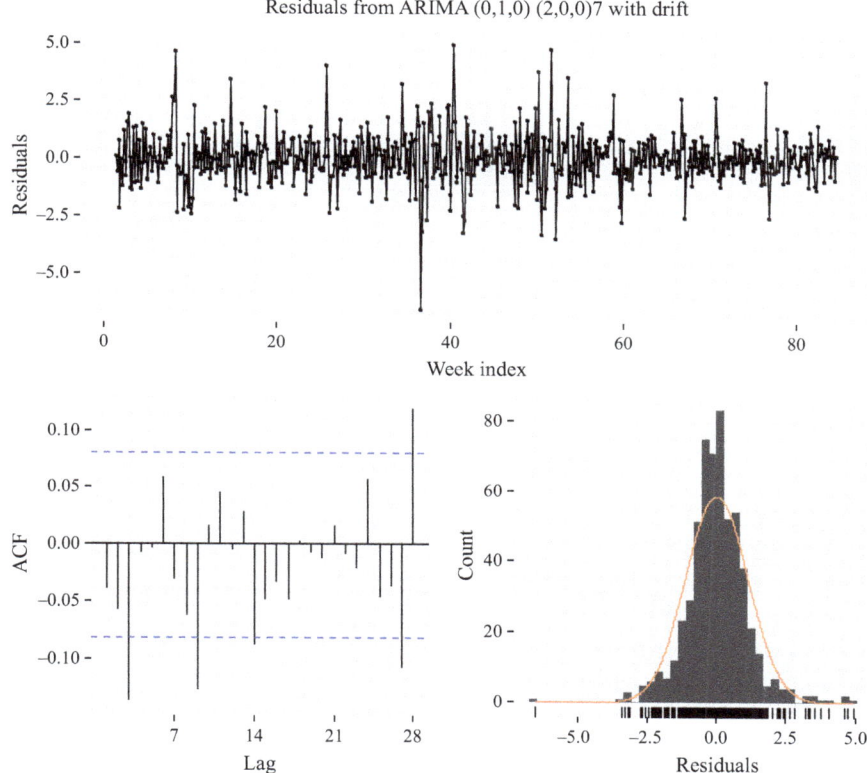

Figure 3.6   Tests on the residuals of the model ARIMA (0,1,0) (2,0,0)$_7$

the critical region, it means that the null hypothesis is unacceptable. In this case, the test statistic is 1.0787, which is larger than the critical value. Therefore, the undifferenced series is nonstationary and needs to be differenced.

After the first-order differencing, the KPSS test statistic is 0.0132 which is still not within the critical region indicating that the differenced series is still nonstationary. The autocorrelation diagram and partial autocorrelation diagram of the differenced series are shown in Figure 3.8. It can be found that the long-term trend after the first-order differencing is in general eliminated, but there is a peak at the seventh seasonal order indicating that a seasonal differencing with a period of 7 days is required. After such a seasonal differencing, the KPSS test statistic is 0.0083 which is within the critical range. Thus, the series after the second differencing is stationary. In the whole process of stabilizing time series, there are two steps of differencing, namely, a regular differencing and a seasonal differencing. Therefore, the *d* parameter of the non-seasonal part of the ARIMA model is one and the *d* parameter of the seasonal part is one. The established model is ARIMA (*p*, 1, *q*) (*P*, 1, *Q*)$_7$ seasonal multiplicative model. Lying on this, other model settings could be further determined.

48 *Traffic information and control*

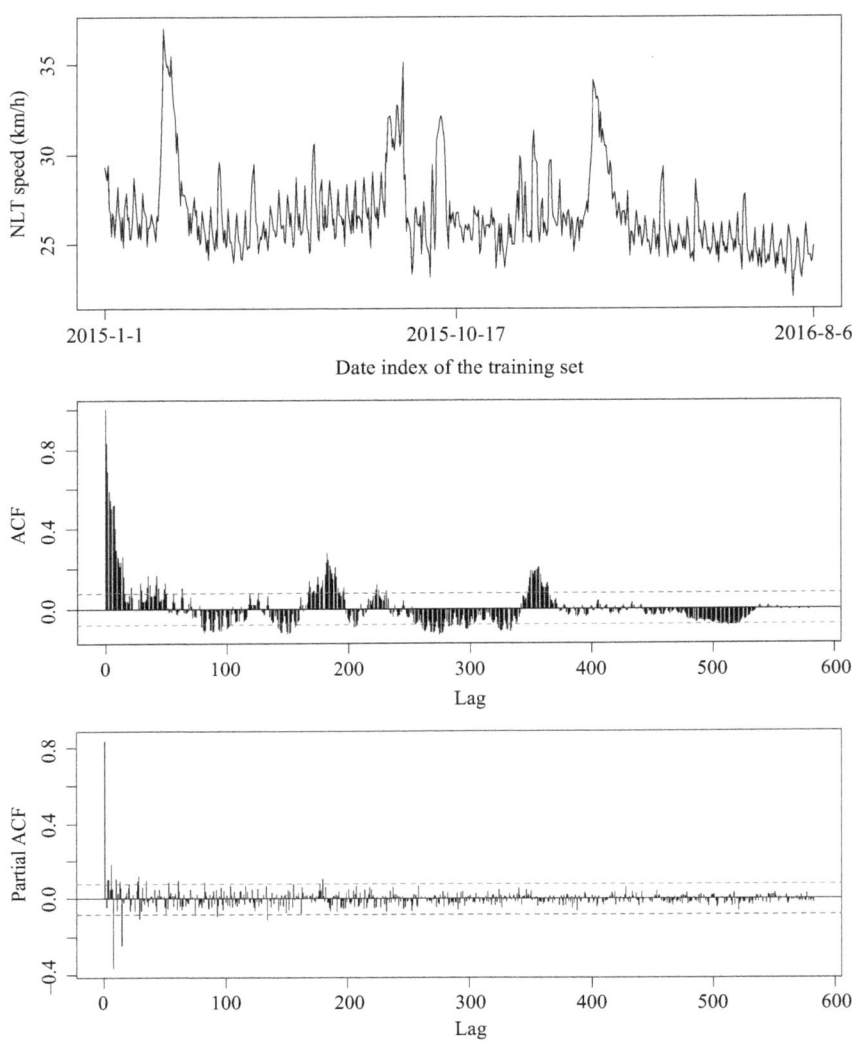

Figure 3.7   The illustration of the training set and its ACF and PACF plots

Figure 3.9 shows the ACF and PACF plots of the NLT speed time series after the seasonal differencing. It can be found from the ACF diagram that there are peaks at the third and seventh orders of the lag, and the ACF values are in the significant range after the seventh order of the lag. Therefore, the value of the q parameter should be 3 or 7. It can be found from the PACF that there are significant peaks at the third and seventh order of the lag, so the value of the p parameter should be 3 or 7. For the values of P and Q, see the number of spikes when the lag order is 7, 14, 21, and 28. There are two spikes in the ACF diagram, so

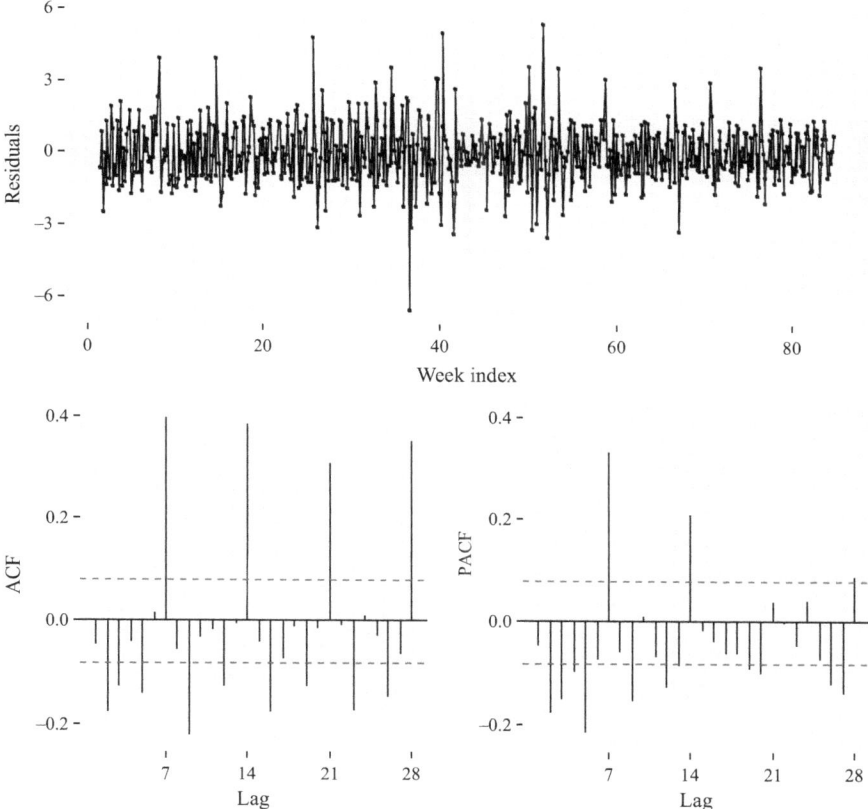

Figure 3.8 The illustration of the residuals and its ACF and PACF plots after the first-order differencing

Q is 3. There are four spikes in the PACF diagram, so P is 4. According to the previous analysis, all possible ARIMA models are listed here, i.e., ARIMA (3,1,7) (4,1,3)$_7$, ARIMA (3,1,3) (4,1,3)$_7$, ARIMA (7,1,3) (4,1,3)$_7$, ARIMA (7,1,7) (4,1,3)$_7$, ARIMA (3,1,0) (4,1,3)$_7$, and ARIMA (0,1,3) (4,1,3)$_7$. Establishing the previous models, respectively, the modeling performance criteria are shown in Table 3.4.

Among the earlier candidate models, the model ARIMA (3,1,7) (4,1,3)$_7$ is preferable in favor of the AIC and AICc values, while the model ARIMA (0,1,3) (4,1,3)$_7$ is preferable in favor of the BIC value. The following analyses will take the former model as an example. The estimated parameters of the model are shown in Table 3.5. Figure 3.10 illustrates the prediction results of the model ARIMA (3,1,7) (4,1,3)$_7$ and Figure 3.11 illustrates the properties of modeling fit.

Figure 3.12 is the standardized residual diagram of the model ARIMA (3,1,7) (4,1,3)$_7$ as well as the ACF and PACF plots of the residual. For all lag orders, the ACF values have passed the test and are all exhibited white-noise features, indicating that

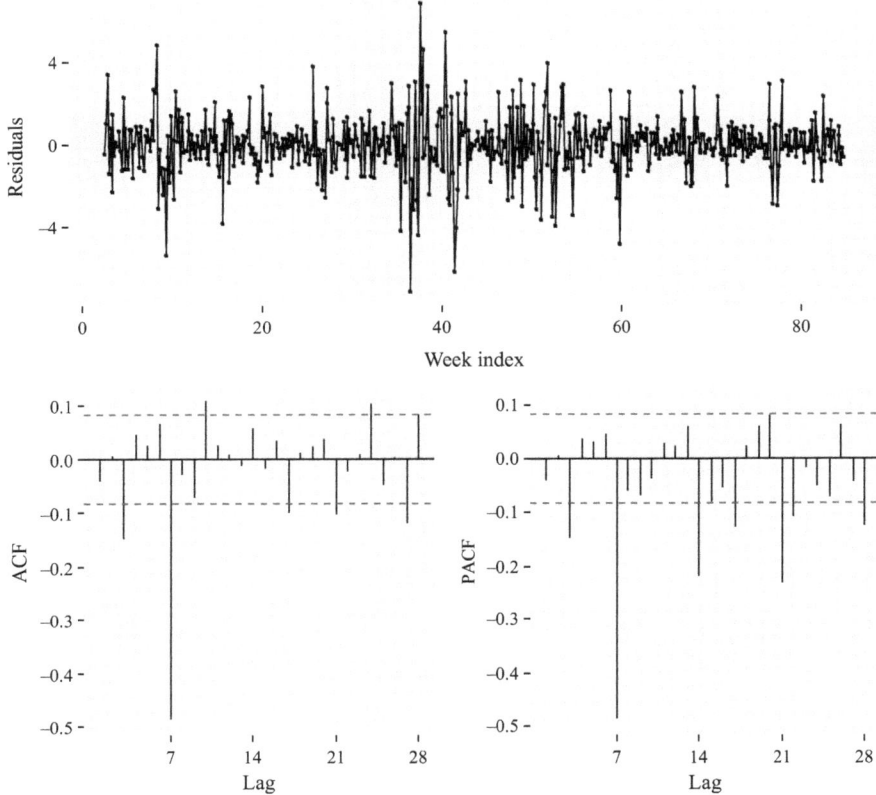

Figure 3.9 The illustration of the residuals and its ACF and PACF plots after the seasonal differencing

the residual does not have features of autocorrelation. Based on the distribution of residuals, most of them are concentrated in the range of −0.5 to 0.5. The Ljung–Box test shows that the *p*-value is 0.1155, which is greater than 0.05 indicating that the residual series is white noise.

3.4.5 The analysis based on a hybrid ARIMA–SVM Model

The ARIMA model has advantages in dealing with linear problems, while the SVM model has advantages in dealing with nonlinear problems [40–43]. Therefore, a combination of the two approaches could have substantial improvements in terms of fitting to the NLT speed time series. The combined model is called the hybrid ARIMA–SVM model in this chapter. Assuming that the original time series is Y_t, it can be divided into a linear autocorrelation part A_t and a nonlinear residual B_t, i.e., $Y_t = A_t + B_t$. First, the ARIMA model is used to predict the original series. The

Table 3.4 The modeling performance criteria for each ARIMA model

ARIMA model	AIC	AICc	BIC
$(3,1,7)\ (4,1,3)_7$	1,699.26	1,700.49	1,777.67
$(3,1,3)\ (4,1,3)_7$	1,721.05	1,721.8	1,782.04
$(7,1,3)\ (4,1,3)_7$	1,705.89	1,707.12	1,784.3
$(7,1,7)\ (4,1,3)_7$	1,706.65	1,708.48	1,802.49
$(3,1,0)\ (4,1,3)_7$	1,721.74	1,722.21	1,769.66
$(0,1,3)\ (4,1,3)_7$	1,720.79	1,721.26	1,768.71
auto.arima	1,801.64	1,801.71	1,819.11

Table 3.5 The estimated coefficients for the model ARIMA $(3,1,7)\ (4,1,3)_7$

Parameter	Coefficient	Parameter	Coefficients
AR1	0.2252	MA6	−0.0668
AR2	0.1648	MA7	−0.3926
AR3	−0.1966	SAR1	−0.9487
MA1	−0.3785	SAR2	−0.0813
MA2	−0.2081	SAR3	0.4211
MA3	0.0661	SAR4	0.0498
MA4	−0.0223	SMA1	0.4270
MA5	0.0023	SMA2	−0.6101
		SMA3	−0.8169

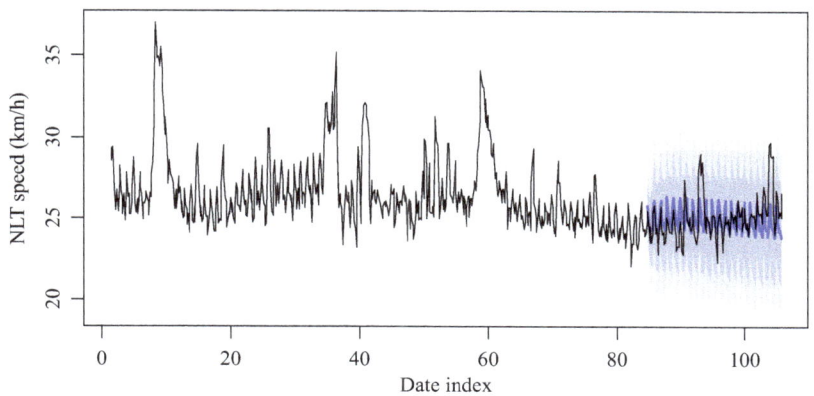

Figure 3.10 The prediction results of the seasonal ARIMA model

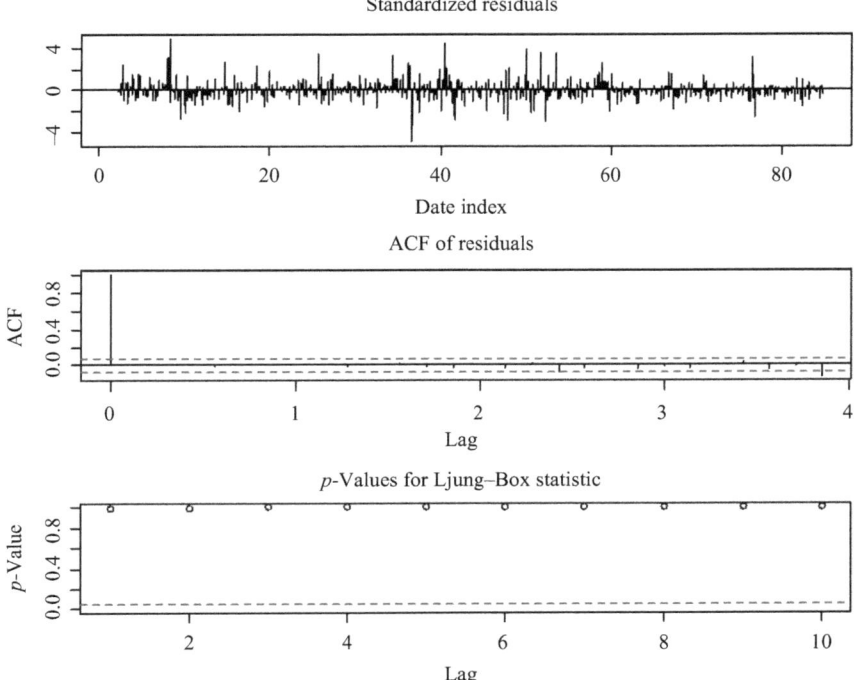

Figure 3.11 The property of modeling fit of ARIMA (3,1,7) (4,1,3)$_7$

predicted value is A'_t, and the residual is B_t; then, the residual B_t is used as the input of the SVM model which outputs the predicted residual B'_t; finally, the linear-part prediction A'_t and the nonlinear-part prediction B'_t are combined to get the final ARIMA–SVM prediction Y'_t, i.e., $Y'_t = A'_t + B'_t$.

In the part of the ARIMA model, the seasonal ARIMA (3,1,7) (4,1,3)$_7$ model was chosen. In the part of the SVM model, the selection range of the γ parameter was set between 0.0001 and 0.1, and the selection range of cost value was set between 10 and 10,000. After searching for the combinations, the optimal settings of γ and cost of the SVM model were obtained. Under the condition of $\gamma = 0.001$ and cost $= 1,000$, the model error rate reaches the lowest, i.e., 0.4220. Table 3.6 reports the settings and modeling performance of those SVM models.

After determining the model settings, the residual of the ARIMA model is predicted. In the SVM model, the residual serves as the response variable, weekend, long vacation, OEPN travel restriction, temperature, AQI, and vehicle ownership serve as independent variables. Finally, the predicted residual and the predicted value from the seasonal ARIMA model were combined as shown in Figure 3.13.

It can be found that the prediction of the hybrid ARIMA–SVM model is closer to the original time series than the prediction of the seasonal ARIMA model because

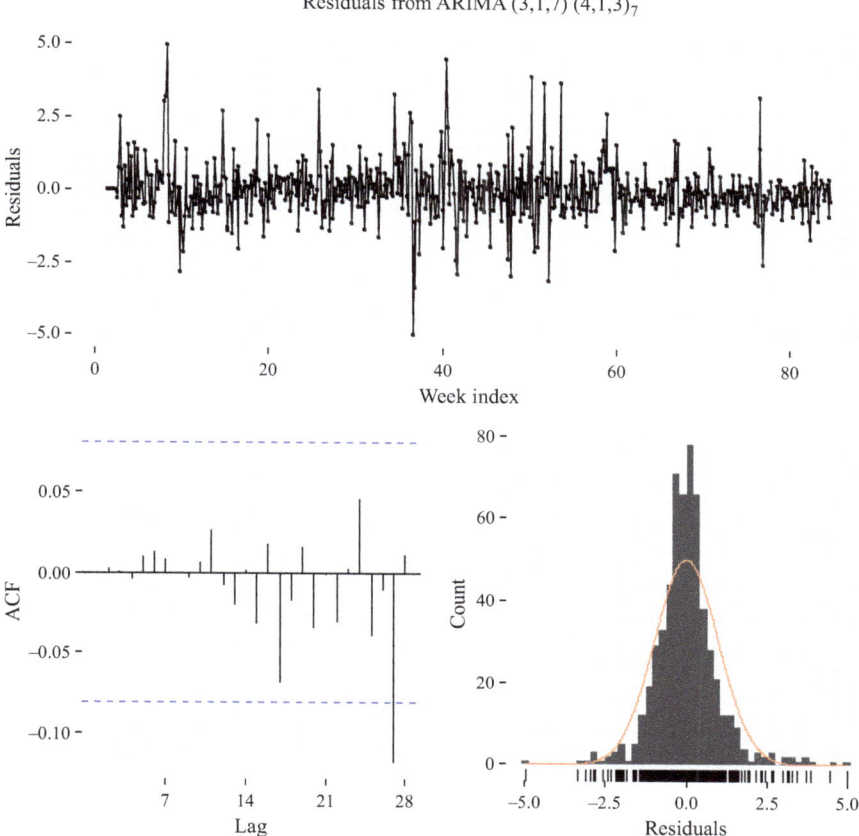

Figure 3.12 The test on the residual of the model ARIMA (3,1,7) (4,1,3)$_7$

the hybrid model not only considers the seasonality effect but also adds nonlinear features and the influence of exogenous factors. Especially, for the long-term pattern, the precision of the hybrid ARIMA–SVM model is substantially higher than that of the seasonal ARIMA model.

3.5 Conclusions

This chapter proposed a macroscopic traffic performance indicator, i.e., the NLT speed seeking to comprehensively assess the degree of traffic congestion of a city or region. Based on the data within the sixth ring road of Beijing, China between 2015 and 2016, several influential factors of the NLT speed were investigated and the factor with the largest influence is long vacation. Moreover, weekend, OEPN travel restriction, and temperature are also highly correlated with the NLT speed.

54 Traffic information and control

Table 3.6 The settings and modeling performance of the SVM models

No.	γ	Cost	Error	Dispersion
1	0.0001	10	1.428	0.922
2	0.001	10	0.547	0.269
3	0.01	10	0.441	0.156
4	0.1	10	0.430	0.169
5	0.0001	100	0.554	0.265
6	0.001	100	0.486	0.191
7	0.01	100	0.453	0.120
8	0.1	100	0.670	0.370
9	0.0001	1,000	0.510	0.199
10	0.001	1,000	0.422	0.129
11	0.01	10,00	0.571	0.280
12	0.1	1,000	0.795	0.588
13	0.0001	10,000	0.471	0.166
14	0.001	10,000	0.450	0.106
15	0.01	10,000	1.266	1.662
16	0.1	10,000	1.055	0.612

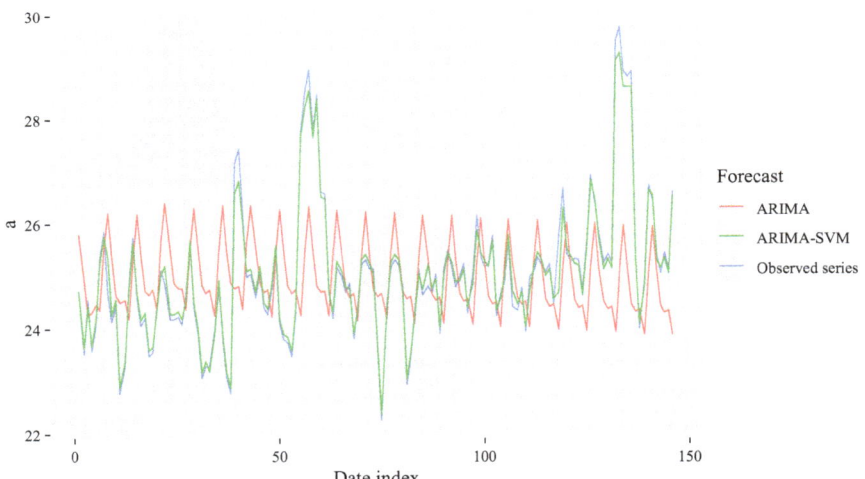

Figure 3.13 The comparison between the hybrid ARIMA–SVM and the seasonal ARIMA models in terms of the predicted NLT speed

Then the NLT speed time series was analyzed using several exponential smoothing and ARIMA models. For the exponential smoothing methods, the optimal model is the state-space ETS model, but the prediction of the exponential smoothing method is not stable in such a complicated condition. In contrast, a seasonal ARIMA model was chosen as it can provide better performance. A hybrid ARIMA–SVM model was also adopted to improve the seasonal ARIMA in terms of prediction accuracy.

There are several avenues for future studies. First, the time span of this study is from January 1, 2015 to December 31, 2016, which is limited in understanding the evolvement of urban traffic conditions for a broader sense. Additional data before or after this study period could be incorporated to perform more comprehensive analyses. Second, several influential factors were adopted in this study which enhanced the prediction ability of the model. However, there are not only those factors that affect the NLT speed but also some important factors such as macroscopic socioeconomic characteristics that were not considered. In the future, with the enrichment of exogenous factors, the model could exhibit a better performance of prediction. Third, in the process of modeling, the impact of various behaviors had not been fully considered. For example, the long vacation of the spring festival has 7 consecutive days. In fact, after the end of this vacation, the traffic status usually does not resume back to the normal level. Such social phenomenon should be considered in the modeling process, which would provide potential references for designating urban traffic policies.

References

[1] de Oliveira E. L., da Silva Portugal L., and Porto Junior W. 'Determining Critical Links in a Road Network: Vulnerability and Congestion Indicators'. *Procedia – Social and Behavioral Sciences*. 2014;162:158–167.

[2] Eisele W. L., Zhang Y., Park E. S., Zhang Y., and Stensrud R. 'Developing and Applying Models for Estimating Arterial Corridor Travel Time Index for Transportation Planning in Small to Medium-Sized Communities'. *Transportation Research Record*. 2011;2244:81–90.

[3] Bremmer D., Cotton K. C., Cotey D., Prestrud C., and Westby G. 'Measuring Congestion: Learning From Operational Data'. *Transportation Research Record*. 2004;1895:188–196.

[4] Wang Z., Que L., Liu J., and Guan J. 'Study on the evaluation system and prediction approach on traffic congestion index of RTMS data'. *In Proc. the Seventh International Conference on Traffic and Transportation Studies*. 2010, pp. 903–914.

[5] Bok G. C., Lee S. J., Choe Y. H., Gang J. G., and Lee S. H. 'Development of a Traffic Condition Index (TCI) on Expressways'. *Journal of the Eastern Asia Society for Transportation Studies*. 2009;27(5):85–95.

[6] Thurgood G. S. 'Development of a Freeway Congestion Index Using an Instrumented Vehicle'. *Transportation Research Record*. 1995;1494: 21–29.

[7] Zuo Y., and He Q. 'Intersection approach road congestion index and application for Beijing, China'. *In Proc. the 2nd IEEE International Conference on Information Management and Engineering*. 2010, pp. 263–266.

[8] Taylor J. W. 'Short-Term Electricity Demand Forecasting Using Double Seasonal Exponential Smoothing'. *Journal of the Operational Research Society*. 2003;54(8):799–805.

[9] Kim M. S. 'Modeling Special-Day Effects for Forecasting Intraday Electricity Demand'. *European Journal of Operational Research*. 2013;230(1):170–180.

[10] Sampathirao A. K., Grosso J. M., Sopasakis P., Ocampo-Martinez C., Bemporad A., and Puig V. 'Water Demand Forecasting for the Optimal Operation of Large-Scale Drinking Water Networks: The Barcelona Case Study'. *IFAC Proceedings Volumes*. 2014;47(3):10457–10462.

[11] Grosso J. M., Ocampo-Martínez C., Puig V., and Joseph B. 'Chance-Constrained Model Predictive Control for Drinking Water Networks'. *Journal of Process Control*. 2014;24(5):504–516.

[12] Dong Z., Yang D., Reindl T., and Walsh W. M. 'Short-Term Solar Irradiance Forecasting Using Exponential Smoothing State Space Model'. *Energy*. 2013;55:1104–1113.

[13] Schrank D., Eisele B., and Lomax T. 'TTI's 2012 urban mobility report'. *Texas A&M Transportation Institute*. Dec. 2012, TX, USA, Tech. Rep.

[14] Cohn N. 'TomTom traffic index: Toward a global measure'. *In Proceedings of ITS*. 2014.

[15] Sun J., Wen H., Gao Y., and Hu Z. 'Metropolitan Congestion Performance Measures Based on Mass Floating Car Data'. *Computational Sciences and Optimization*. 2009;2:109–113.

[16] Wen H., Hu Z., and Sun J. 'Road traffic performance monitoring in Beijing'. *Presented at the 6th Advanced Forum on Transportation of China*. 2010, Beijing, China.

[17] Geng J., Long R., Chen H., and Li Q. "Urban Residents" Response to and Evaluation of Low-Carbon Travel Policies: Evidence from a Survey of Five Eastern Cities in China'. *Journal of Environmental Management*. 2018;217(1):47–55.

[18] Liu Z., Li R., Wang X., and Shang P. 'Effects of Vehicle Restriction Policies: Analysis Using License Plate Recognition Data in Langfang, China'. *Transportation Research Part A-Policy and Practice*. 2018;118:89–103.

[19] Ghosh B., Basu B., and O'Mahony M. 'Multivariate Short-Term Traffic Flow Forecasting Using Time-Series Analysis'. *IEEE Transactions on Intelligent Transportation Systems*. 2009;10(2):246–254.

[20] Samaneh A. and Diane J. C. 'A Survey of Methods for Time Series Change Point Detection'. *Knowledge and Information Systems*. 2017;51(2):339–367.

[21] Karim F., Majumdar S., Darabi H., and Chen S. 'LSTM Fully Convolutional Networks for Time Series Classification'. *IEEE Access*. 2018;6:1662–1669.

[22] Cleveland R. B. 'STL: A Seasonal-Trend Decomposition Procedure Based on Loess'. *Journal of Office Statistics*. 1990;6(1):3–73.

[23] Sanchez-Vazquez M. J., Mirjam N., George J. G., and Fraser I. L. 'Using Seasonal-Trend Decomposition Based on Loess (STL) to Explore Temporal Patterns of Pneumonic Lesions in Finishing Pigs Slaughtered in England, 2005-2011'. *Preventive Veterinary Medicine*. 2012;104(1):65–73.

[24] Jesús R., Rosario R., Jorge R., Federico F., and Rosa P. 'Modeling Pollen Time Series Using Seasonal-Trend Decomposition Procedure Based on LOESS Smoothing'. *International Journal of Biometeorology*. 2017;61(2):335–348.

[25] Brown R. G. 'Statistical forecasting for inventory control'. *Journal of the Royal Statistical Society: Series A (General)*. 1960;123(3):348–349

[26] Holt C. E. *Forecasting Seasonals and Trends by Exponentially Weighted Averages*. 2004, Carnegie Institute of Technology, Pittsburgh, USA.

[27] Winters P. R. 'Forecasting Sales by Exponentially Weighted Moving Averages'. *Management Science*. 1960;6:324–342.

[28] Billah B., King M. L., Snyder R. D., and Koehler A. B. 'Exponential Smoothing Model Selection for Forecasting'. *International Journal of Forecasting*. 2006;22(2):239–247.

[29] Koehler A. B., Snyder R. D., Ord J. K., and Beaumont A. 'A Study of Outliers in the Exponential Smoothing Approach to Forecasting'. *International Journal of Forecasting*. 2012;28(2):477–484.

[30] Abdullah M., Zaharim A., Zain A. F. M., Sabirin A., Bahari S. A., and Habib S. N. A. A. 'Forecasting of Ionospheric Delay Obtained From GPS Observations Using Holt-Winter Method'. *European Journal of Scientific Research*. 2009;37(3):471–480.

[31] Odame O., Atinuke A., and Orelwoapo L. 'Using Holt Winter's Multiplicative Model to Forecast Assisted Childbirths at the Teaching Hospital in Ashanti Region, Ghana'. *Journal of Biology, Agriculture and Healthcare*. 2014;4(9):83–88.

[32] Rob H. and George A. *Forecasting: Principles and Practice*. 2014, Bowker Saur Pharo, London.

[33] Williams B. M. 'Multivariate Vehicular Traffic Flow Prediction: Evaluation of ARIMAX Modeling'. *Transportation Research Record*. 2001;1776:194–200.

[34] Lippi M., Bertini M., and Frasconi P. 'Short-Term Traffic Flow Forecasting: An Experimental Comparison of Time-Series Analysis and Supervised Learning'. *IEEE Transactions on Intelligent Transportation Systems*. 2013;14(2):871–882.

[35] Holens G. A. 'Forecasting and selling futures using ARIMA models and a neural network'. *ProQuest Dissertations & Theses Global A&I: The Sciences and Engineering Collection*. 1997.

[36] Alex J. S. and Bernhard S. 'A Tutorial on Support Vector Regression'. *Statistics and Computing*. 2004;14(3):199–222.

[37] Christopher J. C. B. 'A Tutorial on Support Vector Machines for Pattern Recognition'. *Data Mining and Knowledge Discovery*. 1998;2(2):121–167.

[38] Chang C. and Lin C. 'LIBSVM: A Library for Support Vector Machines'. *ACM Transactions on Intelligent Systems and Technology*. 2011;2(3):27.

[39] Hyndman R. J. and Yeasmin K. 'Automatic Time Series Forecasting: The Forecast Package for R'. *Journal of Statistical Software*. 2008;27(3):1–22.

[40] Pai P. and Lin C. 'A Hybrid ARIMA and Support Vector Machines Model in Stock Price Forecasting'. *Omega-International Journal of Management Science*. 2005;33(6):497–505.

[41] Nie H., Liu G., Liu X., and Wang Y. 'Hybrid of ARIMA and SVMs for Short-Term Load Forecasting'. *Energy Procedia*. 2012;16:1455–1460.
[42] Wang Y., Li L., and Xu X. 'A piecewise hybrid of ARIMA and SVMs for short-term traffic flow prediction'. *In International Conference on Neural Information Processing*. 2017, pp. 493–502.
[43] He Y., Zhu Y., and Duan D. 'Research on hybrid ARIMA and support vector machine model in short term load forecasting'. *In Sixth International Conference on Intelligent Systems Design and Applications*. 2006;1, pp. 804–809.

Chapter 4

Short-term travel-time prediction by deep learning: a comparison of different LSTM-DNN models

Yangdong Liu[1], Lan Wu[2], Jia Hu[1], Xiaoguang Yang[1] and Kunpeng Zhang[2]

Predicting short-term travel time with considerable accuracy and reliability is critically important for advanced traffic management and route planning in Intelligent Transportation Systems (ITS). Short-term travel-time prediction uses real travel-time values within a sliding time window to predict travel time one or several time step(s) in future. However, the nonstationary properties and abrupt changes of travel-time series make challenges in obtaining accurate and reliable predictions. Recent achievements of deep learning approaches in classification and regression shed a light on innovations of time series prediction. This study establishes a series of Long Short-Term Memory with Deep Neural Networks (LSTM-DNN) layers using 16 settings of hyperparameters and investigates their performance on a 90-day travel-time dataset from Caltrans Performance Measurement System (PeMS). Then competitive LSTM-DNN models are tested along with linear regression, Ridge and Lasso regression, ARIMA and DNN models under ten sets of sliding windows and predicting horizons via the same dataset. The results demonstrate the advantage of LSTM-DNN models while showing different characteristics of these deep learning models with different settings of hyperparameters, providing insights for optimizing the structures.

4.1 Introduction

The short-term prediction of traffic flow parameters, including flows, speeds, densities and travel time, composes an important major part in the development and application of ITS, forming the foundation of Advanced Traffic Management Systems and providing essential supports for travel guidance and route planning of travelers. A time series of traffic flows or travel time generally has similarity among its inner subseries

[1]The Key Laboratory of Road and Traffic Engineering, Ministry of Education, Tongji University, Shanghai, China
[2]College of Electrical Engineering, Henan University of Technology, Zhengzhou, China

corresponding to the same weekdays (e.g., Monday in last week to Monday in next week), weekends and same quarters. Despite the periodicity and seasonality, however, these time series are always being affected by various external factors, including traffic crashes, fluctuation in traffic demand and abrupt changes of weather conditions, which results in nonstationary properties and brings difficulties to the prediction.

The short-term travel-time prediction refers to predicting travel time of a certain road section or a route for a short period of time, from several minutes to hours. Consider a time series of travel time $X = \{x_1, x_2, \ldots, x_T\}$, short-term prediction aims to predict $X_m = \{x_{T+1}, \ldots, x_{T+m}\}$ in next time step(s) $\{T + 1, \ldots, T + m\}$, while $m \geq 1$. Here m, representing the number of time steps to predict, is generally addressed as forecasting horizon or prediction horizon. Prediction horizon can contain one time step or several time steps of prediction. One-step prediction is mostly adapted in short-term time series prediction because the accuracy is relatively more possible to be guaranteed. In general, the more time steps for prediction, the more possible the accuracy to decrease. Full part or a subpart of travel-time series X is used to do prediction. Online or real-time prediction can then be achieved when a sliding time window of a certain length moves along the time axis and processes a subpart or a section of a time series at each time step. The short-term travel-time prediction in this study is real time.

4.2 Traffic time series estimation with deep learning

To now, an amount of different kinds of approaches have been studied and applied to short-term traffic flow parameters prediction. These methods can generally be separated into two groups: 1) statistical methods, and 2) machine learning methods. In statistical methods, autoregressive integrated moving average (ARIMA) model is mostly common used for doing traffic flow parameters prediction. [1] and [2] used ARIMA models for freeway travel time prediction. Several variants of ARIMA models have been proposed since the first attempt of using ARIMA for freeway travel time prediction in [1]. Reference [3] proposes H-ARIMA+ model to incorporate more patterns of data within peak hour and spatial influence from traffic events when considering the possible failure in predicting during peak hours and in effect of events as the traffic flow parameters values are only treated as generic time series. In [4], A spatial-temporal ARIMA (STARIMA) model is addressed for integrating spatial and temporal information to predict traffic flow. Reference [5] presents an Bayesian network approach for traffic flow prediction. A Bayesian dynamic linear model is proposed for dynamic short-term travel time prediction in [6]. In [7], a local online kernel method for Ridge regression is developed for predicting urban travel times. For machine learning methods, the most popular are artificial neural networks (ANN) and support vector regression (SVR), a regression method based on support vector machine (SVM). An SVR based travel time prediction method is proposed in [8] and ANN based methods can be seen in [9] and [10]. Some approaches also incorporate statistical and machine learning methods and traffic flow theories to conduct modeling and predicting of travel time data. For instance, Reference [11] utilize bottleneck identification algorithm from traffic flow theories and incorporate clustering, stochastic congestion maps, online congestion searching methods to conduct prediction of freeway travel time using both historical and real-time data.

On the other hand, deep learning methods as a branch of machine learning studies have recently drawn massive attentions from worldwide researchers separated in various fields, since a big breakthrough for training deep neural networks (DNN) was proposed in [12]. In general, the existing deep learning methods can be classified into four categories: Recurrent Neural Networks (RNN), Convolutional Neural Networks (CNN), and Generative Adversarial Networks (GAN).

4.2.1 Recurrent neural network

Deep learning methods form a certain amount of state-of-the-art solutions to many classification and regression problems in several different subjects, in which RNN is now often being used for modeling sequential data, such as time series data and video or audio data, which is specialized for processing time series data by introducing memory to retain information. The modeling h in a fully connected layer can be formulized as

$$h = f(x_t) \tag{4.1}$$

where x_t is the input of the fully connected layer at time t.

The calculation procedure of RNN can be formulized as

$$a_t = W_{a1}h_{t-1} + W_{a2}x_t + b_a \tag{4.2}$$

$$h_t = \tanh(a_t) \tag{4.3}$$

$$o_t = W_o h_t + b_o \tag{4.4}$$

$$\widehat{y}_t = \sigma(o_t) \tag{4.5}$$

where W_* denotes the weight matrix, b_* indicates the bias, σ is the sigmoid function and x_t is the input of an RNN layer at time t. The internal structures of a fully connected layer and RNN are shown in Figure 4.1.

However, RNN is difficult to train due to the exploding and the vanishing gradient problems. In this condition, sophisticated recurrent hidden units, such as

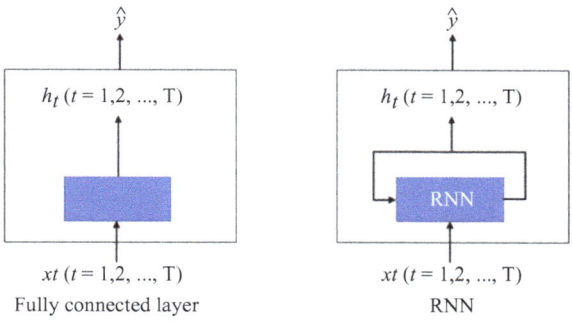

Figure 4.1 Internal structures of a fully connected layer and RNN

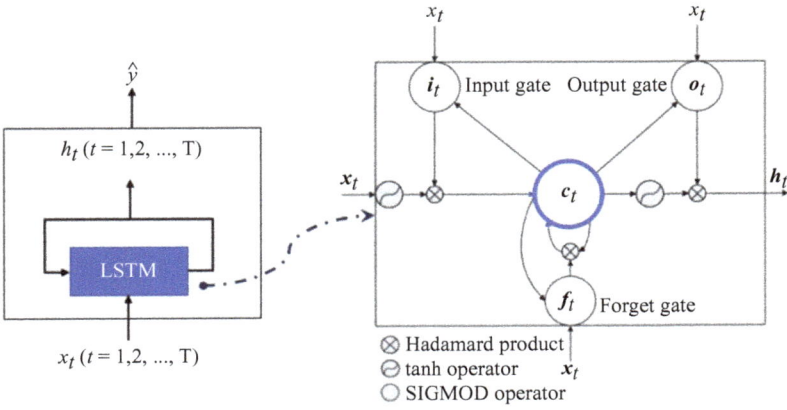

Figure 4.2   Internal structure of an LSTM unit

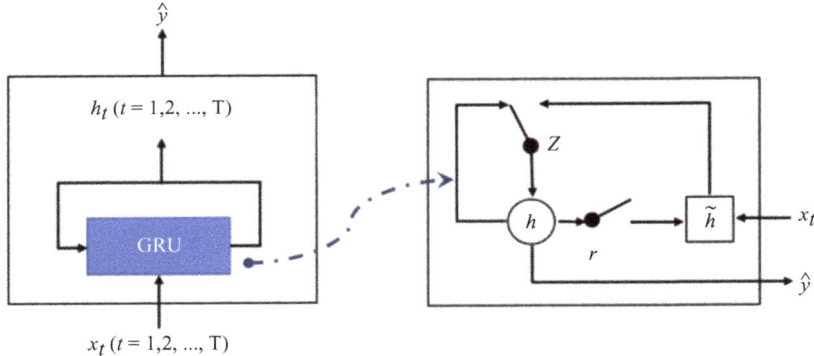

Figure 4.3   Internal structure of a GRU

LSTM and Gated Recurrent Unit (GRU), are proposed to address the difficulty of training RNN. The calculation procedure of LSTM [13] can be formulized as

$$i_t = \sigma(W_{xi}x_t + W_{hi}h_{t-1} + W_{ci} \odot c_{t-1} + b_i) \quad (4.6)$$

$$f_t = \sigma(W_{xf}x_t + W_{hf}h_{t-1} + W_{cf} \odot c_{t-1} + b_f) \quad (4.7)$$

$$c_t = f_t \odot c_{t-1xf} + i_t \odot \tanh(W_{xc}x_t + W_{hc}h_{t-1} + b_c) \quad (4.8)$$

$$o_t = \sigma(W_{xo}x_t + W_{ho}h_{t-1} + W_{co} \odot c_t + b_o) \quad (4.9)$$

$$h_t = o_t \odot \tanh(c_t) \quad (4.10)$$

where W_* denotes the weight matrix, b_* indicates the bias, σ is the sigmoid function and x_t is the input of an RNN layer at time t. The operator $\odot$ denotes the element-wise vector product. The internal structure of an LSTM unit is shown in Figure 4.2.

The calculation procedure of GRU can be formulized as

$$r_t = \sigma(W_{xr}x_t + W_{hr}h_{t-1} + b_r) \tag{4.11}$$

$$z_t = \sigma(W_{xz}x_t + W_{hz}h_{t-1} + b_z) \tag{4.12}$$

$$\tilde{h}_t = \tanh(W_{xh}x_t + W_{hh}(r_t \odot h_{t-1}) + b_h) \tag{4.13}$$

$$h_t = z_t \odot h_{t-1} + (1 - z_t) \odot \tilde{h}_t \tag{4.14}$$

where W_* denotes the weight matrix and b_* indicates the bias. The operator $\odot$ denotes the element-wise vector product. σ is the sigmoid function and x_t is the input of a GRU layer at time t. The output of each GRU layer is the hidden state at each time step. The internal structure of a GRU is shown in Figure 4.3.

LSTM and GRUs are now widely adapted as alternatives of simple RNNs to model or learn complicated information contained within sequential data. A number of studies on time series and other sequential data that applied deep learning approaches indicated the advantage property of LSTM or GRU neural networks over other deep learning methods. In the transportation research area, the early attempt of using deep learning methods is [14] that presents a Deep Belief Networks (DBN) method to predict freeway traffic flow. Reference [15] proposes a Stacked Autoencoder model for freeway traffic flow prediction. In [16], LSTM networks are used for the first time for traffic flow parameters prediction and are successively be experimented in several studies such as [17]. References [18] and [19], respectively, proposes a DBN model and a DNN model for traffic speed and traffic flow prediction. Reference [20] develops an error-Feedback Recurrent Convolutional Neural Networks structure, incorporates spatial and temporal relationship to conduct traffic speed prediction for ring roads in Beijing, China. Reference [21] presented a GRU-based multitask learning model, MTL-GRU, to predict the network-wide short-term traffic speed with the consideration of spatiotemporal correlations at a network-wide level.

4.2.2 Convolutional neural networks

Although sequential data modeling is synonymous with RNN, many researches claim that CNN-based models can achieve a comparable performance on a variety of tasks such as audio synthesis, word-level language modeling and machine translation [22–24]. In practice, CNN is often utilized over a multidimensional problem. For example, a two-dimensional (2D) image I as input, the calculation procedure of CNN, can be formulated as

$$S(i,j) = (I * K)(i,j) = \sum_m \sum_n I(m,n)K(i-m,j-n) \tag{4.15}$$

where $*$ is the convolution operation, K is a 2D kernel.

Since convolution is commutative, the earlier equation can be equivalently written as

$$S(i,j) = (K * I)(i,j) = \sum_m \sum_n I(i-m,j-n)K(m,n) \tag{4.16}$$

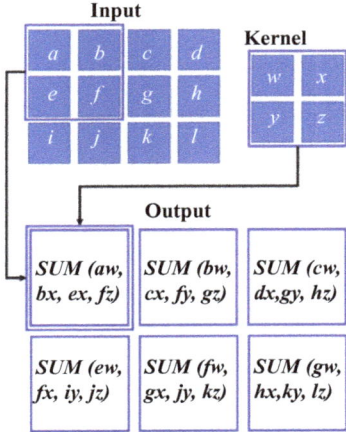

Figure 4.4 An example of 2D convolution

Usually, the latter formula is more straightforward to implement, because there is less variation in the range of valid values of m and n. Figure 4.4 for an example of convolution is applied to a 2D tensor.

Compared with RNN, CNN extracts data information for fixed size contexts, which makes CNN less common for time series modeling due to the lack of memory for a long sequence [25]. However, recent work manages to utilize CNN for sequence modeling by preprocessing sequential data or modifying CNN structures [23]. In the transportation research area, as for the former practice, [26] converted traffic speed dynamics to images, based on which CNN was employed to predict the network-wide traffic speed in Beijing, China. Based on images converted from the citywide crowd flow data, [27,28] utilized a residual CNN to predict the citywide crowd flow with the consideration of spatial–temporal correlations at a network level. As for the latter one, [29] proposed temporal CNN to predict passenger demand. To consider network-wide spatial–temporal correlations, a multitasking learning framework was introduced to capture passenger demand dynamics with trajectory data from Chengdu, China and New York City, respectively.

In general, RNN demonstrates its superiority to model sequential data and capture their temporal correlations, while CNN has been proved to be effective to capture spatial correlations [31]. Inspired by the desirable performance of RNN and CNN in traffic state prediction, many researches tried to combine RNN and CNN in a ConvRNN-based framework to carry out the prediction. Typically, in a ConvRNN model, CNN and RNN layers are employed to capture spatial correlations and temporal correlations in traffic data, respectively. For example, [32,33] employed Convolutional LSTM (ConvLSTM) to capture spatial and temporal relations for the short-term passenger demand prediction

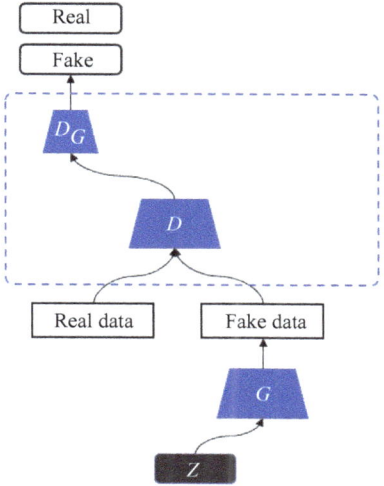

Figure 4.5 Internal structure of a GRU

with trajectory data from Didi Chuxing. Reference [34] proposed an end-to-end deep learning architecture that consists of CNN and LSTM. The proposed ConvLSTM model was capable of capturing the spatial–temporal correlations when predicting short-term traffic flow.

4.2.3 Generative adversarial networks

Most recently, the deep-learning-based GAN [34] draws increasing attention in the scientific community. GAN is a promising approach in the field of image generation, because it can excellently model the complicated probability distribution of a dataset and then generate samples accordingly. As shown in Figure 4.5, a general structure of GAN consists of two models that are alternatively trained to compete with each other. The generator G is optimized to represent the true data distribution p_{real}, by generating samples that are difficult for the discriminator D to differentiate from real samples. Overall, the training procedure is similar to a two-player min–max game with the objective function as

$$\min_G \max_D V(D, G) = \mathbb{E}_{x \sim p_{real}}[\log D(x)] + \mathbb{E}_{z \sim p_z}[\log(1 - D(G(z)))] \quad (4.17)$$

where x is a real image (real travel times in the context of the travel times imputation) from the true data distribution p_{real}, and z is a noise vector sampled from the distribution p_z (e.g., a uniform or normal distribution). In practice, the generator G is modified to maximize $\log(D(G(z)))$ instead of minimizing $\log(1 - D(G(z)))$ to mitigate the gradient vanishing [34]. The training procedure of the original GAN model is given as Algorithm 1.

Algorithm 1 The original GAN model training

---
for number of training iterations **do**
Initialize hyperparameters;
Feed the noise vector z from p_z into D;
Feed x from the true data distribution p_{real} into D;
Maximize $V(D, G)$, update D;
Feed the noise vector z from p_z into G;
Minimize $V(D, G)$, update G.
end for
return G

In the transportation research area, some scholars creatively introduce GANs to estimate traffic states with data from stationary loop detectors [35–39]. As an advanced deep learning paradigm, GAN is found to be capable of well imputing missing traffic data. Specifically, with traffic data from several loop detectors installed along a freeway, [35] proposed a deep Generative Adversarial Architecture to estimate the missing traffic flow and traffic density values. Reference [38] proposed a GAN-based approach to impute traffic flow data with a representation loss to narrow the gap between generative data and real data. However, this study predicted the missed traffic flow values of one detector station by using its own records only. Obviously, these GAN-based methods impute the missed values with traffic data from one or several detectors, which fails to consider spatiotemporal correlations at a network level. Meanwhile, they only impute a single data point for a certain detector during a time interval, which does not satisfy the need to impute multiple data points. Thus, it is still at an early stage, although the GAN-based method provides a novel way for traffic time series estimation.

4.3 The LSTM-DNN models

Despite the studies cited earlier using deep learning methods for traffic time series prediction, these studies did not go deeper into studying the optimization of structure design and hyperparameter settings of deep learning models in use for traffic data. It is important and essential because the designs of deep learning models are variable for learning different types of data. The design is more flexible when compared to traditionally well-used models and therefore it can also be more difficult to achieve an absolutely optimal effect of model establishment. Hence, this study is aimed to explore how the "accurate" performance could change in quantity with varying settings of LSTM neural networks with deep fully connected layers. We call these models LSTM-DNN deep learning models. In the meantime, we compare these models with commonly used time series regression/prediction models as well.

Specifically speaking,

- we investigate short-term prediction results of travel time using 16 different LSTM-DNN models by arranging and combining three important hyperparameters, including the number of LSTM units and layers, and of deep fully connected layers;
- we select competitive LSTM-DNN models from earlier and test them along with linear models and machine learning models such as linear regression, Ridge and Lasso regression, ARIMA and DNN models in several experiment groups containing ten sets of sliding windows and predicting horizons to further check the accuracy of both deep learning and well-used prediction methods.

Again, the historical travel-time series in current time step T is $X = \{x_1, x_2, \ldots, x_T\}$, the problem is to predict travel time x_{T+m} while $m \geq 1$, indicating either predict value in the next time step or in several time steps ahead. And the input data .. has total length of $k (1 \leq k \prec T)$, which is the length of sliding window for prediction.

As illustrated in Figure 4.6, the proposed LSTM-DNN models are composed of

1. one or several stacked LSTM layers (shown in green blocks): the number of the layers can be one to many, two LSTM layers are stacked as an example in Figure 4.6;
2. Deep fully connected neural layers (DNN layers, shown by gray circle blocks): the number of the layers can be 1 to many, three DNN layers are exemplified in Figure 4.6; and
3. A deep neural layer with one hidden neural with linear activation (shown by blue circle block). The output only has one value of travel time, so only one unit exits in this layer and this layer is not considered as the DNN layer's part mentioned earlier.

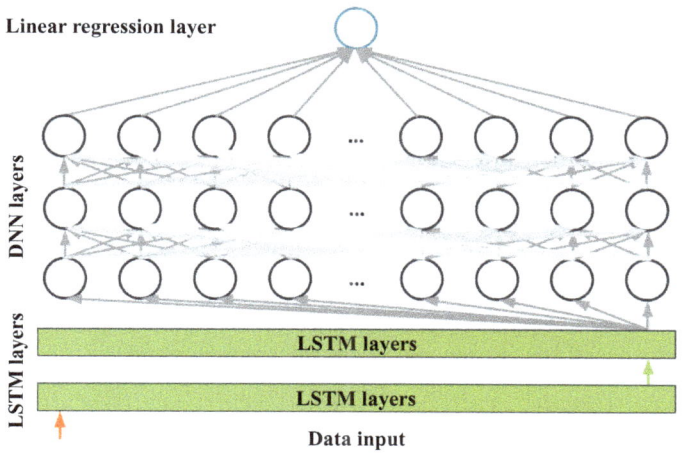

Figure 4.6   The architecture of an LSTM-DNN model

Data are directly input to the first LSTM layer. The orange arrows shown under the bottom LSTM layer only represent where to input the data. The outputs from an LSTM layer cannot be used directly as prediction results because in each time step LSTM yields output with the same length of input data (e.g., have a dimension of time steps×1). Hence, one or several DNN layers should be added above the LSTM layer for transferring output dimension. So how many DNN layers should be added to obtain best predictions becomes a problem to be studied. The size of each stacked LSTM layer can be various and unique. Second, the size of each DNN layer should better be equal but can also be designed with various values and unequal with each other.

The objective function is set as Mean Square Error (MSE) as (4.18), where x_i denotes the actual travel time in training data and $\hat{x}_i$ is the prediction value. We use adaptive subgradient methods developed in [40] to train the proposed models. The entire network can be established by using TensorFlow, a recently developed deep learning framework in Python.

$$MSE = \frac{1}{n}\sum_{i=1}^{n}(x_i - \hat{x}_i)^2 \qquad (4.18)$$

4.4 Experiments

4.4.1 Datasets

The data for this study are obtained from Caltrans PeMS, an online system that provides traffic data collected over 16,000 loop detectors located in California, the United States. We collected travel-time data on Corridor 13 (Alameda I-80) freeway section. The data duration is 90 days from Sep 18 00:00, 2016 to Dec 17 23:59, 2016. Travel-time values are aggregated per 5 min, so there are 288 data points for 1 day and 25,920 points in total. The data were split into training (70%, 63 days), validating (10%, 9 days) in order to train the LSTM-DNN models and testing (20%, 18 days) data for investigate the prediction ability of the models, since testing data are never seen by the models during training. As for using statistical methods, there is no need of validating data so the data were split into training (80%) and testing (20%) data for training statistical models such as linear regression, Ridge regression, Lasso regression and ARIMA.

4.4.2 Evaluation metrics

Mean Absolute Percentage Error (MAPE) and Root Mean Square Error (RMSE) are employed as evaluation metrics for our travel-time prediction models. x_i denotes the actual travel time and $\hat{x}_i$ are the prediction values.

$$MAPE = \frac{1}{n}\sum_{i=1}^{n}\frac{|x_i - \hat{x}_i|}{x_i} \qquad (4.19)$$

$$RMSE = \sqrt{\frac{1}{n}\sum_{i=1}^{n}(x_i - \widehat{x}_i)^2} \qquad (4.20)$$

4.4.3 Hyperparameter settings for LSTM-DNN models

In these experiments, we intended to create a highly challenging prediction mission for deep learning models. After several preliminary experiments, we set the input data length only be 6 time steps (6×5 min = 30 min), and the task is to predict travel time at 12 time steps ahead (60 min ahead). From Section 4.2, we can find that there are three important hyperparameters for LSTM-DNN models: (1) the number of LSTM units, which represents the number of cell states in an LSTM layer and is abbreviated as "len_LSTM" (here for simplification, we make equal size of each LSTM layer for stacking), we set four values increasing with same interval; (2) number of LSTM layers (n_LSTM), we chose only two values—1 and 2 and (3) number of DNN layers (n_DNN), two options—2 and 4—are provided. The summary of hyperparameter settings is presented in Table 4.1.

In fact, various choices can be made toward the selection of these parameters and extremely a wide range of numbers can be used. But to leverage the training complexity and grained accuracy, we only used these values for experiments. Note that each number chosen for len_LSTM can exactly be associated with integer hours, e.g., 36 × 5 min = 3 h and 216 × 5 min = 18 h, assuming that there are connections between number of cell states and hours. Besides, the learning rate, number of epochs and batch size for each time training one single model are equal and are respectively set to be 0.001, 80 and 180 based on experience from several preliminary experiments.

Six individual LSTM-DNN models with different combination of the three hyperparameters have been established and each model is trained eight times separately and tested eight times separately as well. So, 8 groups of 16 experiments (totally 128 times of experiments) have been conducted for comparing prediction results. Each experiment contains a training and a testing phase. The training process for each group is kept the same. The reason for conducting multiple time experiments is that the parameters obtained from a trained deep learning model can be different each time; thus, the prediction results vary. We illustrated MAPEs and RMSEs in testing phases

Table 4.1 Hyperparameter settings

Sliding window length	Prediction horizon	Number of LSTM units (len_LSTM)	Number of LSTM layers (n_LSTM)	Number of DNN layers (n_DNN)
30 min (6 time steps)	60 min (12 time steps)	[36, 108, 216, 288]	[1, 2]	[2, 4]

Note: The bold values imply the best value of metrics in each column for a certain scenario.

of eight group experiments in Figure 4.7. The legend "LSTM1_DNN2" denotes that the number of LSTM layers (n_LSTM) equals to 1 and the number of DNN layers (n_DNN) 2, and so forth. For convenience, we use style LSTM 1+DNN 2 to describe the model structures.

Figure 4.7 indicates that better results cannot always be obtained when making deeper or wider neural networks. For MAPE, it is clear from Figure 4.7(a) that when len_LSTM is 36, combination of LSTM 2+DNN 2 is better than LSTM 2+DNN 4, but when len_LSTM increases to 108, 216 and 288, performance from the latter becomes better as seen from the median levels. More interestingly, when len_LSTM is the biggest (288), LSTM 1+DNN 4 is relatively better than LSTM 2+DNN 2 (or +DNN 4) as the box of distribution is narrower. In overall scope, LSTM 1+DNN 4 with 108 len_LSTM and LSTM 1+DNN 4 with 288 len_LSTM get more centralized low MAPEs and hence are the relatively best two models with the measurement of MAPE. Also, from Figure 4.7(b), a phenomenon should also be paid attention that models with moderate len_LSTM (i.e., 108) are more stable and get more centralized low MAPEs than the others in three out of four structures. When the structure becomes LSTM 2+DNN 2, all four len_LSTM types of models become dramatically unstable, indicating LSTM 2+DNN 2 structure possibly an inappropriate one for prediction tasks, since it lacks robustness.

When comparing with RMSE in Figure 4.7(c) and (d), models with one LSTM layer get better results (more centralized distribution and lower median) than those with two LSTM layers in 128 experiments. And the results from models with 36 and 108 len_LSTM are more stable, despite of outlier models. That exits in all four types of len_LSTM groups. Furthermore, LSTM 1+DNN 4 with 108 len_LSTM gets the best performance and robustness in eight groups. The smallest values of MAPEs and RMSEs are more likely to be obtained from models with 216 and 288 len_LSTM, but correspondently, the larger values have higher probabilities to be seen in these models. And no matter for MAPE or RMSE, when len_LSTM is 36, LSTM 1+DNN 2 model is always better than LSTM 1+DNN 4 according to the box widths and levels, but this situation gets contrary with the increasing of len_LSTM (or at least almost equal). Therefore, LSTM 1+DNN 4 with 108 len_LSTM is overall optimal choice among those models in both precision and robustness.

4.4.4 Comparison between LSTM-DNN models and benchmarks

In order to compare the prediction ability of LSTM-DNN with well-used statistical and machine learning models, we continually conducted 2 groups of totally 10 experiments, in which sliding window length is set to be 12 and 6 (i.e., 60 and 30 min), and prediction horizon is [1, 2, 4, 6, 12] time steps (i.e., [5, 10, 20, 30, 60] min), such that 10 experiments can be conducted with each sliding windows length paired to 1 prediction horizon. In each experiment, 6 types of model are trained and tested, which are respectively 3 types of linear models, an ARIMA model, a DNN model (4 fully connected layers with 288 units in each) and an LSTM-DNN model. The utilized linear models are (1) linear

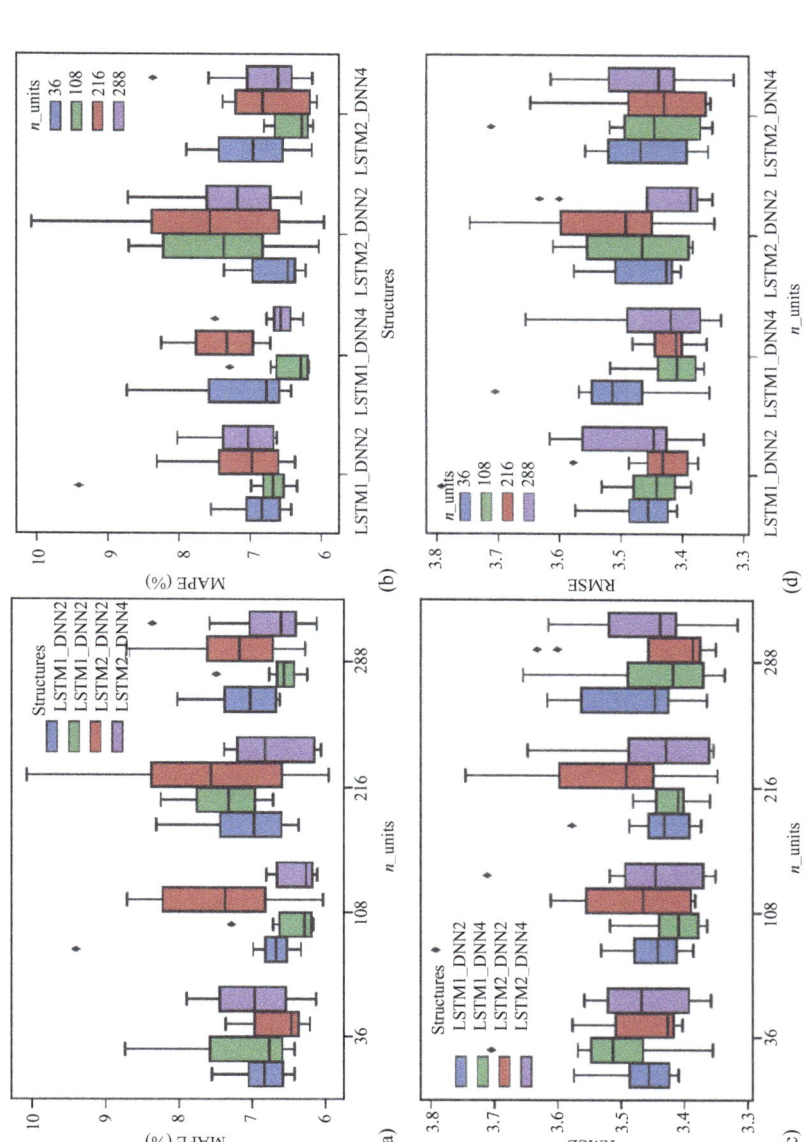

Figure 4.7 Evaluation metrics of prediction on test data (8 groups of 16 experiments): (a) MAPE (%) on test data (by len_LSTM); (b) MAPE (%) on test data (by structures); (c) RMSE on test data (by len_LSTM) and (d) RMSE on test data (by structures)

regression, this is the simplest model in our experiments; (2) Ridge regression and (3) Lasso (least absolute shrinkage and selection operator) regression. The linear models and ARIMA model are trained on training data (80%) and predict on test data (20%). The test data remain unknown for these models until the data are used for testing the models. The LSTM-DNN model we chose here is LSTM 1+DNN 4 with 108 len_LSTM, which is analyzed to be the best model in all LSTM-DNN alternatives in Section 4.4.3. The α for Ridge and Lasso regressions is, respectively, set to 0.5 and 0.05, as this setting can leverage the output of MAPE and RMSE and yield better prediction results on test data in preliminary experiments. All additional hyperparameters of the LSTM-DNN model remain the same with which in Section 4.4.3, LSTM-DNN model is trained six times for each of the ten experiments to get the best result and DNN models also trained several times for each prediction scenario.

From the results shown in Figure 4.8 and Table 4.2, we can clearly see the advantages and drawbacks of our LSTM-DNN model. Both prediction performances in short and long horizons in the deep learning model almost gained every best result among the benchmark models. When the sliding window is narrowing down and prediction horizon stretching, the advantage of the proposed model is dramatically being revealed. For MAPE or RMSE, only one nonoptimal case of LSTM-DNN model occurs when using travel time within 60-min sliding windows to predict 10-min travel time in future. When focusing on ARIMA model, one important condition that needs to be clarified is that the experiments used iterative prediction on test data. ARIMA model can only do one-step prediction directly; if it is required to do multistep prediction, ARIMA model generally takes iterative method, i.e., use the prediction value as feedback to predict the following time step values. To increase the prediction quality, iterative prediction often retrains the trained ARIMA model when a new true value is fed back to it, such that the ARIMA model updates its parameters in every time step. However, for maintaining consistency of the experiments, it is not allowed to use test data to retrain the models. Thus, an ARIMA model conducts i-step prediction only using its own $(i-1)$ step prediction values, which leads to the gradually increasing errors and the huge deviation of prediction in long horizon prediction. Hence, the results of ARIMA are not plotted on Figure 4.8.

When focusing on DNN models, the MAPEs are smaller than the other benchmarks when the horizon is long enough (cases like 60-min for 60-min, and 30-min for >20-min prediction) but DNN models have no advantages in terms of RMSEs. The possible reason is that the causal relation of the time step dimension is partially ignored by DNN units when inputting data into them.

4.5 Conclusion and future work

In this study, we first investigate short-term travel-time prediction effects of 16 LSTM-DNN models by arranging 16 groups of combination for three important hyperparameters, including the number of LSTM units and layers, and of deep fully connected layers (DNN layers) and conducted 128 experiments. We find that adding

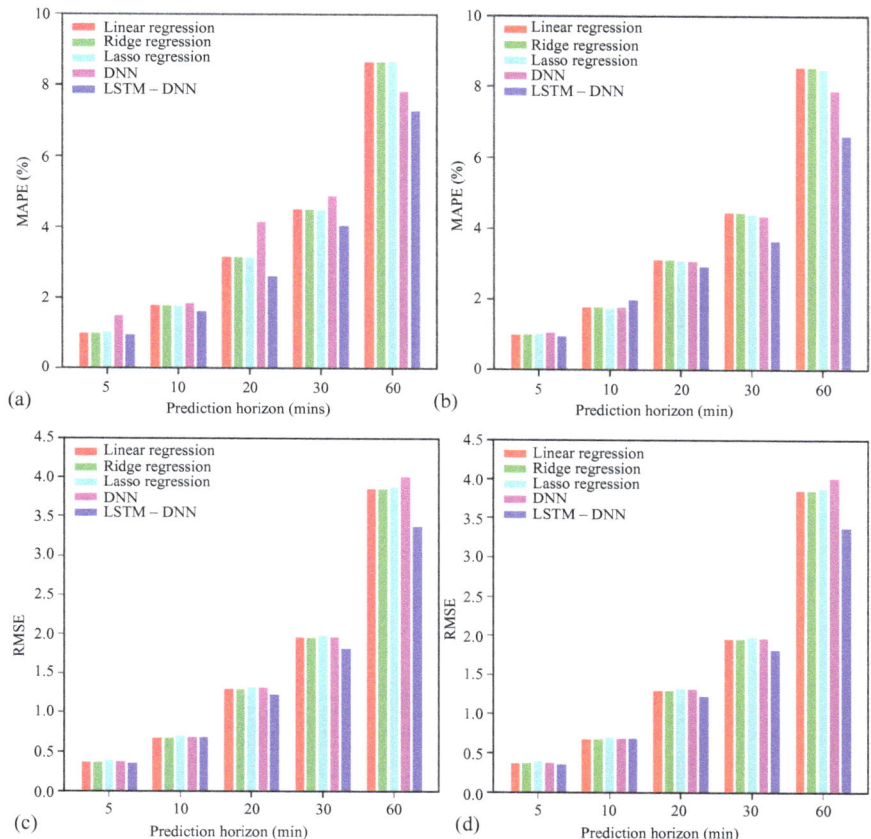

Figure 4.8 Evaluation metrics of prediction on test data: (a) MAPE (%) on test data (60 min as sliding window); (b) MAPE (%) on test data (30 min as sliding window); (c) RMSE on test data (60 min as sliding window); (d) RMSE on test data (30 min as sliding window)

depth and width is not always the best choice to obtain better results. Maintaining a middle scale of deep learning model that appropriate to certain data, in turn, is more possible for generating expected accurate and solid prediction results according to the evaluation metrics of MAPE and RMSE. Second, we choose competitive LSTM-DNN models from earlier and test them along with linear prediction models such as linear regression, Ridge and Lasso regression, ARIMA and DNN models in several experiment groups containing ten combinations of sliding windows and predicting horizons to further check the accuracy of both deep learning and well-used statistical and machine learning methods. The results indicate the strong power of our LSTM-DNN models when the sliding window is narrow and the prediction horizon becomes far ahead, both MAPEs and RMSEs are lower than benchmark methods under these scenarios.

Table 4.2 Evaluation metrics of prediction on test data

Prediction horizon (min)	MAPE (%), sliding window = 60 min					MAPE (%), sliding window = 30 min				
	5	10	20	30	60	5	10	20	30	60
Linear regression	1.010187	1.781746	3.157644	4.513085	8.637864	1.006194	1.773854	3.122103	4.457734	8.528927
Ridge regression	1.010173	1.781706	3.157573	4.513007	8.637823	1.006172	**1.773804**	3.122005	4.457624	8.528827
Lasso regression	1.048166	1.793461	3.165147	4.504833	8.671028	1.028043	1.769604	3.107643	4.407132	8.498129
DNN	1.493290	1.861104	4.151781	4.876978	7.820807	1.065725	1.778982	3.084992	4.358483	7.877626
ARIMA	1.208255	2.484198	5.562116	9.407818	23.736445	1.130722	2.236957	4.790908	7.990758	20.229357
LSTM-DNN	**0.967767**	**1.621363**	**2.616827**	**4.057407**	**7.264099**	**0.961468**	1.997938	**2.920018**	**3.655459**	**6.600965**
Prediction horizon (min)	RMSE, sliding window = 60 min					RMSE, sliding window=30 min				
	5	10	20	30	60	5	10	20	30	60
Linear regression	0.367289	0.677546	1.298362	1.954813	3.853114	0.366843	**0.677064**	1.296032	1.952448	3.858874
Ridge regression	0.367294	0.677550	1.298375	1.954831	3.853134	0.366848	0.677065	1.296040	1.952459	3.858887
Lasso regression	0.404177	0.711253	1.340855	1.996254	3.887231	0.397264	0.703925	1.329985	1.984265	3.884927
DNN	0.459849	0.731222	1.427575	2.161596	4.173776	0.377937	0.687101	1.319625	1.968736	4.011477
ARIMA	0.402316	0.792926	1.687567	2.735245	4.004502	0.388855	0.750618	1.554574	2.497897	5.806814
LSTM-DNN	**0.362902**	**0.655964**	**1.211798**	**1.771794**	**3.064785**	**0.365456**	0.688558	**1.234101**	**1.822048**	**3.364138**

Note: The bold values imply the best value of metrics in each column for a certain scenario.

The results from the comparison of LSTM-DNN models also reveal the complexity and difficulty in the optimization for deep learning prediction models. We mainly focus on three important hyperparameters related to the structure of LSTM-DNN models but several more factors also have impacts on model structure optimization for producing optimal prediction results, such as optimizing methods selection, and the setting of learning rate, epoch amount, batch size and optional regularization for training a model. We mainly employed a set of appropriate default values for these hyperparameters in order to simplify the comparison, emphasize the key aspects for building models and reduce time cost. It should be a huge work for optimization when considering changes of all these hyperparameters, which is still an unsolved problem in deep learning area. Nevertheless, the further work should be focusing on more available structures of deep learning model, increasing experiments, adding more well-used models for comparison and optimizing the structure of deep learning models under different prediction scenarios. Besides, specific prediction effects should be evaluated aside from the evaluation metrics, that is, to look deeper into the shapes of predicted time series to check how well a model can accommodate abrupt changes to be predicting.

References

[1] M. S. Ahmed and A. R. Cook, "Analysis of freeway traffic time-series data by using Box-Jenkins techniques," *Transportation Research Record*, vol. 722, pp. 1–9, 1979.

[2] D. Billings and J.-S. Yang, "Application of the ARIMA models to urban roadway travel time prediction – A case study," *IEEE Int. Conf. on Systems, Man and Cybernetics*, pp. 2529–2534, 2006.

[3] B. Pan, U. Demiryurek, and C. Shahabi, "Utilizing real-world transportation data for accurate traffic prediction," *IEEE 12th Int. Conf. on Data Mining*, pp. 595–604, 2012.

[4] P. Duan, G. Mao, C. Zhang, and S. Wang, "STARIMA-based traffic prediction with time-varying lags," *IEEE 19th Int. Conf. on Intelligent Transportation Systems (ITSC)*, pp. 1610–1615, 2016.

[5] W. Zheng, D.-H. Lee, and Q. Shi, "Short-term freeway traffic flow prediction: Bayesian combined neural network approach," *Journal of Transportation Engineering*, vol. 132, pp. 114–121, 2006.

[6] X. Fei, C.-C. Lu, and K. Liu, "A Bayesian dynamic linear model approach for real-time short-term freeway travel time prediction," *Transportation Research Part C: Emerging Technologies*, vol. 19, pp. 1306–1318, 2011.

[7] J. Haworth, J. Shawe-Taylor, T. Cheng, and J. Wang, "Local online kernel Ridge regression for forecasting of urban travel times," *Transportation Research Part C: Emerging Technologies*, vol. 46, pp. 151–178, 2014.

[8] C.-H. Wu, J.-M. Ho, and D.-T. Lee, "Travel-time prediction with support vector regression," *IEEE Transactions on Intelligent Transportation Systems*, vol. 5, pp. 276–281, 2004.

[9] J. V. Lint, S. Hoogendoorn, and H. V. Zuylen, "Accurate freeway travel time prediction with state-space neural networks under missing data," *Transportation Research Part C: Emerging Technologies*, vol. 13, pp. 347–369, 2005.

[10] R. Li and G. Rose, "Incorporating uncertainty into short-term travel time predictions," *Transportation Research Part C: Emerging Technologies*, vol. 19, pp. 1006–1018, 2011.

[11] M. Yildirimoglu and N. Geroliminis, "Experienced travel time prediction for congested freeways," *Transportation Research Part B: Methodological*, vol. 53, pp. 45–63, 2013.

[12] G. E. Hinton, "Reducing the dimensionality of data with neural networks," *Science*, vol. 313, pp. 504–507, 2006.

[13] S. Hochreiter and J. Schmidhuber, "Long short-term memory," *Neural Computation*, vol. 9, pp. 1735–1780, 1997.

[14] W. Huang, G. Song, H. Hong, and K. Xie, "Deep architecture for traffic flow prediction: Deep belief networks with multitask learning," *IEEE Transactions on Intelligent Transportation Systems*, vol. 15, pp. 2191–2201, 2014.

[15] Y. Lv, Y. Duan, W. Kang, Z. Li, and F.-Y. Wang, "Traffic flow prediction with big data: A deep learning approach," *IEEE Transactions on Intelligent Transportation Systems*, vol. 16, pp. 865–873, 2015.

[16] X. Ma, Z. Tao, Y. Wang, H. Yu, and Y. Wang, "Long short-term memory neural network for traffic speed prediction using remote microwave sensor data," *Transportation Research Part C: Emerging Technologies*, vol. 54, pp. 187–197, 2015.

[17] Y. Tian and L. Pan, "Predicting short-term traffic flow by long short-term memory recurrent neural network," *IEEE Int. Conf. on Smart City/SocialCom/SustainCom (SmartCity)*, pp. 153–158, 2015.

[18] Y. Jia, J. Wu, and Y. Du, "Traffic speed prediction using deep learning method," *IEEE 19th Int. Conf. on Intelligent Transportation Systems (ITSC)*, pp. 1217–1222, 2016.

[19] H. Yi, H. Jung, and S. Bae, "Deep neural networks for traffic flow prediction," *IEEE Int. Conf. on Big Data and Smart Computing (BigComp)*, pp. 328–331, 2017.

[20] J. Wang, Q. Gu, J. Wu, G. Liu, and Z. Xiong, "Traffic speed prediction and congestion source exploration: A deep learning method," *IEEE 16th Int. Conf. on Data Mining (ICDM)*, pp. 499–508, 2016.

[21] K. Zhang, L. Zheng, Z. Liu, and N. Jia, "A deep learning based multitask model for network-wide traffic speed predication," *Neurocomputing*, DOI: https://doi.org/10.1016/j.neucom.2018.10.097, 2019.

[22] A. C. D Oord, D. Sander, H. Zen, *et al.*, "WaveNet: A generative model for raw audio," *arXiv:1609.03499*, 2016.

[23] S. Bai, J. Z. Kolter, and V. Koltun, "An empirical evaluation of generic convolutional and recurrent networks for sequence modeling," *arXiv:1803.01271*, 2018.

[24] S. Bai, J. Z. Kolter, and V. Koltun, "Trellis networks for sequence modeling," *arXiv:1810.06682*, 2018.

[25] J. Schmidhuber, "Deep learning in neural networks: An overview," *Neural Networks*, vol. 61, pp. 85–117, 2015.

[26] X. Ma, Z. Dai, Z. He, J. Ma, Y. Wang, and Y. Wang, "Learning traffic as images: A deep convolutional neural network for large-scale transportation network speed prediction," *Sensors*, vol. 17(4), p. 818, 2017.

[27] J. Zhang, Y. Zheng, and D. Qi, "Deep spatio-temporal residual networks for citywide crowd flows prediction," *Proc. 31st AAAI Conf. Artif. Intell.*, pp. 1655–1661, 2017.

[28] J. Zhang, Y. Zheng, J. Sun, and D. Qi, "Flow prediction in spatio-temporal networks based on multitask deep learning," *IEEE Transactions on Knowledge and Data Engineering*, 2019.

[29] K. Zhang, Z. Liu, and L. Zheng, "Short-term prediction of passenger demand in multi-zone level: Temporal convolutional neural network with multi-task learning," *IEEE Transactions on Intelligent Transportation Systems*, DOI: https://doi.org/10.1109/TITS.2019.2909571, 2019.

[30] L. Deng, and Y. Dong, "Deep learning: Methods and applications," *Foundations and Trends® in Signal Processing*, vol. 7(3–4), pp. 197–387, 2014.

[31] J. Ke, H. Zheng, H. Yang, and X. (Michael) Chen, "Short-term forecasting of passenger demand under on-demand ride services: A spatio-temporal deep learning approach," *Transportation Research Part C: Emerging Technologies*, vol. 85, pp. 591–608, 2017.

[32] X. Zhou, Y. Shen, Y. Zhu, and L. Huang, "Predicting multi-step citywide passenger demands using attention-based neural networks," *Proc. Elev. ACM Int. Conf. Web Search Data Min.*, pp. 736–744, 2018.

[33] Y. Liu, H. Zheng, X. Feng, and Z. Chen, "Short-term traffic flow prediction with Conv-LSTM," *2017 9th International Conference on Wireless Communications and Signal Processing (WCSP)*, 2017.

[34] I. J. Goodfellow, J. Pouget-Abadie, M. Mirza, et al., "Generative adversarial nets," *Proceedings of the 27th International Conference on Neural Information Processing Systems*, Vol. 2. MIT Press, Montreal, Canada, pp. 2672–2680, 2014.

[35] Y. Liang, Z. Cui, Y. Tian, H. Chen, and Y. Wang, "A deep generative adversarial architecture for network-wide spatial-temporal traffic-state estimation," *Transportation Research Record*, vol. 2672(45), pp. 87–105, 2018.

[36] Y. Lin, X. Dai, L. Li, and F.-Y. Wang, "Pattern sensitive prediction of traffic flow based on generative adversarial framework," *IEEE Transactions on Intelligent Transportation Systems*, vol. 20(6), pp. 2395–2400, 2018.

[37] Y. Lv, Y. Chen, L. Li, and F.-Y. Wang, "Generative adversarial networks for parallel transportation systems," *IEEE Intelligent Transportation Systems Magazine*, vol. 10(3), pp. 4–10, 2018.

[38] Y. Chen, Y. Lv, L. Li, and F.-Y. Wang, "Traffic flow imputation using parallel data and generative adversarial networks," *IEEE Transactions on Intelligent Transportation Systems*. Retrieved from http://doi.org/10.1109/TITS.2019.2910295, 2019.

[39] K. Zhang, N. Jia, L. Zheng, and Z. Liu, "A novel generative adversarial network for estimation of trip travel time distribution with trajectory data," *Transportation Research Part C: Emerging Technologies*, vol. 108, pp. 223–244, 2019.

[40] J. Duchi, E. Hazan, and Y. Singer, "Adaptive subgradient methods for online learning and stochastic optimization," *Journal of Machine Learning Research*, vol. 12, pp. 2121–2159, 2011.

Chapter 5

Short-term traffic prediction under disruptions using deep learning

Yanjie Dong[1], Fangce Guo[1], Aruna Sivakumar[1] and John Polak[1]

5.1 Introduction

Rapid urbanisation, population growth and increase in vehicle use have created both social and environmental problems, such as road traffic congestion, poor air quality and environmental degradation. Over the past five decades, considerable effort has been devoted to address these problems, initially through building new roads, widening existing roads, providing traffic information and multimodal (public transport) options. However, over the years, with a saturation of demand and the shortage of road space, recent emphasis has shifted from building new infrastructure to using existing infrastructure more efficiently. Intelligent Transport System (ITS) applications are therefore becoming an increasingly important alternative to avoid or mitigate traffic congestion.

Accurate and robust short-term prediction of traffic variables (e.g. traffic flow, occupancy, speed, travel time and traffic state) is one of the essential components in ITS applications of Advanced Traveller Information Systems and Urban Traffic Control, which require proactive traffic information to improve services effectively. Traffic prediction is an estimation process that aims to identify further traffic data of speed, flow and travel time [1]. The provision of traffic information in the short-term is key to helping traffic operators understand the current traffic conditions, develop more sophisticated strategies to mitigate traffic problems (such as congestion and incidents) and provide real-time traffic advice to drivers (e.g. congestion warnings and efficient travel routes). Therefore, the accuracy and reliability of the traffic prediction model is of paramount importance.

Accurate prediction of traffic conditions is therefore a key component of urban traffic management systems that help traffic operators and engineers to monitor current road networks and the operational performance of the traffic facilities. Road traffic conditions are strongly correlated with traffic variables (e.g. traffic speed,

[1]Urban Systems Lab, Department of Civil and Environmental Engineering, Imperial College London, London, UK

flow and travel time) of relevant links and corridors. The factors influencing traffic conditions can be categorised into two groups: traffic demand and traffic supply. Traffic demand influences the number of vehicles or travellers using a road network; traffic supply reflects the available capacity of the road facility and infrastructure.

Traffic demand factors are expressed in terms of seasonal effects, network effects, population characteristics and traffic information [2]. Traffic supply reflects the capacity of the road facility and is affected by planned and unplanned events, weather conditions, road geometry and dynamic traffic management [3]. In practice, most of these factors overlap and are dependent on each other [3]. For example, some factors affect not only traffic supply but also traffic demand and vice versa; adverse weather conditions may change travellers' routes, modes and departure times as well as reduce the road capacity.

In general, urban traffic congestion states can be categorised into Recurring Congestion (RC) and non-RC. A large number of studies in the literature have focused on traffic state estimation and traffic-data-based prediction of recurrent congestion because most recurrent congestion is caused by high volumes of vehicle flow at specific locations during the peak period, which is easy for network operators to monitor and forecast in advance. Non-recurrent traffic congestion is sometimes caused by unexpected disruptions, such as traffic incidents and accidents, severe weather conditions and planned events, such as sports matches, concerts and road-works. These disruptions may cause sudden changes in traffic patterns and a corresponding reduction of road capacity. Some studies [4] include exogenous weather variables, such as temperature, humidity, visibility and rainfall intensity, for improved accuracy of short-term traffic prediction models. Unfortunately, such weather data are neither easy to access in real time nor easy to link to target traffic corridors. Some research, such as [5,6], directly use conventional machine learning tools and statistical methods to forecast short-term traffic flow and travel time. However, they are unable to detect the change of traffic profiles quickly, and the predicted traffic variables are underestimated during abnormal traffic conditions. Accurate traffic prediction during disruption conditions is still a challenging problem for transport network managers.

Recently, there has been growing interest in using deep learning tools, such as Convolutional Neural Networks (CNNs), Recurrent Neural Networks (RNNs), Graph Convolutional Networks (GCNs), in traffic engineering [7–10] due to their non-parametric nature and superior performance in many domains, such as image processing, object recognition, text mining and sequence modelling. The applications of advanced deep learning techniques suggest a potential for accurate prediction of traffic variables during abnormal conditions. However, some challenges present themselves in the application of deep learning to predict short-term traffic variables, for example the complexity of representing a traffic road network within a deep learning model, heavy preprocessing requirements, the resulting information loss, poor transferability and the lack of recognition of the heterogeneous nature of traffic network characteristics at different locations.

Against this background, the overall aim of this chapter is to review existing studies of methods in short-term traffic prediction and to propose a new deep-learning-based prediction framework to predict traffic conditions in the short-term. The main contributions of this chapter are as follows:

1. We propose a novel deep-learning-based architecture, which is able to learn the spatial and temporal traffic propagation rules on a network, based on a combination of a GCN and depth-wise separable One-Dimensional (1D) CNN model.
2. The proposed model considers the heterogeneity of different parts of the traffic network and understands the importance of adjacent nodes for a given location (either a junction or a section of road) in short-term traffic prediction problems.
3. The proposed prediction model, which is calibrated using generic traffic data, can provide reliable one-step and multistep ahead results under a wide range of traffic conditions.
4. We present a straightforward and scalable solution for converting road networks into a suitable graph structure for deep learning.

This chapter is structured as follows: Section 5.2 reviews existing studies in short-term traffic prediction under different traffic conditions and provides a discussion of machine learning and deep learning techniques in short-term traffic prediction. Section 5.3 proposes a Temporal Spatial Traffic Graph Attention network (TS-TGAT) to capture spatial–temporal features and predict both one-step and multistep ahead traffic speed data. Section 5.4 uses real-world traffic data collected from London to evaluate the proposed model presented in Section 5.3 under both normal and disruptive traffic conditions. After testing the proposed prediction model, Section 5.5 summarises the main findings of this study and suggests further research directions.

5.2   Literature review

Existing literature on short-term traffic prediction models, spanning over the last four decades, covers a broad range of methods, including advanced statistical approaches and machine learning tools. Many alternative approaches can be used to categorise these models based on the factors that may influence final performance. From the point of view of traffic conditions, short-term traffic forecasting models in the literature are summarised principally into two categories: traffic prediction under normal conditions and prediction concerned with abnormal traffic conditions.

5.2.1   Traffic prediction under normal conditions

Various methods of short-term traffic prediction have been proposed using recurrent traffic patterns to predict traffic profiles during normal conditions. The main assumption made in these studies is that there is no occurrence of planned or unplanned events in both historical and currently observed datasets. For example, Autoregressive Integrated Moving Average (ARIMA) time-series models [11] have been widely used to predict future variables using similar daily or weekly patterns from historical datasets under normal traffic conditions [12]. Other time-series models in short-term traffic prediction include Historical Average (HA) algorithms [13],

Exponential Smoothing methods [14] and Seasonal ARIMA (SARIMA) [6,15]. Such time-series methods that use identified traffic patterns from historical data to predict future traffic variables can provide acceptable results under normal free-flow traffic conditions. The accuracy of time-series models, however, deteriorates when data patterns are disrupted during unexpected events [16].

Since the 1990s, machine learning tools as data-driven black-box methods have been widely used to predict traffic under normal conditions due to their higher accuracy and reliability. A large number of short-term traffic prediction models using machine learning methods can be found in the literature, ranging from parametric to non-parametric models, such as k-Nearest Neighbour (kNN) (e.g. [17]), Neural Networks (NNs) (e.g. [18,19]), Support Vector Machine (SVM) (e.g. [20]), Random Forests (e.g. [21]) and hybrid methods (e.g. [22–24]). The disadvantage of machine learning methods is that they are not easy to implement to a large network when the spatial and temporal features are difficult to capture; this is particularly true in urban areas [1].

5.2.2 Traffic prediction under disrupted conditions

5.2.2.1 Traffic characteristics under disrupted conditions

In urban areas, most disrupted traffic conditions are caused by planned and unplanned events resulting in road blockage or closure of lanes [25–27]. During disruptions, the available road capacity is lower than the traffic demand. Therefore, traffic congestion and queues may occur at a location incident or accident occurrence [28]. Many existing studies use *queuing theory* to understand traffic processes that are interrupted by accidents and incidents (e.g. [29,30]). Among queuing models, deterministic arrivals and departures with a single server (D/D/1) is the simplest queuing model [31]. Under abnormal traffic conditions, a road segment occupied by a stopped vehicle can be considered as a *server* [32]. The service starts when an individual vehicle joins this link and ends when this vehicle passes the end of the link. In this body of research, traffic stream characteristics and diagrams are used to explain the nature of traffic flows and the impacts on capacity. Figure 5.1 is a fundamental diagram used to explain the relationships among traffic variables during disruption. For more details refer to [31,33,34].

In the applications of delay analysis to implement Traffic Incident Management, some actions are taken after the incident, such as using Variable Message Signs, which will change the vehicle queuing diagram. More complex delay analysis using queuing theory in these circumstances can be found in [31].

5.2.2.2 Traffic prediction under disrupted conditions

Much effort has been put into addressing traffic prediction under normal traffic conditions. Disrupted traffic conditions such as non-recurrent traffic congestion, which might be caused by planned events such as road works or unplanned events such as incidents or accidents, cannot be neglected. In the last two decades, there has been an increasing focus on prediction under abnormal traffic conditions for real-world ITS applications. Short-term prediction is arguably more important

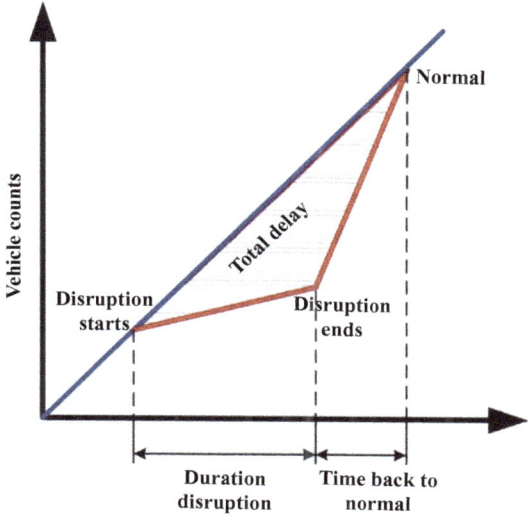

Figure 5.1 Vehicle queuing-capacity-time diagram

during abnormal conditions due to the uncertainty regarding how the traffic state will evolve into the future. Abnormal traffic conditions and disruptions are mainly caused by two factors: severe weather and traffic incidents and accidents [26].

Because historical weather information such as visibility, temperature and humidity can be easily collected, some researchers have included weather factors into short-term traffic prediction models under abnormal traffic conditions. For example, Huang and Ran [35] develop a traffic prediction model based on NNs. Both traffic speed and weather information are used as inputs to their prediction model. Similarly, Samoili and Dumont [36] also directly added weather information to their prediction model as an explanatory variable to the traffic forecasting framework. These traffic prediction models will not work, however, when the traffic abnormality is caused by traffic-related factors such as incidents and accidents.

Increasingly, there are several studies focused on developing traffic prediction models that consider the impacts of traffic-related factors (i.e. traffic incidents and accidents) on road traffic conditions. For example, Tao *et al.* [37] present NN-based models to predict travel times on a highway corridor when an incident occurs. Sun *et al.* [38] use a Bayes network algorithm with Principle Component Analysis to forecast traffic flow in the short term, where the conditional probability of a traffic state on a given road can be calculated, given the states of topological neighbours on a road network. The proposed method is designed to deal with traffic incident and accident conditions in principle, however, given the difficulties of flow data collection for abnormal conditions, the proposed algorithm was not evaluated using abnormal traffic flow data. In order to overcome the limitations of previous studies in evaluation, Castro-Neto *et al.* [26] developed an SVM model to predict traffic flow during normal days and abnormal conditions, such as holidays and traffic incidents. Although the studies described here have developed models to predict traffic variables during abnormal traffic conditions, it should be noted that most of these studies were implemented for freeways and motorways.

To extend the implementation of traffic prediction during disruptions to arterial roads, Guo *et al.* [39] used the *k*NN-based algorithm to predict traffic flow in central London during traffic incidents. The authors compared the prediction results of the *k*NN models against RNNs and Time Delay Neural Networks [40]. Although the results showed that the *k*NN method outperforms other competitors under traffic incident conditions, the proposed model overestimates traffic flow after the occurrence of incidents. In order to quickly detect changes of traffic profiles caused by traffic incidents and accidents, a two-stage prediction structure with Singular Spectrum Analysis (SSA) for data smoothing and statistical/machine learning tools for forecasting was proposed in [16,41]. Comparisons of prediction accuracy with and without the proposed framework show that the SSA method as a data smoothing step before the application of machine learning or statistical prediction methods could improve the final traffic prediction accuracy, especially under abnormal traffic conditions. However, the improvement resulting from the smoothing stage was expected to be lower as the temporal granularity of data decreases and the pre-processing is not necessary for traffic data with a low sampling frequency.

In order to improve prediction accuracy under a wide range of traffic conditions using traffic data with different sampling frequency, Wu *et al.* [42] use regime shift strategies to predict traffic under different conditions. For prediction under abnormal conditions, a boosting element is applied; when traffic conditions are back to free-flow, the boosting element is disabled. The regime shift strategy was tested using traffic flow data collected on highways and the prediction results indicate that the proposed method is more computationally effective than traditional online learning models under abnormal traffic conditions. More recently, Guo *et al.* [22] proposed a fusion-based framework to improve the accuracy of traffic prediction based on a combination of multiple stand-alone predictors that work accurately under different traffic conditions.

5.2.2.3 Summary

The previous subsections have discussed the characteristics of traffic disruptions and reviewed existing studies on short-term traffic prediction during disruptions. Table 5.1 shows a comparison summarising the key features of the literature reviewed in this section. The existing applications of machine learning tools for short-term traffic prediction under normal and incident conditions are mainly focused on a single corridor or single sensor problems. During disruption conditions, the road capacity of a blocked link could reduce, resulting in traffic congestion in the entire area rather than a single link. Hence, understanding the spatial–temporal relationship among sensors that are in close proximity to each other can help to improve prediction accuracy during abnormal traffic conditions. None of the conventional machine learning tools is easy to implement during both normal and abnormal traffic conditions, learning the complex spatial–temporal relationships to

Table 5.1 Categorisation of available literature in existing studies of traffic prediction during abnormal conditions

Author	Context	Prediction step	Traffic variable	Exogenous variable	Method	Traffic condition	Data source
Huang and Ran [35]	Corridor in urban areas	One and multistep	Flow, speed	Weather	NN	Normal and adverse weather condition	US Department of Transportation
Tao et al. [37]	Corridor of freeways	One-step	Travel time	Weather and incident	NN	Normal, incident and adverse weather	Archived Data Management System (ADMS)
Vlahogianni et al. [43]	Urban area	One-step	Volume and occupancy	N/A	NN	Various traffic conditions	
Castro-Neto et al. [26]	Corridor of freeways	One-step	Flow	N/A	Online-SVR	Normal and abnormal	Freeway Performance Measurement System (PeMS)
Guo et al. [39]	Corridor in urban areas	One and multistep	Flow	N/A	kNN	Normal and abnormal	UK Transport for London
Wu et al. [42]	Freeways	One-step	Flow	N/A	Online boosting regression	Normal and abnormal	
Guo et al. [16]	Corridor in urban areas	One and multistep	Flow	N/A	SARIMA, Grey Model	Normal and abnormal	UK Transport for London
Guo et al. [41]	Corridor in urban areas	One-step	Travel time	N/A	Smoothing and kNN	Normal and abnormal	UK Transport for London
Guo et al. [22]	Corridor in urban areas	One and multistep	Flow and travel time	N/A	Fusion of machine learning	Normal and abnormal	UK Transport for London

accurately and robustly predict traffic variables remains a challenge. Moreover, traffic prediction under disrupted conditions is important but challenging, and there is limited research especially for urban arterial roads. One of the objectives of this chapter is to address this research gap. The following section will review and discuss the applications of deep learning techniques to short-term traffic prediction problems.

5.2.3 Review of traffic prediction using deep learning techniques

5.2.3.1 Introduction

As discussed in Section 5.2.1, a large number of conventional machine learning methods to forecast traffic data can be found in the literature. Despite advances in these machine-learning-based traffic prediction models, a challenging problem is the implementation of machine learning methods for large-scale networks to understand both spatial and temporal relationships [9], especially for urban arterial networks [44].

In order to better understand complex non-linear spatial–temporal relationships within historical traffic profiles, deep learning techniques have been applied to short-term traffic prediction under normal traffic conditions. For example, a prediction architecture using Long Short-Term Memory (LSTM) Neural Networks was developed by Ma *et al.* [45] to capture dynamic traffic patterns for short-term traffic speed prediction using data collected from microwave detectors on expressways. The results show that the proposed method outperforms conventional statistical and machine learning models, such as ARIMA, SVM, Kalman filter and NN, for both one-step and multistep prediction. Although the proposed LSTM model can better understand temporal correlations of a road network, it cannot explain the spatial features well. In order to capture spatio-temporal dependencies in a prediction model, Ma *et al.* [46] introduced a CNN method to predict traffic speed data, as collected by GPS units installed in taxis in Beijing. The fundamental idea from the paper is that spatio-temporal relationships of traffic profiles can be transferred to images which are used as inputs to the CNN. The proposed method with CNN was shown to produce more accurate prediction results than the LSTM. On the other hand, an LSTM-based method was used to predict traffic volumes from three locations in Beijing [47]. The results of this study indicate that deep learning tools can predict traffic variables more accurately, especially for multistep ahead prediction. As a pure image processing method, CNN-based model is not computational efficient, which limits its real-time application. In order to address this limitation, Zhang *et al.* [48] proposed a hybrid deep learning framework called Attention Graph Convolutional Sequence-to-Sequence model to predict traffic speed at a road network level. The prediction results of the proposed framework are more accurate than ARIMA, NN, *k*NN and Support Vector Regression (SVR) and more computational efficient than CNN, especially for multistep prediction.

It can be seen from the papers reviewed earlier that deep learning techniques can better extract spatio-temporal features in a traffic network, which, in turn, can improve the accuracy of multistep ahead prediction under normal traffic conditions. More recently, a methodology for deep learning with transfer learning was proposed to predict traffic flow under insufficient data conditions [49]. In the proposed three-stage framework, both RNNs and LSTM were selected for prediction, and transfer learning techniques were used to improve the transferability of the deep learning tools.

Deep learning is clearly becoming a popular technique for short-term traffic prediction. The main advantages of deep learning include structural flexibility and accuracy. Similar to a conventional machine learning tool, deep learning as a data-driven method does not require extensive theoretical physical rules for modelling traffic [2].

In general, there are two main elements in the application of deep learning tools to short-term traffic prediction. They are data representation and the design of model architecture to capture spatiotemporal traffic features. The following two subsections will review these two design elements in detail.

5.2.3.2 Data representation in traffic prediction using deep learning

The main challenge of traffic prediction using deep learning is to accurately and efficiently transfer traffic datasets into readable inputs for the deep learning tools.

For traffic data collected from corridors, vectors can be simply generated by location and time information. For example, Polson and Sokolov [9] predict short-term traffic flow speed using data collected from 21 road segments for a corridor in Chicago and represent the traffic speed by vectors $[x^1_{1,t-t_{lag}}, \ldots, x^1_{t,}; \ldots; x^{21}_{t-t_{lag}}, \ldots, x^{21}_{t,}]$, where $x^i_{t,}$ is the cross-sectional traffic flow speed at location i at time t.

For complex network structures, there are the following three widely used methods to represent input traffic data.

- *Graph representation*: Li *et al.* [7] proposed a Diffusion Convolutional Recurrent Neural Network (DCRNN), where spatial dependence, temporal dependence and scheduled sampling can be modelled. In order to represent network-level traffic data, the authors adopt a weighted directed graph structure, where each node is a traffic sensor and the weights on the edges represent the proximity between sensor pairs measured by road network distance. Then the DCRNN was built by replacing matrix multiplication in Gated Recurrent Unit (GRU) with the diffusion convolution. However, they implemented random walks on the graph model traffic diffusion process. In addition, only proximity was used to describe node relationships, whereas in reality there are many other factors, such as speed limits/traverse time and type of roads (e.g. major and minor types). Yu *et al.* [50] proposed a graph convolution network, where the authors used a graph method to structure traffic data, $\mathcal{G} = (\mathcal{V}, \mathcal{E}, \mathcal{W})$, where $\mathcal{V}$ is a set of nodes representing observations from N monitoring stations; $\mathcal{E}$ is a set of edges representing connectedness between stations and $\mathcal{W}$ is a weighted adjacency matrix. There are two widely used approaches to generalise CNNs to structured data formats.

- *Linkage network*: Wang et al. [51] proposed a linkage network to represent the traffic network, where the nodes on the linkage network represent road segments from the real world and can carry properties, such as speed limit, number of lanes and length. The edges of the linkage network represent linkages of the road network, for example turning left, right or going straight. In addition, there is a propagation module based on GRU to learn the propagation of traffic flows along with different nodes and their impacts on traffic. The proposed model was tested with taxi GPS data and the results show that the model outperforms ARIMA, SVR, Gradient-Boosted Regression Tree and Graph Recurrent Neural Network models.
- *Images*: Ma et al. [8] converted traffic time and space matrices into images with every pixel representing a value in a matrix. The converted images can be fed into a CNN network for traffic prediction. The image matrix can be written as $M = [m_{11}, \ldots, m_{1N}; \ldots; m_{Q1}, \ldots, m_{QN}]$, where N is the length of time intervals, Q is the length of road sections and m_{ij} is the average traffic speed on section i at time j. The image matrix is similar to the one used for data representation of corridors; spatial information, such as upstream and downstream, can be easily represented as an image format. However, the converted images cannot learn dependency and relationships at network levels. Moreover, the image representation method cannot be easily transferred from one location to another, reducing computational efficiency significantly.

5.2.3.3 Spatio-temporal features in traffic prediction using deep learning

Polson and Sokolov [9] applied deep learning to predict traffic speed within the next 5-min interval. They developed a model with a hierarchical sparse vector autoregressive technique as the first deep layer to capture spatio-temporal characteristics of traffic data and other layers were used to model nonlinear relationships. In the design of the network structure and weights, stochastic gradient descent was implemented for network optimisation and a random search approach was used for parameter optimisation. Zhao et al. [47] predicted one-step and multistep ahead traffic flow using a DCRNN network, which can identify spatial relationships in a diffusion process characterised by a bidirectional random walk and capture temporal dependency using an encoder–decoder architecture with a scheduled sampling technique. Yu et al. [50] proposed a Spatio-Temporal Graph Convolutional Network to predict short-term traffic speed under typical conditions using traffic data collected from Beijing and California. The input traffic network was created as a graph structure. In order to extract spatial features, the graph convolution was employed on graph-structured inputs directly; the authors used entire convolutional structures on the time axis for temporal feature identification. Zhang et al. [52] developed a deep-learning-based model for traffic flow prediction. Within the proposed architecture, different timestamps were selected to analyse temporal properties, such as trend, closeness and period, and the convolution method was used to capture spatial dependencies. Liao et al. [53] proposed a deep spatiotemporal residual network to exploit spatial and temporal features in forecasting hotspot traffic speed.

Deep learning methods have become quite popular in short-term traffic prediction given their ability to more easily capture spatial and temporal dependencies than conventional machine learning tools, which has also been demonstrated in other traffic domains, such as incident detection [54,55], congestion detection [56], traffic condition prediction [57] and accident prediction [58,59]. Although different network structures have been used to exploit spatial and temporal features as discussed earlier, the main disadvantages of these methods are the reduction of computational efficiency and robustness in capturing spatial dependencies during a sudden change in traffic flow profiles at a network level as occurs during disruptions.

5.2.4 Summary

As reviewed in Section 5.2.3, deep learning enables the prediction process to adapt more easily to normal traffic conditions and to learn the spatial–temporal relationship among sensors in a network, which is key to effective traffic prediction during disruptions. Although the benefits of deep learning tools in short-term traffic prediction are being explored in the literature, the application of deep learning for traffic forecasting during intermittent disruptions is still at an early stage of research and empirical experience is limited. Very few researches have looked at the issue of traffic prediction under disruption specifically. On the algorithm side, many existing studies require heavy preprocessing in order to fit the transport prediction problem into the rigid format requirements of deep learning models, which results in the loss of information, limited transferability and scalability. Examples of these include converting traffic flows into images before using CNN models [8] or converting traffic network into matrices [55]. Other researchers may have started using a more flexible structure to represent a traffic network, such as the graph model; however, the treatment of each node as being homogeneous does not reflect the real-world context of a traffic network. In the real world, each node (e.g. a junction or a traffic measuring site on the road) has unique characteristics which affect its capacity and has different impacts to its adjoining neighbour nodes. These issues need to be addressed before a deep learning model can be more widely used in short-term traffic prediction.

5.3 Methodology

In this section, we propose a TS-TGAT as a flexible, scalable, and high-performance model to solve the traffic prediction problem. The following subsections define the problem formulation and introduce each component of the proposed prediction model.

5.3.1 Traffic network representation on a graph

The network-level traffic speed forecast is a classic time-series prediction with strong spatial propagation and it is well suited to be represented as a graph model.

A directed graph can be defined as $G(\mathsf{N}, \mathcal{E})$, where:

- N represents a set of N nodes (or vertices);

- $\mathcal{E}$ is a set of edges which directionally connect a pair of nodes, $e_{ij} \in \mathcal{E}$ where e_{ij} connects origin node i to destination node j and $(i,j) \in \mathsf{N}$.

Hence, the traffic network in our model is converted to a graph as follows:

- nodes: represent traffic monitoring stations (can be loop detectors, traffic counters or cameras);
- edges: define connections between two nodes.

The edges between each pair of nodes are bidirectional (represented by two separate edges), even though many roads may be unidirectional. This is due to the nature of traffic flow propagation, depending on whether the traffic is free-flow or congested [60].

5.3.2 Problem formulation

Given the graph $G(\mathsf{N}, \mathcal{E})$ with a total of N nodes and M edges, and traffic speed data $X \in \mathbb{R}^{N \times T}$, the objective of traffic prediction at time t is to find a function f which predicts the most likely traffic speed for the next step based on previous $k(k \geq 1)$ time steps of data:

$$\widehat{X}_{t+1}^{N} = f(X_{t-k}^{N}, X_{t-k+1}^{N}, \ldots, X_{t}^{N}) \tag{5.1}$$

If multistep prediction is required, for example, to forecast traffic for the next h steps ($h \geq 1$), (5.1) will be applied recurrently to use the predicted traffic speed as inputs for the prediction of the next time step.

The function f can be obtained by minimising a loss function $\|\cdot\|_p$ which measures the discrepancy between the predicted and observed values in the training data set for all nodes:

$$\mathcal{L} = \sum_{n=1}^{N} \|\widehat{X}_{T_{train}}^{n} - X_{T_{train}}^{n}\|_p \tag{5.2}$$

The notations used in this chapter are given in Table 5.2.

5.3.3 Model structure

In this section, we present the overall architecture of the proposed TS-TGAT model, as shown in the following graph in Figure 5.2.

There are a number of components in this structure, for example the historical input data for the past k time periods are first fed into a 1D CNN block, where the temporal dependencies are processed. The outputs are then used as inputs to a TGAT block, where the multi-attention graph convolution operation is carried out to process the traffic network spatial dependencies. The outputs from the TGAT block is then fed into a non-linear activation function, before going through the same block again. The predicted traffic conditions for the next time step are the outputs from the final block. As the number of layers (TGAT blocks) increases, the nodes on the graph are able to process the information further and further away (will be further discussed in

Table 5.2 Notations

Notations	Description
G	A graph
$\mathbb{N}$	A set of nodes
ε	A set of edges
$e_{n_i n_j}$	Directional edge connecting node n_i to n_j
N	The total number of nodes in $\mathbb{N}$
n_i	Node i in $\mathbb{N}$
T	Total number of time steps available for each node
T_{train}	Total number of time steps used in the training process
t	Time step t in T
X	Observed value (traffic speed)
$\widehat{X}$	Predicted value (traffic speed)
k	k steps of input data used in prediction
$\|\|\ \|\|_p$	L-p normalisation
H, h	Feature embedding in the hidden layer
W	Learnable weights
$A, \widehat{A}$	Adjacency matrix of a graph
$D, \widehat{D}$	Degree matrix of a graph
I	Identify matrix
c_{ij}	A normalisation constant for edge (n_i, n_j) used in graph propagation
l	A layer in the graph model
σ	Activation function
λ	A scaling hyper-parameter for weights regularisation
$\mathcal{L}$	Loss function
K	Total number of attention heads
F	Total number of input features

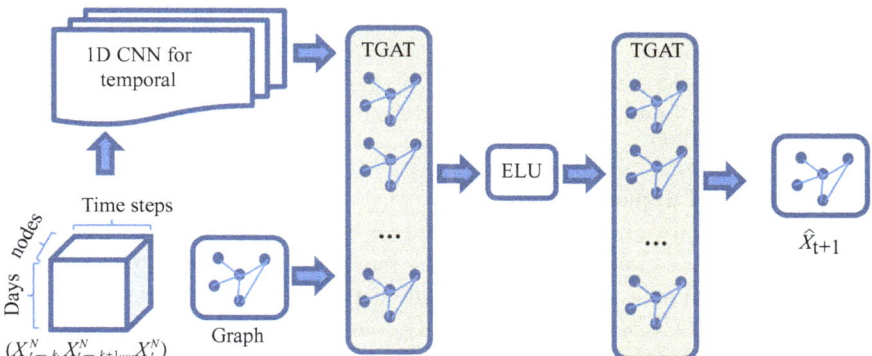

Figure 5.2 Overview of the proposed model structure

Section 5.3.3.2). However, there is a trade-off between the overall performance and the computational cost as the model becomes increasingly difficult to train as it grows deeper. In our model, a two-layer graph convolution is used, as shown in Figure 5.2. Each component in the proposed model framework will be discussed in the following subsections.

5.3.3.1 Temporal dependencies

RNNs such as LSTM and GRU have been a popular choice for time-series analysis. However, RNN models are slow to train due to the sequential forward and backward training processes. CNNs have been found to have the same or better performance compared to RNN in some time-series prediction problems [61], and more importantly, can run in parallel to improve training speed significantly. Inspired by [62], we implement a depth-wise separable 1D CNN model for modelling temporal dependencies.

The difference between depth-wise separable CNN and normal CNN is that the convolutions are not summed across channels. In our traffic network graph, it means each node i would have its own filter weights, convoluted on its own features.

Therefore, for a given input vector $X^{N \times D \times T' \times F}$, where N is the total number of nodes, D is the total number of days, T' is the number of time steps per day ($T' \times D = T$, where T is the total number of time steps) and F is the number of features which in our research is 1 (i.e. traffic speed only), replication padding will be carried out initially to keep the size of vector unchanged after convolution operation:

$$H^{N \times D \times (T'+k-1) \times F} = \text{ReplicationPad}\left(X^{N \times D \times T' \times F}\right) \tag{5.3}$$

Assuming a filter size of $1 \times k$, the 1D temporal convolution operation at time $t \in T'$ for node n can be obtained using the following formula:

$$h_t^n = \sum_{i=0}^{k-1} \left(W_i^n H_{t-k+1+i}^n + b^n\right) \tag{5.4}$$

where $W^{N \times 1 \times k}$ is the weight vector and b^n is the bias. Equation (5.5) can be calculated in parallel for all the inputs data and the outputs from temporal convolution would be $H^{N \times D \times T' \times F}$.

5.3.3.2 Spatial dependencies

Spatial dependencies are important in network-level traffic forecast, due to the nature of traffic flows. When traffic is under free-flow condition, traffic prediction of a given location on a road network usually depends on the characteristic of this location itself (such as speed limit and road geometry) and on the upstream traffic, such as traffic flows. However, when traffic is congested, then the traffic condition on the current site usually depends more on the downstream traffic due to capacity constraints [60].

One of the key advantages of the TS-TGAT model compared to some of the location-specific models reviewed in Section 5.2.3 is that it can learn spatial dependencies for different traffic propagation conditions through the graph NN learning process.

Given the graph $G(\mathsf{N}, \mathcal{E})$ defined in Section 5.3.1 and an $N \times F$ feature matrix X, an adjacency matrix $A^{N \times N}$ can be generated as a matrix representation of G, where the entry A_{ij} indicates if node i is connected to node j. Therefore, the propagation of a graph network can be expressed as

$$H^{l+1} = AH^l W^l \tag{5.5}$$

where H^{l+1} is the hidden state at the $(l+1)$th layer, and $H^0 = X$ and W^l is the weight at layer l with shape $F \times C$ and C is the size of the hidden state.

There are many methods to carry out the operation in (5.5), including spectral graph convolution [63] and Chebynets [64]. In the proposed prediction model, a GCN [65] is selected due to its capability of producing good performance while significantly reducing the number of parameters and the avoidance of computing costly eigen-decomposition compared with other methods.

In the Kipf and Welling's GCN model, the propagation can be written as

$$H^{l+1} = \sigma\left(\widehat{D}^{-1/2} \widehat{A} \widehat{D}^{-1/2} H^l W^l\right) \tag{5.6}$$

where $\widehat{A} = A + I$, which is to resolve the issue that A is not self-connected. After adding the identity matrix, features from both the node itself and the node's neighbours will now be aggregated in the GCN operation. $\widehat{D}$ is a degree matrix of $\widehat{A}$ and $\widehat{D}^{-1/2} \widehat{A} \widehat{D}^{-1/2}$ is a normalisation trick to keep the scale of feature vectors similar between each propagation. Equation (5.6) can be written in a vector form:

$$h_{n_i}^{l+1} = \sigma\left(\sum_j \frac{1}{c_{ij}} h_{n_j}^l W^l\right) \tag{5.7}$$

where n_i is the node i, j is the node i's neighbouring nodes, including itself, c_{ij} is a normalisation constant for edge (n_i, n_j) and $h_{n_j}^l$ is the feature embedding of node n_j on layer l.

The propagation rule in (5.7) gets applied to all the nodes in the graph and that is how a node's neighbouring feature embedding is propagated into its own features at the next layer.

5.3.3.3 Attention mechanism

With the existing temporal and spatial dependencies learning described in Sections 5.3.3.1 and 5.3.3.2, there is one important issue unresolved which is related to the transport network specifically. In the spatial propagation process (Equation (5.7)), for any given node, the weight matrix is shared by all its neighbours. While this might work well for some problems where nodes are homogeneous (such as the pixels on an image), for network-level traffic prediction, this weight sharing does not accurately reflect the real-world situation. For example, in a situation where a given node is connected to two other nodes, including one major road and one minor road, one would expect the traffic on the major road to have higher impacts on the given node than the minor road. Hence, the weight vector should reflect this by having different values for the neighbour nodes.

This is achieved through the implementation of Graph Attention Network [66], which replaces the statically normalised convolution with attention mechanism in the GCN model.

The GAT model propagation is defined based on the following four equations:

$$z_{n_i}^l = W^l h_{n_i}^l \tag{5.8}$$

$$e_{n_i n_j}^l = \sigma_e \left(\omega_{n_j}^l \left(z_{n_i}^l \| z_{n_j}^l \right) \right) \tag{5.9}$$

$$a_{n_i n_j}^l = \frac{\exp\left(e_{n_i n_j}^l\right)}{\sum_{r \in N(i)} \exp\left(e_{n_i n_k}^l\right)} \tag{5.10}$$

$$h_{n_i}^{l+1} = \sigma \left(\sum_{j \in N(i)} a_{n_i n_j}^l z_{n_j}^l \right) \tag{5.11}$$

Equation (5.8) is a lower layer linear transformation of feature embedding for all the nodes in the graph, the weight W is shared at the layer level. Equation (5.9) calculates an additive attention score for a given edge (n_i, n_j), by first concatenating (the '$\|$' operation) the feature embedding of origin and destination nodes, and multiplying with a trainable weight vector $\omega_{n_j}^l$, then through a non-linear activation function σ_e where the LeakyReLU is used [66]. Note that we employed a node-specific weight vector rather than using the same weight shared across all nodes as in the original article. Equation (5.10) is essentially a softmax transformation to normalise the attention scores (weights) on each node's incoming edges. Equation (5.11) aggregates the feature embedding from its neighbours based on the normalised attention weights, before going through a non-linear activation function, where ELU is used in our model.

In addition, in order to consider the situation where spatial dependencies might change in different traffic states, a multi-head attention structure is used. According to the authors in [67], '*multi-head attention allows the model to jointly attend to information from different representation subspaces at different positions, whereas with a single attention head, averaging inhibits this*'.

$$h_{n_i}^{l+1} = \|_{k=1}^{K} \sigma \left(\sum_{j \in N(i)} a_{n_i n_j}^k W^k h_{n_i}^l \right) \tag{5.12}$$

where K represents the number of attention heads, and each head has its own weight parameters. The output embedding is the concatenation of the outputs from K attentions.

Note that the multi-head is applied to all layers apart from the last. The output from the last layer is the prediction result, as this is a regression problem:

$$\widehat{x}_{n_i} = h_{n_i}^{l_{last}} = \sigma\left(\sum_{j \in N(i)} \alpha_{n_i n_j}^{l_{last}-1} W^{l_{last}-1} h_{n_i}^{l_{last}-1}\right) \quad (5.13)$$

5.3.3.4 Loss function and parameter optimisation

As discussed in Section 5.3.2, the parameters can be estimated by minimising a loss function, which is given as follows:

$$\mathcal{L} = \sum_{t=1}^{T}\sum_{i=1}^{N}\left(\widehat{x}_{n_i}^t - x_{n_i}^t\right)^2 + \lambda \sum_{i=1}^{N}\sum_{l} W_l^{i2} \quad (5.14)$$

where $\widehat{x}_{n_i}^t$ and $x_{n_i}^t$ are the predicted and observed traffic speed on node n_i at time t, respectively. We adopted an L2 squared loss with regularisation; where λ is the hyper-parameter defining how much weight is to be given to the regularisation term. In our model, we use the most widely used parameter value $\lambda = 0.01$.

During the weight updating process, an optimisation method needs to be defined. Resilient backpropagation (Rprop), which is a gradient-descent-based algorithm [68], is used in our model. The reasons we choose Rprop over other optimisation algorithms, such as stochastic gradient descent or Adam, are 3-fold: first, it is one of the fastest weight update mechanisms [69]; second, Rprop is agnostic to gradient magnitudes, which makes it really versatile; third, it adapts the step size dynamically for each weight independently which means it needs a lot less parameter tuning. A detailed explanation of the Rprop algorithm can be found in [69].

5.3.4 Quantification of prediction accuracy

The metrics used to evaluate prediction accuracy in this research are: Mean Squared Error (MSE) and Mean Absolute Error (MAE). MAE is one of the most widely used criteria in short-term traffic prediction to calculate the average of the absolute difference between true and predicted values. Compared to the MAE, the MSE tends to add additional weight to larger absolute errors. Both MSE and MAE are computed as shown next:

- MSE:

$$\text{MSE} = \frac{1}{T \cdot N}\sum_{t=1}^{T}\sum_{i=1}^{N}\left(\widehat{x}_{n_i}^t - x_{n_i}^t\right)^2 \quad (5.15)$$

- MAE:

$$\text{MAE} = \frac{1}{T \cdot N}\sum_{t=1}^{T}\sum_{i=1}^{N}|\widehat{x}_{n_i}^t - x_{n_i}^t| \quad (5.16)$$

5.4 Short-term traffic data prediction using real-world data in London

The structure of the proposed model for short-term traffic prediction has been presented in the previous section. This section presents the real-world data collected for this study, discusses the model training and testing strategy, evaluates the proposed prediction structure for its performance analysis based on different traffic situations and then discusses the results. As reviewed in Section 5.2, the robustness of a prediction model could be improved by better understanding the spatial and temporal dependencies of a network to enable models to predict traffic under a wide range of conditions. The prediction accuracy of the proposed model under both normal and disruption conditions is evaluated in this section.

5.4.1 Traffic speed data

All the traffic speed data used in this study was obtained from Webtris by Highways England (webtris.highwaysengland.co.uk). The Webtris platform has provided both historical and real-time traffic information for 15-min periods. Traffic data from the Webtris platform are collected by Inductive Loop Detectors (ILDs) from the MIDAS system (https://www.midas-data.org.uk). MIDAS is a sensor-based network along UK motorways and is designed to collect traffic data (e.g. traffic flow, speed and occupancy) on road networks.

More than 6,000 ILDs have been deployed in London to provide near real-time traffic data. Given the comprehensive spatial and temporal coverage, ILDs have been widely used in the application of traffic estimation and prediction in London [40,70]. The ILD traffic speed data used in this research are from the area between M25 Junction 2 (J2) and the mid-point of the Dartford Tunnel. The selected area of the network suffers frequent congestion and incidents due to very heavy traffic flow to cross the River Thames. A map of the selected area of the M25 is shown in Figure 5.3. The direction of traffic is anticlockwise, from the south to the north.

5.4.2 Preparation for the prediction model
5.4.2.1 Traffic speed data preprocessing

The traffic speed data were obtained for 4 years from 1 January 2015 to 31 December 2018. A total number of 23 ILDs were identified initially in the section of the study area, which covers the main motorway sections, on-slip roads and off-slip roads. However, traffic data of 9 ILD sites had to be removed due to poor data quality or due to duplicate sites, and the resulting 14 ILDs were chosen to evaluate the prediction accuracy of the proposed models under different traffic conditions.

In addition, the data availability for each day from all the 14 ILDs has been reviewed. Given 15-min intervals, there should be 96 records per day if data are complete. The structure of our model requires data to be available from all nodes at the same time, which means for a given date to be included in the cleaned dataset, the data available from each of the 14 sites should be satisfactory. The threshold for the maximum number of missing records per day for one site is 10 (i.e. approximately

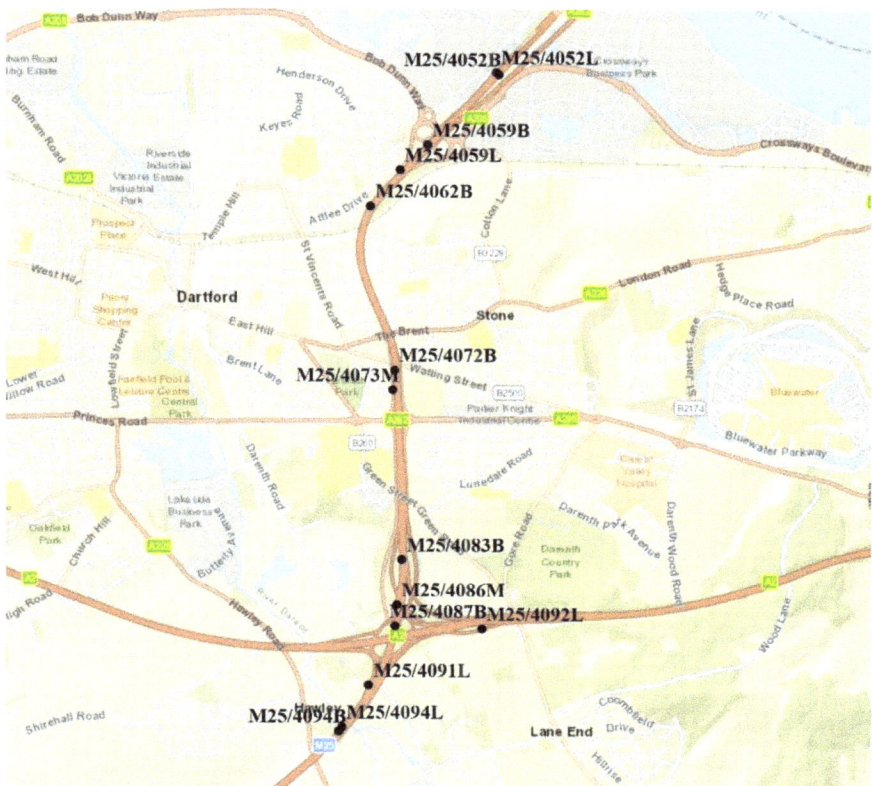

Figure 5.3 Location of selected ILDs on the M25 road (Source: OpenStreetMap)

10.5 per cent of the total 96 records per day), which means if any site has more than 10.5 per cent of its data missing for a given date, then this date will be excluded.

There are a total of 253 days from 14 ILD sites left after the cleaning process. The locations of each ILD site are shown in Figure 5.3. The distance between the furthest sites is approximately 5 km, and the selected network includes some important interchanges, such as A2, A296 and A206.

After the data selection and filtering process, any missing data points are patched based on the average values of the nearest available data points both before and after the missing data. The preprocessed data then form a matrix of dimensions $14 \times 253 \times 96 \times 1$, which means that there are 14 nodes (N), for each node there are data from a total of 24,288 time steps (T), and 1 feature input (speed).

Descriptive analyses of characteristics at the chosen locations are reviewed in this subsection. Figure 5.4 is a scatter plot of speed flow relationships for all the 14 sites from the 253 selected days, whereas Figure 5.5 presents time-series speed plots where each line represents a day.

98 *Traffic information and control*

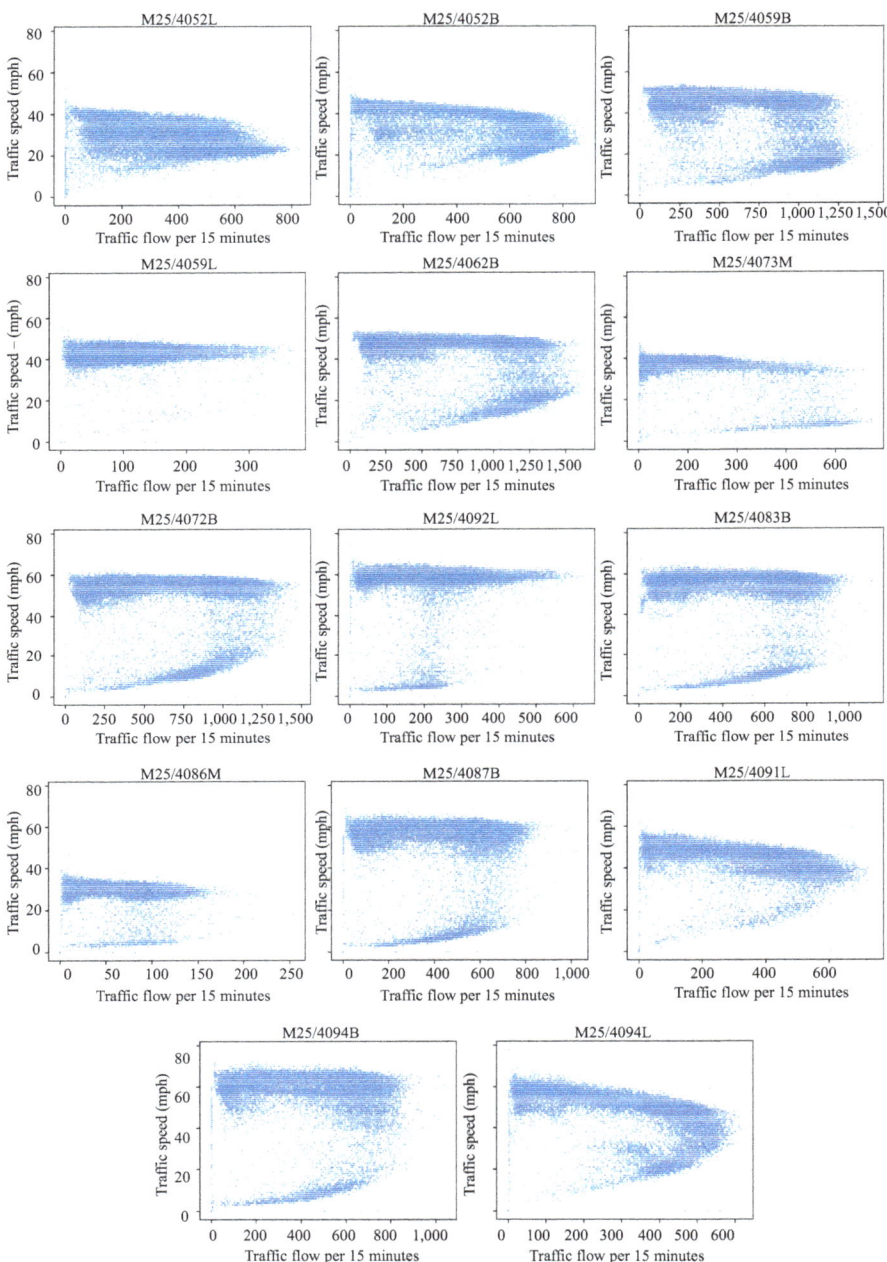

Figure 5.4   Plots of speed-flow for all the 14 sites from 253 selected days

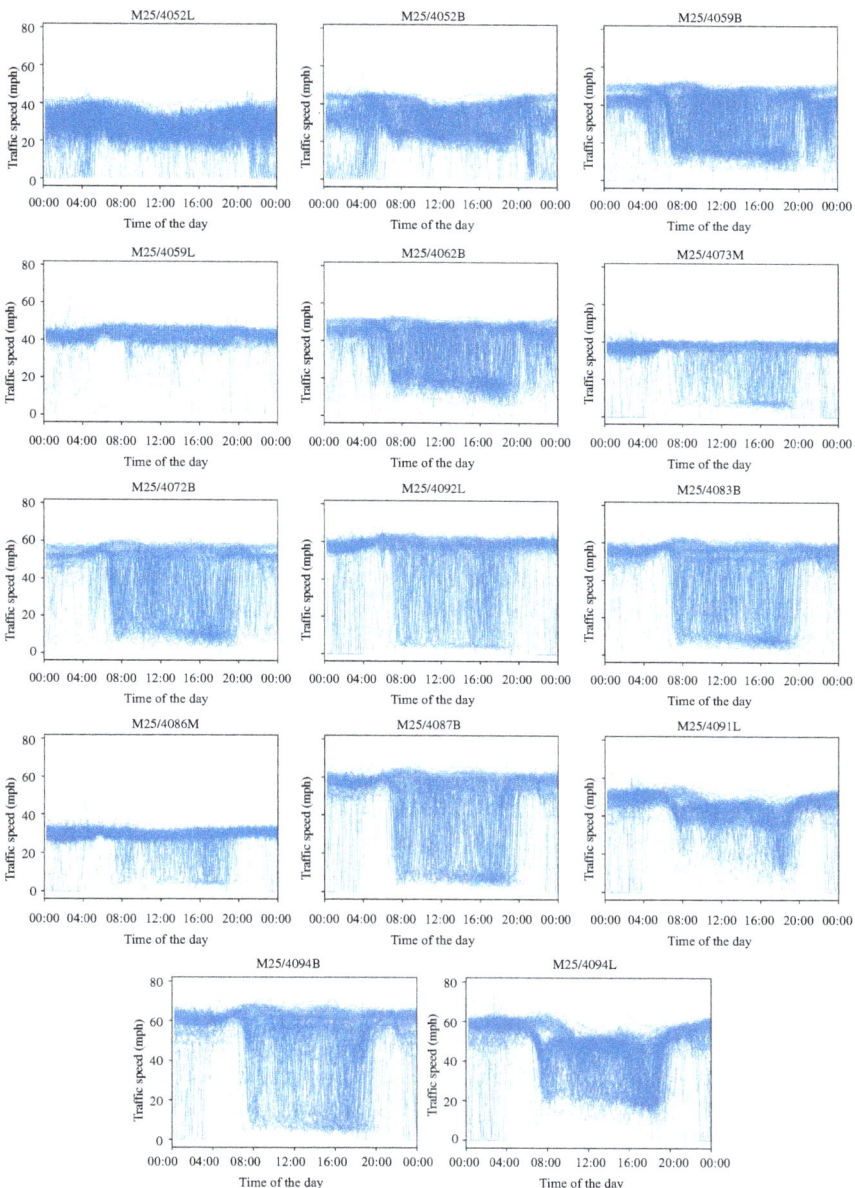

Figure 5.5   Time-series speed plots speed-flow for all the 14 sites

A number of observations can be concluded based on Figures 5.4 and 5.5:

- It can be identified that some sections of the study area use Variable Speed Limit, such as near the sites M25/4052L, M25/4052B, M25/4059B, M25/4072B and M25/4094B, due to the separable speed clusters under a non-congestion situation.
- The 'incomplete' shape of speed-flow scatter plots means that some of the sites suffer from vehicles blocking back from downstream junctions before the maximum capacity is reached. These sites include, for example, M25/4073M, M25/4083B, M25/4087B and M25/4094B.
- The time-series speed plots show that there is a lot of congestion during the daytime hours (between 7 a.m. and 7 p.m.) for the section between the M25 J2 (M25/4094B) to the M25 J1a (M25/4059B); however, it is clear that there is strong day-to-day variability in terms of the start and end time of congestion, because these times are not clustered on the plots.
- During the night time, it is clear that the Dartford tunnels were sometimes closed for maintenance (near M25/4052L and M25/4052B) as the speed readings are zero. This would have knock-on effects on the upstream sites.

In general, data quality is acceptable as there are no outliers, such as extremely large and negative speeds. However, there are a few notable issues:

- Traffic speed data are collected from ILDs, which requires a vehicle to pass over an ILD to generate speeds. For any periods with zero traffic flow, the speed reading would be zero, which generates the challenge of identifying the specific reasons causing a zero-speed phenomenon. For example, under both extremely congested and road-closed conditions, the speed record is zero; however, these congested and road-closed situations are very different from the perspective of short-term traffic prediction.
- When a road is closed for maintenance, there will still be speed readings because these are generated by the maintenance vehicles. These events can be clearly identified on the speed-flow scatter plots, where there are some very low-speed readings even though the flow is low.
- Traffic data with incidents or data with suspicious readings can be identified from the plots.

Although it is possible to remove all the data with the issues mentioned earlier, we keep all the data in our training and testing process; because it is important for the model to learn the relationships and increase the model robustness under all circumstances. In addition, to use the model in the real world, it needs to work well using data with outliers.

5.4.2.2 Graph representation

Based on the discussions in Section 5.3.1, the traffic network in Figure 5.3 can be represented in a graph, as shown in Figure 5.6.

Each node in Figure 5.6 is a traffic count site, with the correspondence between graph node numbers and ILD site IDs given in Table 5.3. The edges are

Short-term traffic prediction under disruptions using deep learning 101

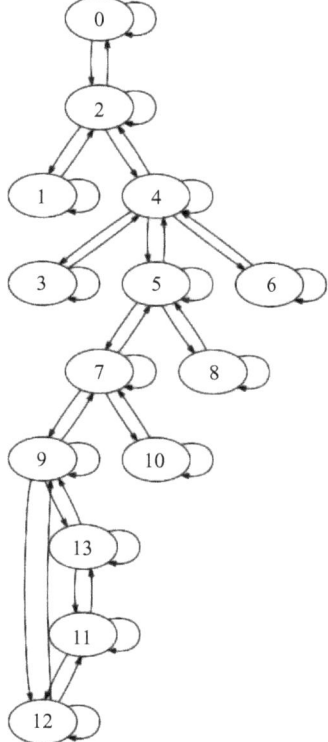

Figure 5.6 Graph representation of the study network

Table 5.3 Correspondence between graph node ID and ILD sensor ID

Node ID	ILD ID
0	M25/4052L
1	M25/4052B
2	M25/4059B
3	M25/4059L
4	M25/4062B
5	M25/4072B
6	M25/4073M
7	M25/4083B
8	M25/4092L
9	M25/4087B
10	M25/4086M
11	M25/4091L
12	M25/4094L
13	M25/4094B

directional connections between the ILD sites. As discussed in Section 5.3.3.2, a self-connection is necessary for each node due to the GCN propagation process. In total, there are 14 nodes and 56 edges on the graph representation of the study network.

5.4.2.3 Baseline methods for comparison

Four comparative methods are used: HA, kNN, Gradient Boosting Decision Tree (GBDT) and LSTM. The parameter settings of each method are presented next:

- HA: the simplest baseline method where the predicted traffic speed at each node is the average speed over all training data at the node level.
- kNN: a simple non-parametric but highly effective method in short-term traffic prediction. In this comparison, it is found that when the number of neighbours equals to 4 the model has the best performance using a grid search method.
- GBDT: a high-performance ensemble-based machine learning method which has been very popular in regression or classification problems. We carried out a grid search method to identify the best performance parameters. The learning rate is set to 0.1, the max depth of the tree is 4, the number of estimators is 100 and the minimum sample for a leaf node is 1.
- LSTM: a popular deep learning model for time-series prediction and has been used by a number of studies in the literature review for short-term prediction. We use a two-layer LSTM with 100 hidden neurons and use a batch training method. The loss function, learning rate and optimiser are kept the same as in the graph model.

5.4.3 Short-term traffic speed prediction under non-incident conditions

5.4.3.1 Model setups

In order to optimise model parameters, we implemented a grid search method to search for the most accurate parameters in the proposed TS-TGAT model. The search results show that the parameters with the most accurate results are 1D CNN kernel size $k = 4$, multi-head numbers $K = 8$ and the size of hidden neurons is 24. As discussed earlier, we are using a two-layer (two blocks in total) model, batch training method with the weight regularisation parameter $\lambda = 0.01$.

The kNN and GBDT methods are implemented based on the scikit-learn package in Python, whereas the LSTM and TS-TGAT models are implemented using Pytorch and Deep Graph Library, an open-source deep learning library.

In order to evaluate the performance of the proposed prediction model, the first 189 days of the preprocessed dataset are used for training the model, which represents 75 per cent of the whole dataset; the remaining 64 days are used for testing, which is 25 per cent of the whole dataset.

5.4.3.2 Prediction results under non-incident conditions

In addition, in order to fully understand the accuracy of the prediction models, both one-step ahead (15 min ahead) prediction and multistep ahead (30, 45 and 60 min

Table 5.4 Performance comparison of different methods under non-incident conditions

	Prediction horizon (min)	HA	kNN	GBDT	LSTM	TS-TGAT
MSE	15	135.40	32.08	26.45	26.14	24.98
	30	135.98	58.87	48.80	50.87	45.66
	45	137.55	81.66	67.68	73.26	64.45
	60	140.02	101.35	83.72	94.74	82.44
MAE	15	7.08	3.22	2.74	2.68	2.68
	30	7.11	4.29	3.87	3.79	3.77
	45	7.17	5.24	4.77	4.61	4.59
	60	7.25	5.77	5.42	5.42	5.30

ahead) prediction results are compared under incident and non-incident traffic conditions.

The prediction results from the different methods are shown in Table 5.4. These are the results based on all 14 ILD sites, using the 64-days testing data, which has never been seen by any of these models.

Five prediction methods were tested under non-incident traffic conditions. As expected, the HA method cannot predict accurately under both one-step ahead and multistep ahead, because it cannot detect the current change and patterns in traffic profiles. The kNN method performs better than the HA method but still worse than other methods, i.e. GBDT, LSTM and TS-TGAT. Moreover, the prediction accuracy of the kNN-based model deteriorates quickly when the temporal prediction horizon increases. The GBDT and LSTM results are quite similar to the prediction accuracy of the TS-TGAT. For example, compared to TS-TGAT, GBDT has very close MSE scores but relatively larger MAE gaps, whereas LSTM has closer MAE scores but relatively larger MSE gaps. This phenomenon suggests that GBDT has performed better than LSTM in reducing the number of larger discrepancies between predicted and observed data points, but LSTM is better in producing predictions that are in general closer to the observed values than GBDT.

TS-TGAT has the most accurate prediction results among the five predictors for one-step, two-step, three-step and four-step ahead, with the smallest losses of MSE and MAE. As expected, the most accurate result is the prediction for the next step (i.e. 15-min). As the prediction horizon increases from 15 min to 1 h, the results deteriorate gradually.

Figure 5.7 shows the scatter plots between predicted and observed traffic speed values using TS-TGAT. It can be seen that for majority of the 14 nodes, the TS-TGAT performs well, as the points are roughly diagonally aligned, which is especially the case when the speeds are either low or high for all nodes. This indicates that the TS-TGAT model can predict accurately under both free-flow and congested conditions. For some nodes, such as Node 2, 4, 5, 8 and 9, the points between free-flow speed and congested speed are scattered more widely. This suggests that the model is struggling to make the correct prediction when traffic is

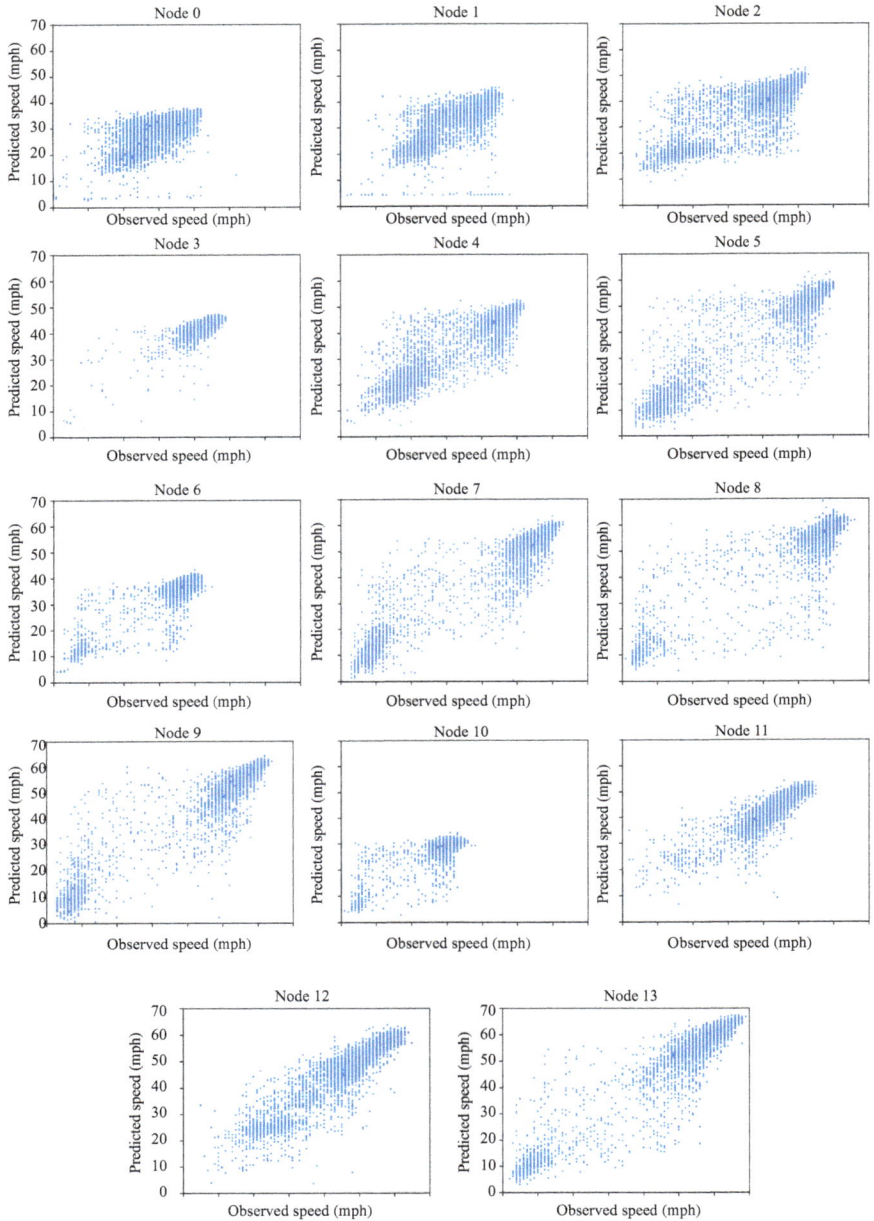

Figure 5.7 TS-TGAT predicted vs observed for the 15-min prediction scale

in an unstable condition, such as the transition from free-flow to congested, or from congested to free-flow. This traffic flow breakdown phenomenon is notoriously difficult to model because of the number of factors involved and the randomness of the process [60].

The model prediction accuracy is further constrained by the granularity of data obtained for this study. The model was designed with a strong emphasis on the spatial and temporal dependencies. However, the 15 min temporal scale with a relatively sparse loop detector network on a motorway might not have strong temporal and spatial connections compared to a more granular dataset, such as 5-min data on a dense urban network. The impact of data granularity on prediction performance should be investigated in future studies.

5.4.4 Short-term traffic data prediction under incidents

As discussed in the previous sections, although the study of deep-learning-based short-term traffic prediction under incidents is relatively limited, the aim of this research is not to train an incident-only prediction model specifically based on traffic data with incidents, but to develop a robust model with flexible frameworks, where the spatial and temporal relationships among different nodes on a graph network can be learned automatically. If the proposed TS-TGAT model is able to learn how traffic behaves on a network, then it should be able to predict the impacts of incidents with good accuracy. This subsection evaluates the prediction accuracy during incidents in more detail.

5.4.4.1 Traffic incident data

Traffic incident information is only used in the prediction stage to identify disruption events. Information about abnormal traffic conditions was directly captured from STATS19, which is Great Britain's official road traffic casualty database that contains police data on road accidents provided by the Department for Transport, including event date, start time, end time, category, location, severity, road surface conditions and other accident-related information.

In total, over 4 years of STATS19 data were collected, from 1 January 2015 to 31 December 2018. The STATS19 data thus collected have been filtered based on the same 253 days that were identified in Section 5.4.2. Because the geospatial coverage of STATS19 is the whole of Great Britain, a map matching process was carried out to identify the incidents that occurred on the road network in our study area (Figure 5.3).

After the map matching process, a manual process was carried out to look through the filtered incidents and their impacts on the network. Only the incidents with wider area impacts (such as causing severe congestion, blocking lanes, blocking-back to up-stream junctions) were selected.

In summary, a total of 8 days were selected, where incidents occurred on our study network and caused significant impacts.

5.4.4.2 Prediction results during disruptions

For prediction during disruptions, we use the same TS-TGAT model trained for prediction under normal conditions to predict the traffic speed data under disruptive conditions. Because the model has already been pre-trained, there is no parameter tuning required in prediction during disruptions.

All the baseline methods were based on the previously trained models as well. A comparison of the results between different methods is shown in Table 5.5. It can be concluded that the TS-TGAT model outperforms other models for traffic prediction under disruptive conditions. The increase of both MSE and MAE values across all methods shows that the performance of all methods worsened when predicting traffic under disruption conditions. However, the gap in MSE values has increased between the TS-TGAT method and the others, which suggests that the TS-TGAT method suffers less compared to the others.

In addition, we randomly choose a date with an accident and compare the results from all methods for traffic prediction under disruption. The chosen date is 27 December 2016, where a crash involving four cars at the M25 J1a (before the Dartford tunnel, near ILD M25/4052B and M25/4052L as shown in Figure 5.3) around 17:00, which caused significant delays, with tailbacks to M25 J3 (which is outside the study area, further south from the last detector M25/4094B). As a result, one of the tunnels had to be closed for a while. Earlier that day, a broken-down lorry between J1a and J1b, and extremely heavy traffic, had already caused lengthy delays in the study area.

Figure 5.8 is a plot comparing the prediction performances of the different algorithms for the next 15 min of this day. In order to avoid over cluttering the plot, only the top three performing algorithms have been presented (TS-TGAT, LSTM and GBDT), whereas the HA and kNN methods have been excluded.

It can be seen that all the three algorithms perform well, with TS-TGAT and LSTM seeming to be better aligned with the observed speed curve. The GBDT tends to overestimate speed slightly at many locations. We believe the performance difference between TS-TGAT and LSTM is caused by the spatial dependencies that the TS-TGAT is able to learn, which results in two effects: first is that it seems to be able to

Table 5.5 *Performance comparison of different methods under disruption conditions*

	Prediction horizon (min)	HA	kNN	GBDT	LSTM	TS-TGAT
MSE	15	123.86	40.36	35.36	34.29	33.24
	30	125.78	75.42	66.30	66.89	61.53
	45	129.33	104.84	89.62	92.90	85.28
	60	133.81	130.18	110.25	116.80	106.53
MAE	15	6.59	3.57	3.12	3.22	3.18
	30	6.68	4.82	4.48	4.49	4.42
	45	6.80	5.98	5.46	5.44	5.33
	60	6.96	6.62	6.21	6.25	6.07

Short-term traffic prediction under disruptions using deep learning 107

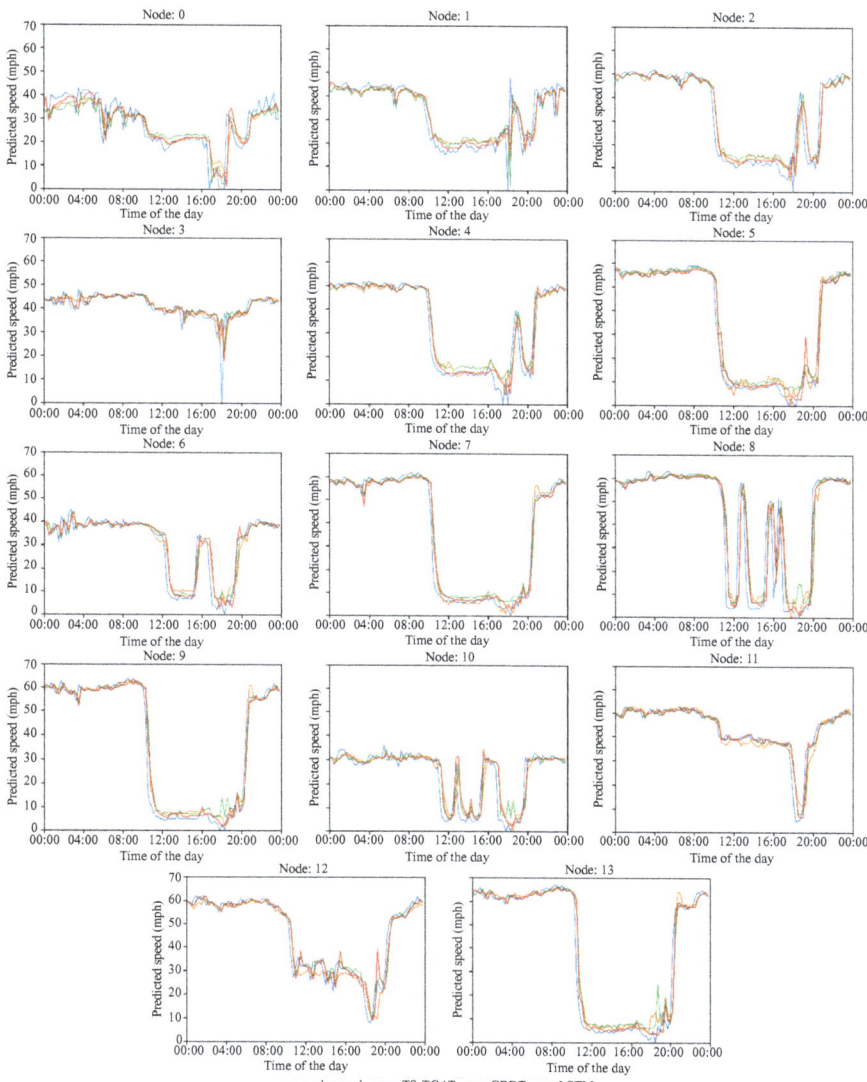

Figure 5.8 Prediction accuracy between the predicted and observed traffic speed

predict the start of speed change slightly faster than the LSTM, when the downstream road starts to become congested. This can be observed, for example, from Node 5 in Figure 5.8. When Node 4 starts to slow down first, TS-TGAT is able to capture this information earlier than LSTM for the speed prediction at Node 5. The second effect is that for prediction on a specific node, TS-TGAT is able to carry information from its neighbours. For example, Node 12, TS-TGAT is able to smooth the speed during the congested periods, due to the fact that its neighbour at Node 13 is severely congested during that period. A similar phenomenon can also be observed on Node 5, where the little spike of speed is caused by the change of speed at downstream Node 4, and TS-TGAT is able to propagate this information down the nodes.

From this analysis of the short-term traffic prediction under incidents, it is clear that the strategy works well: TS-TGAT can be trained using generic data and then used to make predictions for situations where there are disruptions on the network. Although we have faced similar data issues as discussed in the previous subsection, where the spatial and temporal granularity is perhaps not as detailed as we hoped, TS-TGAT is still able to demonstrate its capability and produces superior results compared to the other state-of-art methods.

5.5 Conclusions and future research

In this chapter, we have proposed a novel graph-based model with TS-TGAT to predict short-term traffic speed under both normal and abnormal traffic flow conditions. The novelty of the proposed prediction model is that it can learn both spatial and temporal propagation rules for traffic on a network. Important concepts and improvements are introduced to the model, for example node-level attention weights, multi-head attention and depth-wise separable CNN module to take account of the unique and complex interactions between traffic flows and traffic network characteristics.

The proposed prediction model was trained and tested using ILDs on a section of the M25 motorway network just before the Dartford Crossing (between Dartford Tunnel and M25 J2 with all slip roads). In order to make the model generic and reusable, the model was trained using generic data (including both normal and abnormal traffic flow data) and was tested under mixed conditions and disrupted conditions. A selection of baseline methods was used to benchmark the proposed model performance, including HA, kNN, GBDTs and LSTM, some of which are state-of-the-art methods in the problem of short-term traffic prediction. The results have shown that the proposed TS-TGAT method outperforms other benchmarking methods under both normal and abnormal traffic conditions.

Overall, the proposed model with TS-TGAT can be trained with generic traffic data, including both normal and abnormal traffic flow information. It can capture spatial and temporal propagation rules on a given network and make plausible predictions for both disrupted and normal traffic conditions.

A number of potential investigations or improvements can be considered for future research, including the following:

- The proposed prediction model can be tested using a more granular dataset, both spatially and temporally, for example 5-min traffic data within an urban network. In theory, a dataset with stronger connections spatially and temporally would lead to a better performance from TS-TGAT, comparing to other methods that have been tested in this study.
- The proposed model can be compared against a conventional traffic model for short-term traffic prediction, to better understand the strengths and weaknesses of

the proposed method. A conventional traffic model is usually built on equilibrium and queueing theory, which has been used in the transport industry for many decades. If any benefits are clearly and robustly quantified, this could potentially accelerate the implementation of data-driven methods in real-world applications.
- TS-TGAT has strong capability in propagating information through adjacent nodes, as demonstrated in this chapter. It also has some level of understanding about different states of traffic (i.e. congested or free-flow); for example, identifying which nodes have a higher impact on the states of adjoining nodes, and training attention weights based on this. However, a gating mechanism might be a better solution to tell the model about different states of traffic and might give the model more expressive power.

References

[1] Vlahogianni, E.I., Golias, J.C., and Karlaftis, M.G.: 'Short-term traffic forecasting: Overview of objectives and methods' *Transport Reviews*, 2004, **24**(5), pp. 533–557.

[2] Van Lint, J.W.C.: 'Reliable travel time prediction for freeways'. Thesis, Delft University of Technology, Delft, The Netherlands, 2004.

[3] Van Lint, J.W.C., Van Zuylen, H.J., and Tu, H.: 'Travel time unreliability on freeways: Why measures based on variance tell only half the story' *Transportation Research Part A: Policy and Practice*, 2008, **42**(1), pp. 258–277.

[4] Koesdwiady, A., Soua, R., and Karray, F.: 'Improving traffic flow prediction with weather information in connected cars: A deep learning approach' *IEEE Transactions on Vehicular Technology*, 2016, **65**(12), pp. 9508–9517.

[5] Smith, B.L., Williams, B.M., and Keith Oswald, R.: 'Comparison of parametric and nonparametric models for traffic flow forecasting' *Transportation Research Part C: Emerging Technologies*, 2002, **10**(4), pp. 303–321.

[6] Ghosh, B., Basu, B., and O'Mahony, M.: 'Bayesian time-series model for short-term traffic flow forecasting' *Journal of Transportation Engineering-ASCE*, 2007, **133**(3), pp. 180–189.

[7] Li, Y., Yu, R., Shahabi, C., and Liu, Y.: 'Diffusion convolutional recurrent neural network: Data-driven traffic forecasting' in International Conference on Learning Representations (ICLR), 2018.

[8] Ma, X., Dai, Z., He, Z., Ma, J., Wang, Y., and Wang, Y.: 'Learning traffic as images: A deep convolutional neural network for large-scale transportation network speed prediction' *Sensors*, 2017, **17**(4), p. 818.

[9] Polson, N.G. and Sokolov, V.O.: 'Deep learning for short-term traffic flow prediction' *Transportation Research Part C: Emerging Technologies*, 2017, **79**, pp. 1–17.

[10] Dong, Y., Polak, J., Sivakumar, A., and Guo, F.: 'Disaggregate short-term location prediction based on recurrent neural network and an agent-based platform' *Transportation Research Record*, 2019, **2673**(8), pp. 657–668.

[11] Box, G.E.P. and Jenkins, G.M.: 'Time Series Analysis: Forecasting and Control' (Holden-Day, 1970).
[12] Hamed, M.M., Almasaeid, H.R., and Said, Z.M.B.: 'Short-term prediction of traffic volume in urban arterials' *Journal of Transportation Engineering-ASCE*, 1995, **121**(3), pp. 249–254.
[13] Jeffery, D.J., Russam, K., and Robertson, D.I.: 'Electronic route guidance by AUTOGUIDE: The research background' *Traffic Engineering & Control*, 1987, **28**(10), pp. 525–529.
[14] Williams, B.M., Durvasula, P.K., and Brown, D.E.: 'Urban freeway traffic flow prediction – Application of seasonal autoregressive integrated moving average and exponential smoothing models' *Transportation Research Board*, 1998, **1644**, pp. 132–141.
[15] Williams, B.M. and Hoel, L.A.: 'Modeling and forecasting vehicular traffic flow as a seasonal ARIMA process: Theoretical basis and empirical results' *Journal of Transportation Engineering-ASCE*, 2003, **129**(6), pp. 664–672.
[16] Guo, F., Krishnan, R., and Polak, J.: 'A computationally efficient two-stage method for short-term traffic prediction on urban roads' *Transportation Planning and Technology*, 2013, **36**(1), pp. 62–75.
[17] Davis, G.A. and Nihan, N.L.: 'Nonparametric regression and short-term freeway traffic forecasting' *Journal of Transportation Engineering-ASCE*, 1991, **117**(2), pp. 178–188.
[18] Dougherty, M.: 'A review of neural networks applied to transport' *Transportation Research Part C: Emerging Technologies*, 1995, **3**(4), pp. 247–260.
[19] Hodge, V., Krishnan, R., Jackson, T., Austin, J., and Polak, J.: 'Short-term traffic prediction using a binary neural network', in 43rd Annual Universities Transport Studies Group (UTSG) Conference, (2011).
[20] Wu, C.-H., Ho, J.-M., and Lee, D.T.: 'Travel-time prediction with support vector regression' *IEEE Transactions on Intelligent Transportation Systems*, 2004, **5**(4), pp. 276–281.
[21] Leshem, G. and Ritov, Y.: 'Traffic flow prediction using AdaBoost algorithm with random forests as a weak learner', in World Academy of Science, Engineering and Technology, (Citeseer, 2007), pp. 193–198.
[22] Guo, F., Polak, J.W., and Krishnan, R.: 'Predictor fusion for short-term traffic forecasting' *Transportation Research Part C: Emerging Technologies*, 2018, **92**, pp. 90–100.
[23] Jin, X., Zhang, Y., and Yao, D.: 'Simultaneously prediction of network traffic flow based on PCA-SVR', in Liu, D., Fei, S., Hou, Z., Zhang, H., Sun, C. (Eds.): 'Advances in Neural Networks – ISNN 2007' (Springer Berlin Heidelberg, Berlin Heidelberg, 2007), pp. 1022–1031.
[24] Zheng, W., Lee, D.-H., and Shi, Q.: 'Short-term freeway traffic flow prediction: Bayesian combined neural network approach' *Journal of Transportation Engineering*, 2006, **132**(2), pp. 114–121.
[25] Abbas, M., Chaudhary, N.A., Pesti, G., and Sharma, A.: 'Guidelines for Determination of Optimal Traffic Responsive Plan Selection Control Parameters'

(Texas Transportation Institute, The Texas A&M University System, College Station, TX, 2005).

[26] Castro-Neto, M., Jeong, Y.-S., Jeong, M.-K., and Han, L.D.: 'Online-SVR for short-term traffic flow prediction under typical and atypical traffic conditions' *Expert Systems with Applications*, 2009, **36**(3), pp. 6164–6173.

[27] Venkatanarayana, R. and Smith, B.L.: 'Automated Identification of Traffic Patterns' (University of Virginia, Charlottesville, VA, 2008).

[28] Knoop, V.L.: 'Road incidents and networking dynamics effects on driving behaviour and traffic congestion'. Thesis, Delft University of Technology, Delft, The Netherlands, 2009.

[29] Baykal-Gürsoy, M., Xiao, W., and Ozbay, K.: 'Modeling traffic flow interrupted by incidents' *European Journal of Operational Research*, 2009, **195**(1), pp. 127–138.

[30] Daganzo, C.F.: 'Fundamentals of transportation and traffic operations' (Pergamon, 1997).

[31] Martin, P.T., Chaudhuri, P., Tasic, I., and Zlatkovic, M.: 'Freeway Incidents: Simulation and Analysis' (Civil and Environmental Engineering, University of Utah, Salt Lake City, UT, 2011).

[32] Jain, R., Smith, J.M.: 'Modeling vehicular traffic flow using M/G/C/C state dependent queueing models' *Transportation Science*, 1997, **31**, (4), pp. 324–336.

[33] May, A.D.: 'Traffic Flow Fundamentals' (Prentice-Hall, 1990).

[34] Guo, F.: 'Short-term traffic prediction under normal and abnormal conditions'. Thesis, Centre for Transport Studies, Imperial College London, 2013.

[35] Huang, S. and Ran, B.: 'An application of neural network on traffic speed prediction under adverse weather condition', in Proceedings of the Transportation Research Board 82nd Annual Meeting, (2002).

[36] Samoili, S. and Dumont, A.: 'Framework for real-time traffic forecasting methodology under exogenous parameters', in Proceedings of the 12th Swiss Transport Research Conference (STRC), (2012).

[37] Tao, Y., Yang, F., Qiu, Z.J., and Ran, B.: 'Travel time prediction in the presence of traffic incidents using different types of neural networks', in Proceedings of the Transportation Research Board 85th Annual Meeting, (2005).

[38] Sun, S., Zhang, C., and Yu, G.: 'A Bayesian network approach to traffic flow forecasting' *IEEE Transactions on Intelligent Transportation Systems*, 2006, **7**(1), pp. 124–132.

[39] Guo, F., Polak, J., and Krishnan, R.: 'Comparison of modelling approaches for short term traffic prediction under normal and abnormal conditions', in Proceedings of the 13th International IEEE Annual Conference on Intelligent Transportation Systems, (2010).

[40] Krishnan, R. and Polak, J.W.: 'Short-term travel time prediction: An overview of methods and recurring themes', in Proceedings of the Transportation Planning and Implementation Methodologies for Developing Countries Conference (TPMDC 2008), (2008).

[41] Guo, F., Krishnan, R., and Polak, J.: 'The influence of alternative data smoothing prediction techniques on the performance of a two-stage short-term

urban travel time prediction framework' *Journal of Intelligent Transportation Systems*, 2017, **21**(3), pp. 214–226.
[42] Wu, T., Xie, K., Xinpin, D., and Song, G.: 'A online boosting approach for traffic flow forecasting under abnormal conditions', in 2012 9th International Conference on Fuzzy Systems and Knowledge Discovery, (2012), pp. 2555–2559.
[43] Vlahogianni, E., Karlaftis, M., Golias, J., and Kourbelis, N.: 'Pattern-based short-term urban traffic predictor', in Proceedings of the IEEE Intelligent Transportation Systems Conference, (2006), pp. 389–393.
[44] Vlahogianni, E.I., Karlaftis, M.G., and Golias, J.C.: 'Short-term traffic forecasting: Where we are and where we're going' *Transportation Research Part C: Emerging Technologies*, 2014, **43**, pp. 3–19.
[45] Ma, X., Tao, Z., Wang, Y., Yu, H., and Wang, Y.: 'Long short-term memory neural network for traffic speed prediction using remote microwave sensor data' *Transportation Research Part C: Emerging Technologies*, 2015, **54**, pp. 187–197.
[46] Ma, X., Dai, Z., He, Z., Ma, J., Wang, Y., and Wang, Y.: 'Learning traffic as images: A deep convolutional neural network for large-scale transportation network speed prediction' *Sensors*, 2017, **17**(4), p. 818.
[47] Zhao, Z., Chen, W., Wu, X., Chen, P.C.Y., and Liu, J.: 'LSTM network: A deep learning approach for short-term traffic forecast' *IET Intelligent Transport Systems*, 2017, **11**(2), pp. 68–75.
[48] Zhang, Z., Li, M., Lin, X., Wang, Y., He, F.: 'Multistep speed prediction on traffic networks: A deep learning approach considering spatio-temporal dependencies' *Transportation Research Part C: Emerging Technologies*, 2019, **105**, pp. 297–322.
[49] Li, J., Guo, F., Sivakumar, A., and Dong, Y.: 'Transfer learning in short-term traffic flow prediction with deep learning tools', in Proceedings of the 99th Annual Meeting of the Transportation Research Board, (2020).
[50] Yu, B., Yin, H., and Zhu, Z.: 'Spatio-temporal graph convolutional networks: A deep learning framework for traffic forecasting' in International Joint Conferences on Artificial Intelligence (IJCAI), 2018.
[51] Wang, X., Chen, C., Min, Y., He, J., Yang, B., and Zhang, Y.: 'Efficient metropolitan traffic prediction based on graph recurrent neural network' in Proceedings of the Twenty-Seventh International Joint Conference on Artificial Intelligence (IJCAI-18), 2018.
[52] Zhang, J., Zheng, Y., Qi, D., Li, R., and Yi, X.: 'DNN-based prediction model for spatio-temporal data', in Proceedings of the 24th ACM SIGSPATIAL International Conference on Advances in Geographic Information Systems, (2016), pp. 1–4.
[53] Liao, B., Zhang, J., Cai, M., *et al.*: 'Dest-ResNet: A deep spatiotemporal residual network for hotspot traffic speed prediction', in Proceedings of the 26th ACM international conference on Multimedia, (2018), pp. 1883–1891.

[54] Chen, Q., Song, X., Yamada, H., and Shibasaki, R.: 'Learning deep representation from big and heterogeneous data for traffic accident inference', in Thirtieth AAAI Conference on Artificial Intelligence, (2016).

[55] Zhu, L., Guo, F., Krishnan, R., and Polak, J.W.: 'A deep learning approach for traffic incident detection in urban networks', in 2018 21st International Conference on Intelligent Transportation Systems (ITSC), (IEEE, 2018), pp. 1011–1016.

[56] Sun, F., Dubey, A., and White, J.: 'DxNAT—Deep neural networks for explaining non-recurring traffic congestion', in 2017 IEEE International Conference on Big Data (Big Data), (IEEE, 2017), pp. 2141–2150.

[57] Liao, B., Zhang, J., Wu, C., et al.: 'Deep sequence learning with auxiliary information for traffic prediction', in Proceedings of the 24th ACM SIGKDD International Conference on Knowledge Discovery & Data Mining, (2018), pp. 537–546.

[58] Ren, H., Song, Y., Wang, J., Hu, Y., and Lei, J.: 'A deep learning approach to the citywide traffic accident risk prediction', in 2018 21st International Conference on Intelligent Transportation Systems (ITSC), (IEEE, 2018), pp. 3346–3351.

[59] Yuan, Z., Zhou, X., and Yang, T.: 'Hetero-ConvLSTM: A deep learning approach to traffic accident prediction on heterogeneous spatio-temporal data', in Proceedings of the 24th ACM SIGKDD International Conference on Knowledge Discovery & Data Mining, (2018), pp. 984–992.

[60] Hyder Consulting: 'Motorway Travel Time Variability – Final report' (Highways Agency, 2010).

[61] Gamboa, J.C.B.: 'Deep learning for time-series analysis' *arXiv preprint arXiv:1701.01887*, 2017.

[62] Chollet, F.: 'Xception: Deep learning with depthwise separable convolutions', in Proceedings of the IEEE Conference on Computer Vision and Pattern Recognition, (2017), pp. 1251–1258.

[63] Bruna, J., Zaremba, W., Szlam, A., and LeCun, Y.: 'Spectral networks and locally connected networks on graphs' *arXiv preprint arXiv:1312.6203*, 2013.

[64] Defferrard, M., Bresson, X., and Vandergheynst, P.: 'Convolutional neural networks on graphs with fast localized spectral filtering' *Advances in Neural Information Processing Systems*, 2016, pp. 3844–3852.

[65] Kipf, T.N. and Welling, M.: 'Semi-supervised classification with graph convolutional networks' in International Conference on Learning Representations (ICLR), 2016.

[66] Veličković, P., Cucurull, G., Casanova, A., Romero, A., Lio, P., and Bengio, Y.: 'Graph attention networks' *arXiv preprint arXiv:1710.10903*, 2017.

[67] Vaswani, A., Shazeer, N., Parmar, N., et al.: 'Attention is all you need', in Advances in Neural Information Processing Systems, (2017), pp. 5998–6008.

[68] Riedmiller, M. and Braun, H.: 'A direct adaptive method for faster back-propagation learning: The RPROP algorithm', in IEEE International Conference on Neural Networks, (IEEE, 1993), pp. 586–591.

[69] Igel, C. and Hüsken, M.: 'Improving the Rprop Learning Algorithm' (Citeseer, 2000).
[70] Zhu, L., Guo, F., Polak, J.W., and Krishnan, R.: 'Urban link travel time estimation using traffic states-based data fusion' *IET Intelligent Transport Systems*, 2018, **12**(7), pp. 651–663.

Chapter 6
Real-time demand-based traffic diversion
Pengpeng Jiao[1]

At the present time, technologies about modern communication and electronic computer are developing rapidly. Due to these technologies, traffic researchers and managers can achieve real-time traffic guidance management by releasing some traffic guidance information to travelers. There are two ways to publish traffic guidance information. One is to send guidance information to travelers before they leave, named Pre-trip guidance. The other is to provide guidance information to travelers in travel, named En-route guidance.

The Pre-trip guidance includes bus route adjustment, early warning of bad weather, temporary traffic control, notice of large-scale event, etc. The main function of Pre-trip guidance is to provide some guidance about the schedule of travelers, total traffic volume, choice of trip mode and route. However, the lack of directivity of Pre-trip induction makes it hard to guarantee the induction effect. Therefore, the role of Pre-trip guidance in practical application is relatively limited.

The En-route guidance includes real-time traffic information broadcasting on expressway, variable message sign (VMS), traffic diversion guidance, on-board route guidance system and so on. The En-route guidance improves the ability of the transportation system to meet the travel demand from the view of real-time operation of the system.

The traffic diversion guidance is a common measure to solve traffic congestion, especially nonrecurrent traffic congestion. First, traffic information is collected by information acquisition system and monitoring device of the traffic diversion guidance. Second, the collected information is analyzed by the information analysis system of the traffic diversion guidance. Finally, when the traffic diversion guidance predicts congestion in the upstream road sections, it releases traffic information and diverts vehicles from the congested section to other parallel paths to complete travel [1].

This chapter introduces basic concepts about traffic information guidance and traffic diversion and shows how they can be used to realize transportation real-time demand. The organizational structure of this chapter is as follows: the first section introduces the behavior selection model under the guidance of information, the

---

[1]Department of Transportation Engineering, Beijing University of Civil Engineering and Architecture, Beijing, China

second section introduces the optimization of traffic diversion strategy, the third section introduces the research on Dynamic O–D Estimation (DODE) and the fourth section puts forward the dynamic traffic diversion model based on dynamic traffic demand estimation and prediction.

6.1 Model of path choice behavior of driver under guidance information

After the travel information and diversion guidance information is released by the traffic administrator, the driver will make the choice of diversion according to his own experience and the trust degree of diversion guidance information. The difficulty of building an individual selection model about path choice behavior is very big, and there are two main reasons cause this situation. On the one hand, the compliance of driver with diversion guidance strategy has a great influence on the optimization effect of diversion measures. On the other hand, the individual choice of driver is influenced by many factors, such as experience, current situation of diversion, personality characteristics and so on. Therefore, the feasible method in the study is to establish an analysis model for the driver's overall choice behavior.

The establishment of an analysis model for the driver's overall choice behavior needs to consider the complex composition of drivers. The complexity includes three aspects: first of all, the individual characteristics of drivers are different, including driving experience, age, gender, personality characteristics and personal preferences; second, drivers have different travel characteristics, including travel destination and nature, travel distance and travel time; finally, the number of intersections driver passed by, traffic control modes, traffic safety and road interference factors need to be discussed. Therefore, it is necessary to establish a systematic classification framework considering various factors, to analyze the mechanism of driver choice behavior in depth, and to produce the optimal diversion strategy. The interaction mechanism between driver's choice behavior and shunt-induced information is shown in Figure 6.1.

As shown in Figure 6.1, driver makes initial travel planning routes according to personal factors (travel demand, travel preferences and decision-making criteria), weather and road conditions. In the course of driving on the road, the traffic flow state changes at any time. Traffic administrator distributes traffic information, congestion reports, accident detection data and corresponding diversion guidance strategies to drivers through VMS and road traffic broadcasting. After the driver makes the choice of route diversion or non-diversion, the traffic demand on the road network redistributes and the traffic flow operation state changes again. The online traffic monitoring system can send the change of the traffic flow operation state to the control center to complete the interaction mechanism.

As early as the 1970s, Allsop [2] started the study about the impact of traffic management and control measures on dynamic driver's choice behavior. In Intelligent Transportation System (ITS), the robustness of real-time dynamic model is directly related to the driver's choice behavior. McDonald *et al.* [3] pointed out that the

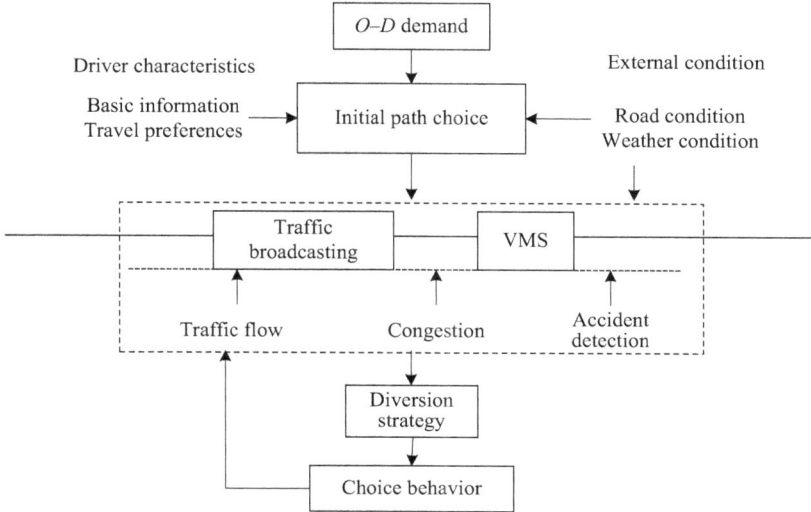

Figure 6.1 Driver selection behavior management mechanism with diversion

compliance of driver with various diversion guidance strategies has a direct impact on the control effect. In the process of ITS development in recent decades, with the improvement of modern computing speed and the improvement of communication technology, the dynamic and volatility of modern traffic is becoming more and more serious.

According to the driving time and distance, driver's behavior can be divided into two categories: within-day and day-to-day. The intraday traffic behavior in some local urban areas is within-day travel, while the inter-day traffic behavior in interurban areas is day-to-day travel. According to the analysis of the influencing factors of driver's choice behavior, the models of path choice behavior of driver are as follows.

6.1.1 Discrete probability selection model

Random utility theory and expected utility theory can build discrete selection models according to the characteristics of selected limbs, and the optimal limb selection index is determined according to the optimal limb selection index. Cascetta *et al.* [4] apply the theory of stochastic utility to the evaluation of passenger travel path scheme. According to the distribution of random error items, a stochastic probability distribution model is formed, which is mainly composed of Probit model and Logit model. The characteristics and influencing factors of drivers' choice behavior can be obtained by Revealed Preference (RP) and State Preference (SP). Abdel-Aly and Peeta *et al.* studied the Logit model based on random utility theory. Abdel-Aly [5] designed an SP survey for commuters on weekdays and built up a model of path choice behavior of driver under traffic management information. Peeta extends Abdel-Aly's selection model and specifies traffic control information as VMS guidance information. Khattak and Palma [6] and Peeta et al. [7] established an orderly Probit model to analyze the impact of weather conditions on commuters' path choice behavior during rush hour on workday. Xu et al. [8] put forward a probability distribution model of passenger flow considering multipath travel selection to solve the problem of network passenger flow distribution. The model uses Probit model that obeys normal distribution to describe the behavior of passenger travel path selection. With the increase of the travel schemes, Probit model is difficult to calculate and has poor practical application, while Logit model is easier to explain and apply [9]. It is found that the different path selection probabilities between the same *O–D* pair of multiple Logit models are determined by the absolute difference between the path utility values, which leads to the occurrence of unreasonable phenomena. With further research, Logit models have also exposed its own defects, such as "Independence from Irrelevant Alternatives" and "utility absolute error determines the selection probability." Yan [10] put forward a model of urban rail transit route selection based on nested Logit, which is more suitable for the large-scale road network. The model divides the passenger route selection into two stages, first select the transfer times, then select the route scheme. In view of the influence of effective path overlap between the same *O–D* pair on path selection, Ling [11] elaborated the probability form of path selection and the determination method of model parameters for path size modification, put forward the perception coefficient of different *O–D* pairs for path size modification Logit-type passenger flow distribution model considering length, and compared and analyzed the polynomial Logit-type passenger flow distribution model through the example experiment. Through the model and path scale, the result of Logit model is modified.

6.1.2 *Prospect theory model*

Path choice behavior of driver has subjectivity and individuality. Considering that the choice-making environment is complex and changeable, the predictive information of random utility theory and expected utility theory will deviate from the actual behavior of drivers. To improve this inherent deficiency of random utility theory and expected utility theory, Bogers and Van Zuylen [12] introduced prospect theory. The model research based on prospect theory shows that, between the path with short travel distance but large travel uncertainty and the path with long travel distance but relatively stable travel, drivers will avoid risks and choose the path planning with strong robustness. Based on the model research framework of prospect theory, Peeta and Yu [13] introduced a theoretical model of travel choice under the condition of prior information. Connors and Sumalee [14] used CPT to transform the actual travel time and probability into perceived value and perceived probability through nonlinear transformation, established a general network equilibrium model, and again verified that the travel path selection behavior based on prospect theory is feasible and effective. This theoretical model can analyze and simulate the choice behavior of drivers under different information conditions. Rasouli and Timmermans [15] prove that social psychology and decision science can also be applied to travelers' path choice behavior in uncertain environment. In order to effectively reflect the perception process of uncertain decision makers,

Chen [16] proposed a decision-making method based on Dempster Shafer theory and prospect theory. Gao [17] compares the utility maximization-based model with the prospect-based model and proves that the non-compensation heuristic model is an alternative to the traditional utility maximization method. Hasuike *et al.* [18] proposed a fuzzy interaction method between travel path selection and traveler satisfaction under uncertain traffic conditions. Luca and Pace [19] use cumulative prospect theory to simulate the behavior of travelers in a dangerous path selection and use different tools to collect data. The static data collected by network driving simulator and travel simulator are compared and analyzed.

6.1.3 Fuzzy logic model

In the decision-making process of driver's route selection, the understanding of route attribute is fuzzy, often with subjective preference consciousness. The previous model cannot deal with these problems well. Foreign scholars began to use fuzzy logic and approximate reasoning generated by subjective judgment to explain path selection behavior [20]. Chakroborthy and Kikuchi [21] first attempt to use fuzzy logic to build a path selection process model (no information) to sense travel time. In this model, travel time is defined as a fuzzy number. For the network composed of two paths, the approximate reasoning method is used to obtain the path preference of each driver, and the network loading mechanism is studied. However, the application of this method is limited to two optional paths. Based on the theory of fuzzy set and approximate reasoning, Lotan and Koutsopoulos [22] established three kinds of path selection models under the condition of traffic information. Ridwan [20] put forward the route selection model based on Fuzzy intention in 2004 and pointed out two shortcomings of the stochastic utility model: first, the stochastic utility model regards all feasible paths as alternative paths for drivers and does not consider the perceptible path of drivers as only a part of all feasible paths; second, the stochastic utility model assumes that drivers always follow the maximum utility. In fact, the driver does not always and cannot always follow the principle of utility maximization. In the research content of driver's path choice behavior, foreign scholars mainly study the influence of real-time information on driver's path choice behavior. Hounsell and Chatterjee [23] based on the RP questionnaire of drivers about VMS on London expressway expound the principles of VMS setting and qualitatively analyze the influence of VMS information on driver path selection. Hussein [24] established a driver behavior model under the influence of real-time information on a crowded commuter road in Brisbane, Australia, and used the agent principle to approximate the model. Khattak used RP survey and SP survey methods to study the route selection behavior of vehicles in radio communication liaison in Chicago area of the United States and found that the delay time had a greater impact on the route selection behavior of drivers. Markku and Heikki [25] studied the influence of weather and weather forecast on the driver's behavior before and during the trip through SP survey. The driver's behavior here includes departure time selection, travel mode selection, travel path selection and driving speed. It is found that the information of bad weather (ice and snow weather) has a greater impact on the behavior of drivers' travel time and mode before travel and has a greater impact on the driving speed of drivers during travel.

6.1.4 Other models

In addition to the previous models, various heuristic theories and other theories are also widely used in the study of travel choice behavior models. According to LOTUS (Logistic Regression Trees with Unbiased Selection), Lee et al. [26] established a Hybrid Tree model of driver's compliance behavior to variable information and used SP data to study the traffic operation status of an interstate highway in the United States. This study pointed out that the optimization of travel time is not a factor that determines the proportion of driver's compliance. Drivers' familiarity with the traffic conditions of bypasses paths and traffic operation conditions will also influence the proportion of driver's compliance.

Comprehensive analysis shows that there are two main problems need to be solved in driver's choice behavior model: first, research on the internal mechanism of driver's choice decision. The research can help traffic managers understand the driver's intuitive and behavioral responses in the process of driving and formulate a more understandable, easily identifiable and acceptable diversion guidance scheme. Second, choose a reasonable driver characteristics research theory and driver selection characteristics survey program. By ensuring the reliability of the initial data, an optimized simulation model can be established to assist traffic managers to generate a reasonable diversion strategy, and ensure the driver's compliance rate, and to make the $O–D$ demand distribution of the road network fit the optimal demand direction of the traffic manager's system.

6.2 Optimization of traffic diversion strategy

Traffic diversion strategy refers to a set of guidance schemes that can achieve diversion objectives, including various control concepts and implementation methods based on traffic flow changes in dynamic road network. Traffic diversion strategy is a methodological system that consists with a theoretical model and a diversion algorithm. Traffic diversion strategy mainly solves the problem of how to guide traffic flow and effectively balance the $O–D$ demand distribution in road network. Essentially, the mechanism of traffic diversion is similar to that of traffic assignment. Most of the theoretical studies of traffic diversion strategies are based on the theory of traffic assignment.

Effective diversion strategy can guide traffic flow around congested sections to other parallel sections in time, thus alleviating the pressure of sections and improving the efficiency of network operation. In order to meet the accuracy requirement of dynamic traffic diversion problem, traffic researchers have done a lot of research on diversion control and route guidance strategy. Researchers hope that the operation state of the expressway system can reach system optimum (SO) or user equilibrium (UE), by continuously optimizing the traffic flow distribution on the road network. These route guidance strategies mainly include responsive strategy and iterative strategy.

6.2.1 Responsive strategy

The reactive travel time is calculated according to the actual traffic state, and then the vehicle traffic assignment strategy is formulated. The control framework of responsive strategy is shown in Figure 6.2.

In the control framework of responsive strategy, the control variables are the traffic status monitored by the system such as travel time, delay or traffic density. These data are sent to the traffic control center as input data. In 1994, Messmer and Mammar first proposed using the responsive strategy to provide guidance management for the traffic flow on the road network, and input the current traffic flow data of the monitoring system online to ensure the real time of the dynamic mathematical model. However, the diversion guidance strategy generated by Messmer and Mammar model [27] is localized, and the applicable network scale is small, that means, only using in a single diversion node or exit ramp. Papageorgiou [28] introduced the optimal control method to solve the traffic assignment and traffic guidance problems. Meanwhile, he introduced the concept of feedback to establish the optimal user conditions. On the basis of macro traffic flow model, a general control framework of traffic flow control is constructed. Lioogendoom and Bovy [29] introduced the classical responsive control strategy into the guidance strategy and carried out simulation experiments. The responsive guidance strategy can be applied to medium-scale road network. In view of the urban expressway network, Wang and Papageorgiou [30] analyzed the operation of responsive guidance strategy and other guidance strategies. The results showed that the travel time of the responsive strategy is relatively deviated but approximate to the actual travel time. As an extension of the single variable responsive guidance strategy, Hawas and Mahmassani [31], Pavlis and Papageorgiou [32] put forward the multivariable responsive strategy, as well as heuristic and advanced feedback control theory to improve the sensitivity of the strategy to demand changes and driver's compliance rate.

Compared with the situation without guidance control, responsive guidance strategy can significantly reduce traffic delay. But it is still difficult to achieve the optimal state of the traffic system because of the inherent locality of responsive guidance control. In addition, the existing guidance settings of responsive guidance strategy cannot provide the traffic operation in the future, which also limits the application of responsive guidance strategy in large-scale traffic network.

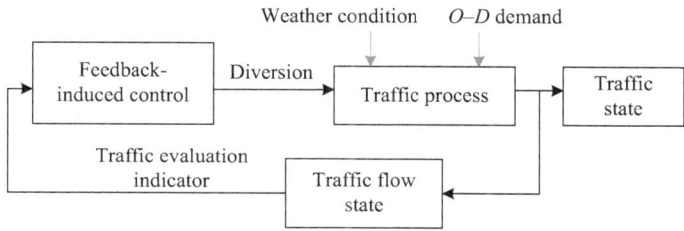

Figure 6.2 The control framework of responsive strategy

To overcome the inherent limitations of classical responsive guidance strategies, Morin [33], Messmer *et al.* [34] and Wang *et al.* [35] proposed predictive responsive guidance strategy. This strategy applies dynamic network traffic flow model in the framework of classical responsive guidance strategy and uses current traffic flow state and future demand forecast to predict the change of future traffic flow. Compared with the classical responsive guidance strategy, the predictive responsive guidance strategy is generally more robust and suitable for the large-scale road network. However, scholars point out that more research and field tests are needed to verify the practicability of this strategy in different road network topologies and traffic conditions, especially in the case of incidental traffic congestion.

6.2.2 Iterative strategy

Corresponding to responsive guidance strategy, another kind of guidance strategy is iterative guidance strategy. In the operational framework of iterative guidance strategy, the current parameters of traffic flow in road network are taken as the initial values of the model. In the guidance framework of iterative guidance strategy, the dynamic network traffic flow model simulates the traffic flow on the road network. The iterative guidance strategy finally predicts the traffic status in the future by updating iteratively between the operational framework and the guidance framework. The control module of the system adjusts the route guidance strategy of users on the road network according to the real-time traffic condition, so as to achieve the control objectives of traffic managers. In addition, iterative guidance strategy is essentially predictive, which can publish pre-guidance strategies according to the predicted traffic state and adjust them in real time. The operational framework of iterative guidance strategy is shown in Figure 6.3.

When the control objective of traffic managers is SO, the mechanism of iterative guidance strategy is to determine the proportion of traffic diversion at traffic diversion nodes. The value of the proportion of traffic diversion is made on the basis of several influencing factors of diversion strategies, including driver's compliance rate, accident and weather conditions. Finally, the diversion information will be distributed to drivers of road network and make the performance

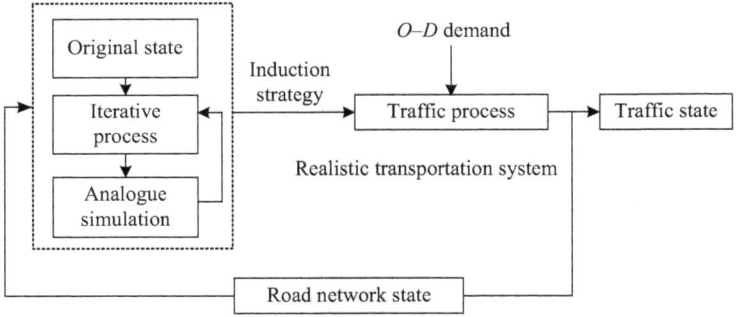

Figure 6.3 The operational framework of iterative guidance strategy

index of the road network reach the optimum level [36–39]. Mahmassani and Peeta [40], Ben *et al.* [41], Wisten and Smith [42] and Wang *et al.* [43] have conducted relevant studies on the UE guidance strategy. The core of UE guidance strategy is to revise the diversion rate of each path in real time and reduce the travel time difference between two alternative paths by running the simulation model iteratively in a given period of time. The direction of iteration is the equilibrium state of road network described by Wardrop equilibrium. He *et al.* [44] put forward a route guidance strategy for a single target road network to achieve the user optimum in the case of high traffic demand. And He *et al.* [45] put forward the mean velocity feedback strategy (MVFS) to achieve user optimization for the asymmetric traffic network. He *et al.* [46] proposed a traffic-condtition-based strategy (TCB). The strategy is based on an estimate of free movement and traffic congestion and is able to approach to the user optimum steadily.

The disadvantage of iterative guidance strategy is that it relies too much on traffic model (such as route selection model), the calculation of iteration process is complicated, the calculation burden is heavy, and it has no advantage for real-time traffic control in the large-scale road network.

6.3 Research on dynamic *O–D* estimation

The *O–D* estimation is a special case of optimization problem. It calculates the *O–D* volume by making the calculated road traffic volume and its observation road traffic volume consistent. Generally, the number of observed sections is far less than the number of unknown *O–D*. The only *O–D* traffic volume cannot be obtained only by observing the traffic volume condition of the section. Some assumptions must be added. For example, it can assume that the *O–D* distribution conforms to the gravity model, or that the current *O–D* matrix can be modified by observing the traffic volume of the road section. Therefore, Cascetta and Nguyen [47] defines the *O–D* estimation problem as the process of determining the *O–D* matrix by effectively combining the observed traffic volume and other information. Since the number of observed traffic volume is generally much smaller than that of *O–D* demand, Cremer and Keller [48] put forward the first series data model in the history of *O–D* estimation theory. The Cremer and Keller model [49] extends the observed traffic data to the traffic data, including current and historical traffic data, thus successfully transforming the original nonstationary problem of *O–D* estimation into a stationary problem. When the selected period increases, the *O–D* estimation even becomes super-statically determinate problem.

According to the scale of road network, the DODE model can be divided into intersection model, expressway model and whole road network model. The main difference between these three models is the consideration of travel time and the assumption of path selection. Intersection model does not need to consider the problem of route selection, and the travel time at intersections is generally much smaller than the observation time and can be neglected. Expressway model does not need to consider the problem of route selection too, but the travel time at expressway section cannot be neglected. As for the whole road network model, the problem of route selection and travel time must be considered at the same time.

6.3.1 Intersection model

The DODE model applied to intersections generally uses the traffic demand or turning split parameters between O–D directly as its state variables. The basic method is to estimate the turning split parameters by using the time series of flow between the import and export sections and then get the O–D matrix.

In the 1980s, Cremer and Keller [50] proposed a non-traffic assignment DODE model, which laid the foundation for the subsequent O–D estimation research. Cremer–Keller model combines short-term traffic flow prediction with time series analysis and establishes a recursive structure of the estimation method, thus establishing a simple O–D estimation model for intersection. In 1989, Nihan and Davis [51] made another progress. Based on the principle of recursive prediction error, Nihan and Davis modified the least square model in the Cremer–Keller model. This model can consider the influence of the past observation on the current state. The disadvantage of this model is that the sensitivity of the parameters to the change of the observation is relatively low. Sherali et al. [52] improved the least square method of O–D estimation problem. Sherali pointed out that the former parameter optimization model does not take the steering ratio as the restriction condition of flow conservation. Besides the shortcomings mentioned earlier, the historical parameter optimization model needs to standardize the natural constraints in the calculation process, which is complex and makes the accuracy of results reduce easily. In order to solve problems on parameter optimization model, Sherali uses the conjugate gradient algorithm to solve the problem of least square method with full constraints. The conjugate gradient algorithm can overcome the overload effect caused by large amount of calculation and high precision, when the O–D estimation model applied in real-time online. As a supplement to the study of parameter optimization model, some scholars also use intelligent algorithm to solve the O–D estimation problem. Jiao et al. [53] used genetic algorithm to solve Cremer and amp; Keller's model and Nihan and amp; Davis's model. In the process of solving these models, he assumed that the steering division parameter in unit time was a constant fixed value.

6.3.2 Expressway model

For the DODE model applied on urban expressway, it is necessary to consider the travel time of vehicles on the section. According to whether the O–D estimation model considers the mapping relationship between the O–D flow and the traffic flow, the O–D estimation model can be divided into the dynamic traffic assignment model and the non-dynamic traffic assignment model. The O–D estimation model based on dynamic traffic assignment needs to explain the relationship between the O–D demand and the traffic flow on the section. The non-dynamic traffic flow assignment model only needs to calculate the turn partition parameters of the main section or on ramp of the urban expressway and does not need to consider the relationship with the traffic flow on the section.

Chang and Wu model [54] is a DODE model based on the non-dynamic traffic assignment. Based on the description equation of macro traffic flow model, Chang and Wu model analyzes the internal causes of O–D demand distribution of expressway under traffic congestion conditions. This model chooses the combination of distribution parameters and distribution proportion as the state vector of the optimization problem. The rule of state evolution is random walk, and the observation equation is the relationship among the section flow, the observed traffic volume, distribution proportion and division parameters of the entrance and exit ramps. Chang and Wu model does not use the prior O–D matrix but uses the extended Kalman filter algorithm to solve the DODE model. Lin and Chang [55] proposed an improved scheme for Chang and Wu model. This model assumes that the travel time of vehicles with the same starting and ending points approximately obeys the normal distribution. By monitoring the departure time of vehicles, the speed change of vehicles with the same travel demand can be predicted. This method can effectively reduce the number of unknown variables and improve the robustness of the model. In order to avoid the influence of the incomplete initial data and the observation error on the accuracy of the O–D estimation, Lin and Chang use the observation error between the initial O–D matrix and O–D matrix with the extension to improve the algorithm and finally form the robust model of the O–D estimation. On the basis of the robust model, Lin and Chang also proposed a method to increase the accuracy of the O–D demand estimation model, by using the deviation between the travel time required by travelers to travel on the road network at the expected speed and the travel time required by drivers to take diversion measures after congestion.

In 2001, Sherali and Park [56] expanded the range of data use on the offline O–D estimation model. They used the road traffic on the whole road network to estimate the path flow or O–D demand and completed the improvement of the parameter optimization model. The Sherali and Park model can obtain the shortest path between specific O–D pairs and optimize the deviation between the assigned value and the actual value of the O–D estimation results.

6.3.3 Network model

The DODE model applied to the whole road network needs to comprehensively consider the impact of the historical period, that is, the travel time factor. The impact of the driver's path choice behavior on the O–D demand also needs to be considered, that is, the path choice factor. Okutani and Stephanedes [57] used the linear KF algorithm to solve the DODE model of the whole road network. This model takes O–D demand as state vector and explicitly considers the influence of random error and observation error on model accuracy. In the Okutani and Stephanedes model, the distribution form of O–D demand is autoregressive process, but the process has not been studied deeply. Besides, this model does not study the acquisition of dynamic traffic assignment matrix considering route selection factors. Generally speaking, Okutani and Stephanedes model is the basis of *O–D* estimation model based on dynamic traffic assignment, which determines the idea and method of combining dynamic traffic assignment and DODE.

In 1991, Bell [58] conducted an in-depth study on the travel time in the DODE model of the road network. Bell model assumes that in a traffic flow model where the travel time of vehicles approximately follows the geometric distribution, the fleet will gradually dissipate. Bell model is the basis of DODE model of the road network, which does not consider the impact of dynamic traffic assignment. Chang and Tao [59] innovatively used the traffic flow through the screen line and the cordon line into the *O–D* estimation model of the road network, which improved the number of effective observation equations. Besides, Chang and Tao considered the influence of the intersection and combined the parameters such as turning proportion and turning flow of the intersection with the DODE model. Finally, they proposed a two-stage estimation framework for solving the model.

Tavana [60] established a two-level optimization model of DODE of road network. The upper layer is the generalized least square model, and the lower layer is the dynamic UE model based on DynaSMART framework. Zhou [61] improved Tavana model. Zhou model used the traffic flow data of multiple normal days on a fixed road section for *O–D* estimation. There are two objectives of this model. One is to minimize the deviation between the actual and estimated traffic flow values on a road section, and the other is to minimize the deviation between the actual and predicted *O–D* estimation values on a road section. Bert (2009) [62] further improved Tavana model with heuristic algorithm. The upper layer of Bert model considers the dynamic propagation characteristics of macro traffic flow and improves the *O–D* model based on the least square method. The lower layer uses the dynamic network loading module of Aimsun software to generate the user demand distribution form under the dynamic UE condition in real time.

6.4 Dynamic traffic diversion model based on dynamic traffic demand estimation and prediction

Despite the promising progress from those integrated diversion control models, there are some researchers also focus on the dealt with real-time diversion control and traffic state estimation and prediction on the basis of rolling time horizon. But there are still some work needs to be done regarding the real-time diversion control, which can best demonstrate its effectiveness in dealing with the following four problems: (1) the coincidental traffic jams. In current techniques, traditional traffic system management and control measures lack effectiveness in dealing with coincidental traffic jams. (2) The phenomenon of congestion transfer. The existing diversion measures are easily to cause congestion transfer for the lack of consideration on the dynamic demand change. (3) Low efficiency of diversion guidance. The information of traffic guidance lacks the prediction of the traffic flow, so that the efficiency of diversion guidance is low. (4) Difficulty in completing the dynamic updating of the traffic diversion guidance strategy. For the existing diversion models, it is difficult to complete the dynamic updating of the traffic diversion guidance strategy according to the traffic detection data of the next period. In response to the previous research needs, this chapter develops a model for integrated diversion control of a DODE, in which dynamic demand change of general road network is employed to coordinately divert traffic under coincidental traffic jams.

6.4.1 DODE model of urban expressway
6.4.1.1 The module of METANET model

In view of the high traffic density and high speed of urban expressway, METANET model was proposed by Papageorgiou, which has a good description effect of free flow, critical traffic state and saturated traffic flow state [63–65]. The METANET model uses the directed graph to represent the actual road network. In the directed graph of road network, the urban expressway sections have the same section properties, and the geometric linearity of the road has no obvious change. The METANET model describes the average behavior of the traffic flow in a certain time and space. Considering the simulation accuracy, the dimensions of time and space of the METANET model need to be discretized as Figure 6.4.

In Figure 6.4, the section m are evenly divided into the same length intervals, and the length of each interval L is generally 500–1,000 m. The whole simulation time is divided into k intervals, and the time of each interval is T. The time of each interval T should satisfy the condition:

$$T < \frac{\min L}{v_{\text{free},m}}$$

which means the time of the vehicle driving through the minimum length interval with the maximum driving speed $v_{\text{free},m}$. Key variables and parameters used in the model presentation are given as follows:

- λ_m: the number of lanes in the section m;
- $\rho_{m,i}(k)$: the average density of each lane in the length interval i and time interval k of the section m, veh/km/lane;
- $v_{m,i}(k)$: the average speed of each lane in the length interval i and time interval k of the section m, veh/km/lane;

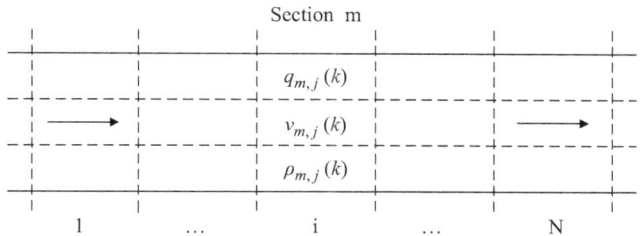

Figure 6.4 Directed graph of urban expressway

- $q_{m,i}(k)$: the traffic flow that leaves the length interval i of the section m during the time interval $[kT, (k+1)T]$, veh/h.

6.4.1.2 The module of DODE model

The O–D demand of urban expressway is equivalent to the traffic flow of on-ramps and exit ramps. As the distance between each pair of on-ramp and exit ramp is so long, the DODE model of expressway should take into account the travel time of traffic flow. To facilitate the establishment of estimation model, the expressway is divided into a number of regions. Each region starts from an on-ramp and ends with an exit ramp, as shown in Figure 6.5.

In this section, the lane-group-based model is extended to incorporate the impacts of multiple detouring traffic flows heading to different downstream on-ramps. Key variables and parameters used in the model presentation are given as follows:

- N: number of subsection, subsection numbers were recorded as $1, 2, \ldots, N$;
- $q_k(i)$: number of vehicles entering the expressway section from the on-ramp i during the time interval k, $i = 1, 2, \ldots, N-1$;
- $y_k(j)$: number of vehicles leaving the expressway section from the exit ramp node j during the time interval k, $j = 2, 3, \ldots, N$;
- $x_{k-m}(i,j)$: number of vehicles depart from the on-ramp i to the exit ramp node j during the time interval k, $i < j$;
- $p_k(i,j)$: percentage of traffic flow in $q_k(i)$ with destination to the exit ramp node j, $i < j$;
- $a_k^m(i,j)$: percentage of traffic flow in $x_{k-m}(i,j)$ arriving at the exit ramp node j during the time interval k, $m = 1, 2, \ldots, M$;
- $m_k(l)$: number of vehicles driving on the expressway section l during the time interval k, $l = 2, 3, \ldots, N-1$;
- $m_k^l(j)$: number of vehicles driving on the expressway section l with destination to the exit ramp node j during the time interval k, $j = l, l+1, \ldots, N-1$;
- $E_{k-m}^k(j)$: percentage of number of detouring vehicles driving across the border of section $j-1$ during the time interval $k-m$ and arriving at the exit ramp node j according to the diversion information during the time interval k;
- $d_k(i)$: percentage of traffic flow in $m_k(j-1)$ detour to the exit ramp node j according to the diversion information.

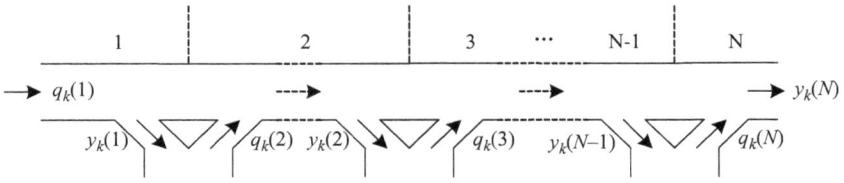

Figure 6.5 Diagrammatic sketch of urban expressway

According to the diagrammatic sketch of urban expressway showed in Figure 6.5, there is only one path for the vehicles on the road, that is, from the upstream or the on-ramp on the expressway, and then runs along to the downstream or the exit ramp. The traffic flow of on-ramp and exit ramp on the expressway section, including the upstream flow of the start section $q_k(0)$, the downstream flow of the end section $y_k(N)$, the inflow of on-ramp $q_k(i)$ and outflow of exit ramp $y_k(j)$, can be obtained by the on-line monitoring system.

Generally, all vehicles have similar behavior characteristic for the traffic flow $q_k(i)p_k(i,j)$. Under the condition of knowing the average travel time $\bar{t}_k(i,j)$ of traffic flow, it can be assumed that the travel time of all vehicles is subject to the normal distribution of average travel time $\bar{t}_k(i,j)$, and the distribution parameters can be determined according to historical data. Thus, the travel time of vehicle $t_k(i,j)$ is obtained. When $\omega\Delta t \leq k\Delta t + t_k(i,j) \leq (\omega+1)\Delta t$, it means that the traffic flow $q_k(i)p_k(i,j)$ can go from on-ramp i within the time interval k to exit ramp node j within the time interval ω. At the same time, the O–D demand $q_k(i)p_k(i,j)$ can affect the traffic flow $y_{\omega-\sigma}(j), \ldots, y_{\omega-1}(j), y_\omega(j), y_{\omega+1}(j), \ldots, y_{\omega+\sigma}(j)$ of exit ramp node j, in the adjacent time interval. And the proportion of the impact on the traffic flow of each exit ramp node j is $a_{\omega-\sigma}^{\omega-\sigma-k}(i,j), \ldots, a_{\omega-\sigma}^{\omega-1-k}(i,j), a_{\omega-\sigma}^{\omega-k}(i,j)$, $a_{\omega-\sigma}^{\omega+1-k}(i,j), \ldots, a_{\omega-\sigma}^{\omega+\sigma-k}(i,j)$. The parameter σ in the formula is the biggest time period of interference, which is decided by the current congestion of the road network. And in this chapter, σ is 5.

Based on the analysis mentioned earlier, the dynamic O–D model for the expressway section is described as follows:

$$J = \min \sum_{k=1}^{K} \sum_{j=1}^{N} \left| y_j(k) - \sum_{m=0}^{M} \sum_{i=1}^{j-1} a_k^m(i,j) q_{k-m}(i) p_{k-m}(i,j) \right| \quad (6.1)$$

6.4.2 Traffic diversion model of urban expressway

6.4.2.1 Simulation of driver's diversion behavior

The traffic diversion strategies are distributed to the drivers through information boards (such as VMS), and information boards are generally set up on the mainline sections of expressway before its exit ramps. When the drivers who are driving on the mainline sections of expressway receive the traffic diversion information, they can choose to detour to the parallel main road or to continue traveling on the mainline section of expressway according to the diversion rate $\beta_k(j)$. The distribution of traffic flow at the exit ramp node j is described as follows:

$$y_k(m,j) = \beta_k(j) m_k(l), \ 0 \leq \beta_k(j) \leq 0.5 \quad (6.2)$$
$$m_k(m,l) = [1 - \beta_k(j)] m_k(l), \ 0 \leq \beta_k(j) \leq 0.5 \quad (6.3)$$

where $y_k(m,j)$ and $m_k(m,l)$ represent the traffic flow driving into exit ramp node j because of traffic guidance and the traffic flow continue driving on the mainline sections of expressway. Since the traffic diversion volume affects arterial traffic flow and tends to form a new equilibrium between the expressway and the arterial traffic network, the upper limit value of the diversion rate in this case can be set to 50%.

6.4.2.2 Influence of diversion on the traffic flow of exit ramp

Traffic diversion strategy based on dynamic O–D is a comprehensive system that involves various aspects of data acquisition and technical support. The basis of traffic diversion strategy is the actual evolution of traffic flow, which can be affected by several disturbance factors like demand fluctuation of general road network, traffic snags and bad weather. Therefore, the model needs to monitor the change of traffic flow on the road network first, including the volume of traffic flow, traffic density, average driving speed, occupancy rate and so on. And then, the real-time traffic flow data need to be processed, including traffic state recognition and short-term traffic flow forecasting. Then we need to process the data of real-time traffic flow, including traffic status recognition, short-term traffic flow prediction and traffic accident detection. After that, in the diversion module, the system evaluates the performance of the road network according to the results of data detection and analysis results. The system will publish the traffic diversion strategy according to the dynamic evolution of current traffic flow. Having received the traffic diversion information, the driver will response to the traffic diversion information according to the validity of information and his experience, that is, to obey the traffic diversion strategy or not. And the diversion rate is decided by the driver's diversion behavior. The driver's diversion behavior affects the traffic flow in the on-ramps and exit ramps, and the dynamic O–D demand changes accordingly. At last, the on-line monitoring system will detect the change of traffic flow and feeds it back to the module of diversion strategy, affecting the diversion behavior once again.

Combined with the analysis of O–D demand estimation and diversion strategy mentioned earlier, it is obviously that the traffic flow on the exit ramp node j in the time interval k includes two parts: O–D volume form on-ramp i to exit ramp node j in the time interval $k - m, k - m + 1, \ldots, k$; traffic flow that detouring to the parallel main road according to the diversion information. Therefore, the traffic flow at the exit ramp node j is described as follows:

$$y_k(j) = \sum_{m=0}^{M} \left[\sum_{i=1}^{j-1} a_k^m(i,j) q_{k-m}(i) p_{k-m}(i,j) + E_{k-m}^k(j) \beta_{k-m}(j) m_{k-m}(j-1) \right]$$

(6.4)

where the O–D distribution rate $a_k^m(i,j)$ can affect the traffic flow which departs from the on-ramp i to the exit ramp node j deeply.

6.4.2.3 Evaluation index of road network performance

Traffic diversion strategy is a service-oriented traffic management for drivers, whose objective is to optimize the performance index of the whole road network and to achieve the optimal system state or user optimal state finally. Therefore, the objective of the traffic diversion model established in this chapter is that the road network performance index is optimal, and Total Time Spent (TTS) of the road network is chosen as the performance index. The formula is shown as follows:

$$J = T \sum_{k=1}^{K} \left\{ \sum_{n=1}^{N} \sum_{i} \rho_{m,i}(k) L_m \lambda_m + \sum_{o} w_o(k) \right\} \quad (6.5)$$

where T represents the time of each time interval, $\rho_{m,i}(k)$ represents the average density of each lane in the length interval i and time interval k of the section m, L represents the length of each length interval, λ_m means the number of lanes in the section m, and w_o means the number of queued vehicles on the starting point o.

6.4.3 Dynamic traffic diversion model based on DODE

Based on the analysis mentioned earlier, a traffic diversion model based on DODE is established. The traffic diversion model based on dynamic traffic demand estimation and prediction consists of the traffic diversion module and the DODE. The traffic diversion module uses the online monitoring O–D volume and road network status to optimize the traffic diversion strategy, which is reflected in the traffic diversion model as the diversion rate and aims to optimize the performance index of road network after diversion. The DODE module gets the O–D volume by minimizing the deviation of the traffic flow in the on-ramps, when the traffic diversion strategy works.

In this process, the DODE module and dynamic traffic diversion module are executed alternately. The framework of the dynamic traffic diversion model based on DODE is shown in Figure 6.6.

The optimal objective of the traffic diversion module is described as follows:

$$J_1 = T \sum_{k=1}^{K} \left\{ \sum_{n=1}^{N} \sum_{i} \rho_{m,j}(k) L_m \lambda_m + \sum_{o} w_o(k) \right\} \quad (6.6)$$

s.t.

$$\rho_{m,l}(k+1) = \rho_{m,l}(k) + \frac{T}{L_m \lambda_m} [\rho_{m,l-1}(k) - \rho_{m,l}(k)] \quad (6.7)$$

$$v_{m,l}(k+1) = v_{m,l}(k) + \frac{T}{\tau} [V(\rho_{m,l}(k)) - v_{m,l}(k)]$$
$$+ \frac{T}{L_m} v_{m,l}(k) [v_{m,l-1}(k) - v_{m,l}(k)] - \frac{vT}{\tau L_m} \quad (6.8)$$

$$V(\rho_{m,l}(k)) = \min \left\{ v_{\text{free},m} \exp \left[-\frac{1}{a_m} \left(\frac{\rho_{m,l}(k)}{\rho_{\text{crit},m}} \right)^{a_m} \right], \eta \cdot v_{\lim} \right\} \quad (6.9)$$

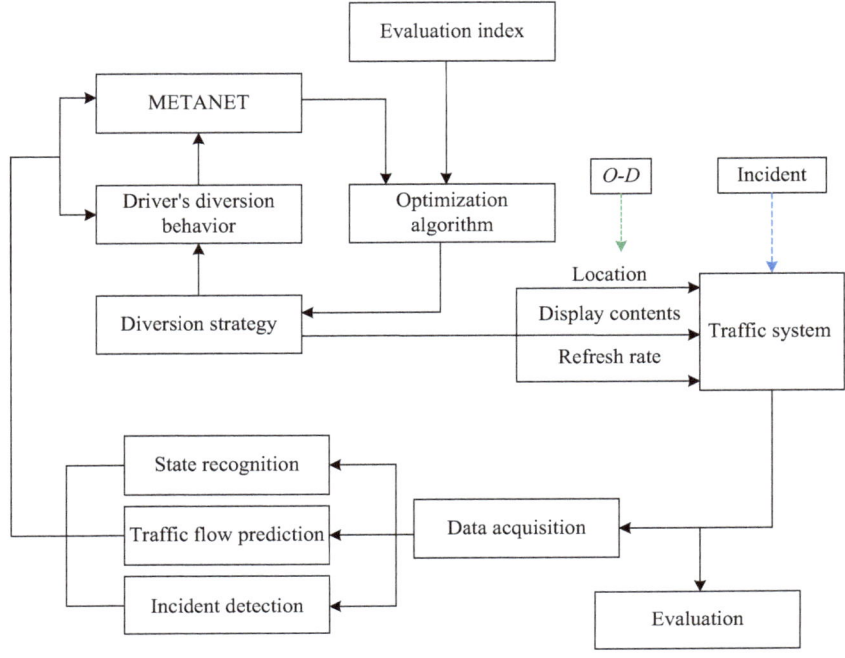

Figure 6.6 Framework of the dynamic traffic diversion model based on DODE

$$w_o(k+1) = w_o(k) + T(d_o(k) - q_{m,o}(k)) \tag{6.10}$$

$$q_{m,o}(k) = \min\left(d_o(k) + \frac{w_o(k)}{T}, Q_{cap,m}, Q_{cap,m}\frac{\rho_{\max,m} - \rho_{m,l}(k)}{\rho_{\max,m} - \rho_{crit,m}}\right) \tag{6.11}$$

when the diversion strategy is implemented on the mainline section of expressway exit ramp node j. The dynamic O–D model for the expressway section is as follows:

$$J_2 = \sum_{k=1}^{K}\sum_{j=2}^{N}\left|y_k(j) - \sum_{m=0}^{M}\left[\sum_{i=1}^{j-1}a_k^m(i,j)q_{k-m}(i)p_{k-m}(i,j) + E_{k-m}^k(j)\beta_{k-m}(j)m_{k-m}(j-1)\right]\right| \tag{6.12}$$

s.t.

$$0 \le p_k(i \cdot j) \le 1, \ i = 1, 2, \ldots, N-1, \ j = 2, 3, \ldots, N$$

$$\sum_{j=i+1}^{N} p_k(i,j) = 1, \ i = 1, 2, \ldots, N-1 \tag{6.13}$$

$$m_k(l) = \sum_{j=l}^{N} m_k^l(j), \forall k, \forall l \tag{6.14}$$

$\beta_k(j) \leq 1, \forall k, \forall j$

$0 \leq a_k^r(i,j) \leq 1, \ 1 \leq j \leq N, \ r = 1, 2, \ldots, k$

$$\sum_{k=r}^{r+u} a_k^r(i,j) = 1, \ 1 \leq i \leq j \leq N \tag{6.15}$$

$$q_k(i) = \sum_{j=i+1}^{N} x_k(i,j), \ i = 1, 2, \ldots, N-1 \tag{6.16}$$

6.4.4 Model solution

This study adopts the genetic algorithm to solve the proposed model, which mainly consists of seven steps:

- Step 1: Description of initial operating conditions.
  According to the objective of the model, this chapter determines several initial operating conditions like the number of controlled ramps, the number of changes in the diversion rate, etc.
- Step 2: Determination about the parameters of genetic algorithm.
  The parameters include the size of the population S, the number of variables, the probability of crossing C_r, the probability of mutation M_r, etc.
- Step 3: Determination of fitness function.
  Plug initial generation population into the objective function $J(s)$ and use formula (6.17) to transform the objective function $J(s)$ into fitness value $F(s)$:

$$F(s) = \begin{cases} C_{\max} - J(s), & J(s) < C_{\max} \\ 0, & J(s) \geq C_{\max} \end{cases} \tag{6.17}$$

  where $J(s)$ is the objective function of diversion model, $F(s)$ is the fitness value of subpopulation s, $C_{\max}$ is the maximum value of the objective function that currently searched.
- Step 4: Operation of selection.
  Use the proportional selection operator to select the fitness individual and define the range of individual fitness with roulette.
- Step 5: Crossover operation.
  Pair all individuals of the $g+1$ generation population randomly and generate a random number $e(e \in [0,1])$ for each pair. If $e < C_r$, this pair needs to operate crossover. The paired individuals composite the $g+2$ generation population.
- Step 6: Executing mutation operation according to the probability of mutation.

134 *Traffic information and control*

Random number $e_m (e_m \in [0, 1])$ is generated for the individual position variable. If $e_m < M_r$, mutation operation will be executed at this position, and the adjusted rates after mutation need to remain within $[0, 1]$.
- Step 7: Judging the stopping condition.

Judge whether the parameter will meet the termination generations of genetic operation.

If the parameter does not meet the termination generations, return to Step 4; and if it satisfies the abort condition, then outputs the optimized operation result.

6.4.5 Case study
6.4.5.1 Experimental design
To illustrate the applicability of the dynamic traffic diversion model based on DODE, a hypothetical corridor, including 10 expressway exits and 29 arterial intersections, was employed in this chapter. Basic layouts of the road network are given in Figure 6.7.

In this hypothetical corridor, an incident occurs on the expressway mainline section (between Exits 6 and 7), and the dynamic traffic diversion model based on DODE determines a set of detour operations. The control method of diversion strategy is to detour the main traffic flow from the upstream ramp of the congestion point to the parallel main road, reducing the local traffic density of the expressway and alleviating the vehicle congestion gradually. The vehicles that run on the parallel main road to detour the congestion points gradually back to the expressway mainline section through the on-ramp on the downstream of the congestion point. Therefore, during the entire experiment period, we only need to focus on several pairs of on-ramps and exit ramps near by the congestion point. The entire experiment period is designed to be 90 min, including the first interval of 15 min for loading traffic flow gradually, the second interval of 60 min with incident and the final interval of 15 min for recovery. The topological relation of the road network is shown Figure 6.8.

The basic data of the hypothetical corridor are shown as Table 6.1.

In order to demonstrate the performance of the model on the different expressway sections, this chapter designed the different traffic demand levels on

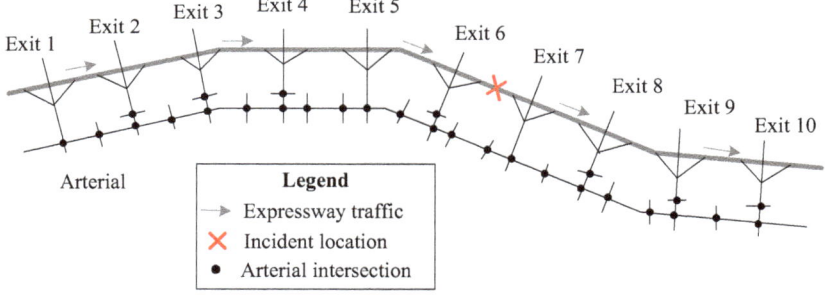

Figure 6.7 *A hypothetical corridor network for case study*

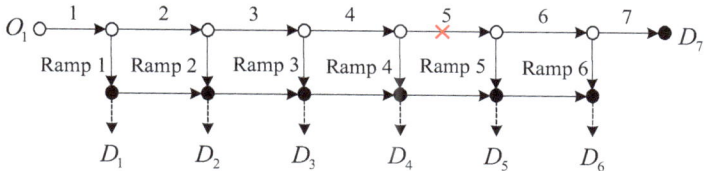

Figure 6.8 Topological relation of the road network

Table 6.1 Basic data of the hypothetical corridor

Section	Distance (km)	Free flow speed (km/h)	Traffic capacity (veh/h/lane)	Jamming density (veh/km)	Number of lanes (lane)
1	1.5	110	2,000	180	3
2	3.0	110	2,000	180	3
3	3.0	110	2,000	180	3
4	3.0	110	2,000	180	3
5	3.0	110	2,000	180	3
6	3.0	110	2,000	180	3
7	1.5	110	2,000	180	3
Ramp 1	1.0	80	1,500	60	1
Ramp 2	1.0	80	1,500	60	1
Ramp 3	1.0	80	1,500	60	1
Ramp 4	1.0	80	1,500	60	1
Ramp 5	1.0	80	1,500	60	1
Ramp 6	1.0	80	1,500	60	1

Table 6.2 O–D volume of loading status

Road network loading	O–D volume							
	O_1D_1	O_1D_2	O_1D_3	O_1D_4	O_1D_5	O_1D_6	O_1D_7	Sum
Basic quantity	96	120	145	166	194	220	1,180	2,121
1.25 times	120	150	181	208	243	275	1,475	2,652
1.50 times	144	180	217	249	291	330	1,770	3,182
1.75 times	168	210	254	291	340	385	2,065	3,712
2.00 times	192	240	290	332	388	440	2,360	4,242
2.25 times	216	270	326	374	437	495	2,655	4,772

hypothetical corridor, and the O–D volume was increased from 2,000 to 4,800 veh/h. Assuming that the time-varying distribution is subordinate to the trapezoidal distribution, and the corresponding load O–D volume is shown in Table 6.2.

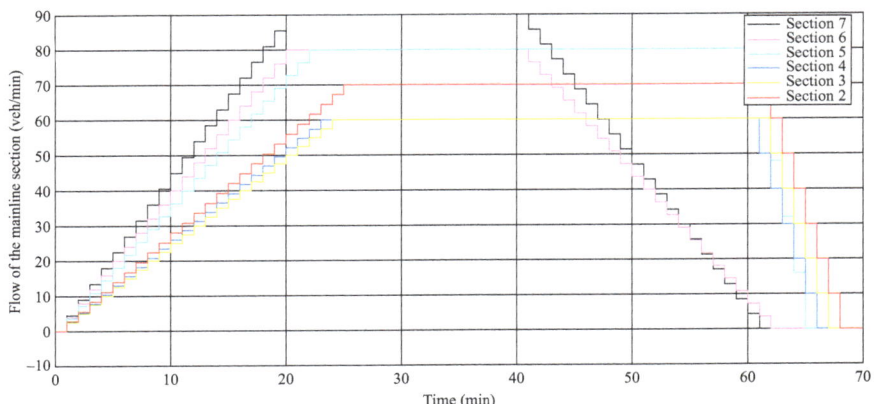

Figure 6.9 Traffic flow of six observed mainline sections

The flow data of the six observed mainline sections 2, 3, 4, 5, 6 and 7 are shown in Figure 6.9.

By designing two scenarios (the normal operation status and the diversion strategy operation status), the result about TTS and O–D estimation can be compared respectively. The GA population size S was set at 100, the maximum number of generation G was set at 200, the crossover probability C_r was set at 0.6 and the mutation probability M_r was set at 0.02.

6.4.5.2 Experimental analysis and results of traffic diversion

The dynamic traffic diversion model based on DODE has two parts: traffic diversion and DODE. Therefore, the results of case study need to be analyzed and evaluated from two aspects. For the traffic diversion, the evaluation index is the TTS of the road network, Total Travel Time (TTT) of the road network, Total Waiting Time (TWT) of the road network and traffic density. For the DODE, two indices are employed to evaluate the coincidence between the dynamic DODE result and the real O–D volume, as next:

MAPE: Mean Absolute Percentage Error

$$\text{MAPE} = \frac{\sum_{k=1}^{n}(|Y(k)-F(k)|/Y(k))}{n} \times 100\%$$

MSE: Mean Square Error

$$\text{MSE} = \frac{\sum_{k=1}^{n}[Y(k)-F(k)]^2}{n} \times 100\%$$

According to the previous analysis of the model, numerical analysis could be carried out for the two scenarios, one was the normal operation status and the other was the

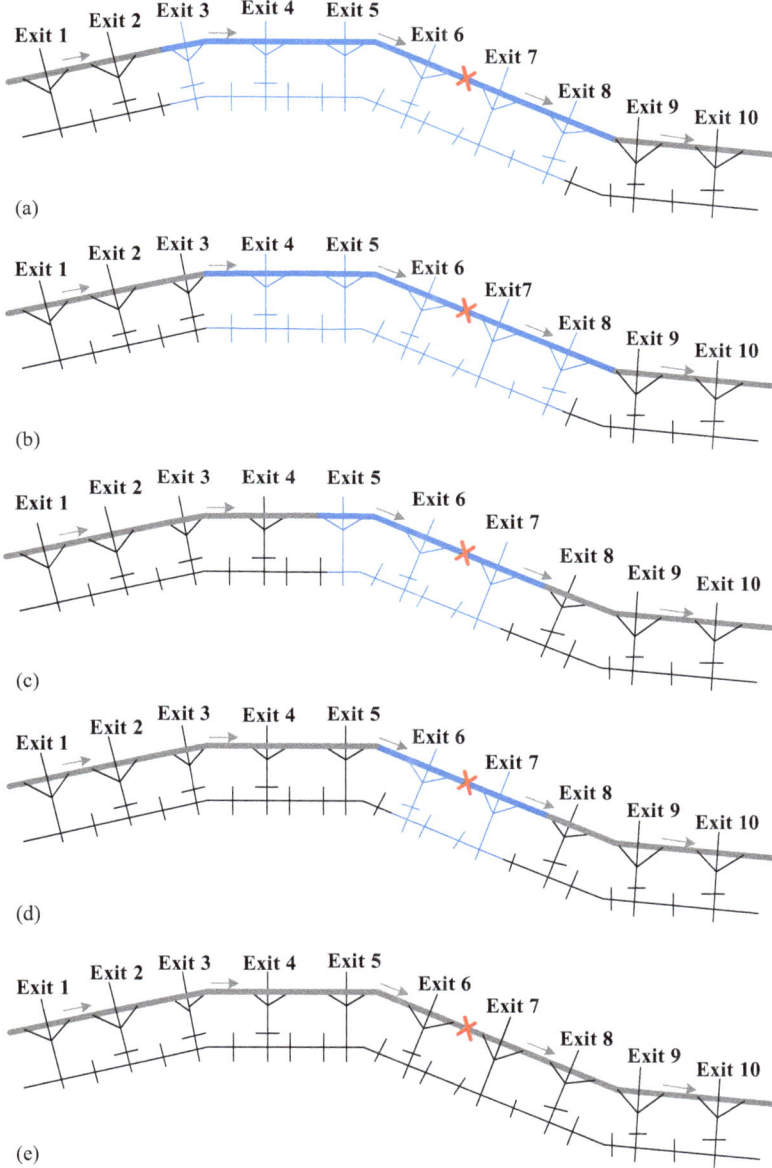

Figure 6.10 The control boundaries generated from the model: (a) the control boundaries-1, (b) the control boundaries-2, (c) the control boundaries-3, (d) the control boundaries-4, (e) the control boundaries-no diversion

Table 6.3 Evaluation index for the traffic diversion model

Control boundary	TTT (veh/h)	TWT (veh/h)	TTS (veh/h)
1	3,031	2,741	5,772
2	3,010(−3.8%)	2,677(−11.1%)	5,687(−7.3%)
3	3,009	2,875	5,884
4	3,048	2,978	6,026
No diversion	3,126	3,007	6,133

diversion strategy operation status. Since existing researches have shown that only diversion strategy implemented on the critical control nodes at the upstream and downstream of incident point could have significant optimization effect, this chapter took four nodes at the upstream and two nodes at the downstream of incident point as critical control nodes. Figure 6.10 presents the variation of the generated control boundaries (shown in the blue color).

The variation of the control boundaries and its impact on evaluation index are summarized in Table 6.3 and Figure 6.11.

Comparison between the results yields the following observations:

1. The TTS of the road network without the diversion measures could grow to a maximum of 6,133 veh/h, which shows that the congestion is serious, and the traffic delays are too large.
2. The diversion measures with control boundaries-2 seem more appropriate for the example road network due to its compact size and shorter distances for detour operations. Most importantly, it can substantially reduce the TTS and TWT of the road network (7.3% and 11.1%, respectively, as highlighted in Table 6.3).
3. The control boundaries-1 and control boundaries-2 have lower traffic density (10–40 veh/km/lane) at most of the time. It can be seen from Figure 6.11 that the red interval represents high traffic density (75–85 veh/km/lane). Compared with not implementing the diversion measures, there is no red interval for the control boundaries-2 (see Figure 6.11), which shows great effect of peak cutting and alleviation of the serious congestion. In real-world applications, the same procedure can be used by the traffic operators to determine the proper control boundaries and achieve the maximal control benefits.

This experimental analysis has also yielded the dynamic diversion rate for every 2 min of different control boundaries. The comparison results, as shown in Figure 6.12, have indicated that

1. With the simulation process proceeding, the traffic flow on the road network is gradually loading, and the diversion rate on each node increasing correspondingly.
2. The closer the diversion rate to the incident, the higher the average diversion rate is. This phenomenon shows that the diversion model based on DODE has taken the distance of detour into consideration and tries to make the most of the

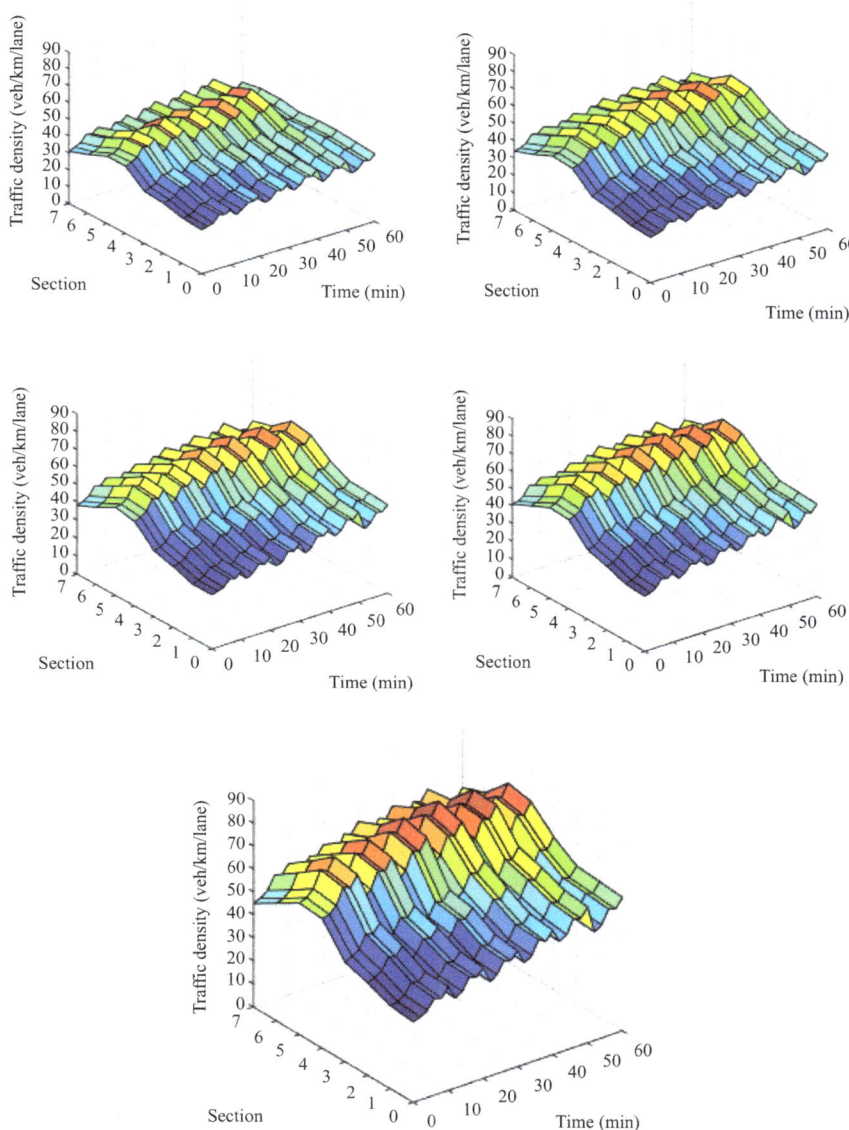

Figure 6.11 Thermodynamic diagram of traffic density: (a) the control boundaries-1, (b) the control boundaries-2, (c) the control boundaries-3, (d) the control boundaries-4, (e) the control boundaries-no diversion

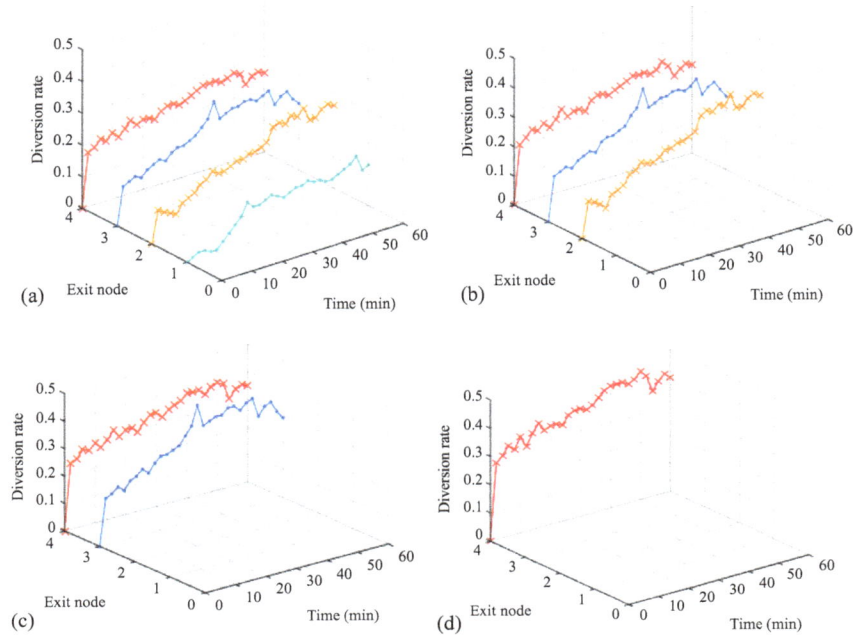

Figure 6.12 Dynamic diversion rate: (a) the control boundaries-1, (b) the control boundaries-2, (c) the control boundaries-3, (d) the control boundaries-4

traffic detour to the parallel road on the nearest upstream node, which can significantly make TTT smaller.

6.4.5.3 Experimental analysis and results of DODE

According to the previous analysis, this chapter used the dynamic diversion rate of control boundaries-2 to calculate the O–D volume. Considering the traffic flow will return to the mainline section of expressway after passing the incident point, this chapter chose the O–D volume of $O_1 D_7$ to check the model (see Figure 6.13).

The line on Figure 6.13 yields the following observations:

1. The distribution of the DODE results of $O_1 D_7$ is trapezoidal distribution basically, which is similar to the known distribution law of the real and historical O–D distribution.
2. The maximum deviation between the real O–D volume and the DODE results is about 10%, the MAPE is 12.31% and the MSE is 14%. All deviation is on the acceptable range (15%), which indicates that dynamic traffic diversion model based on DODE has good accuracy and applicability.

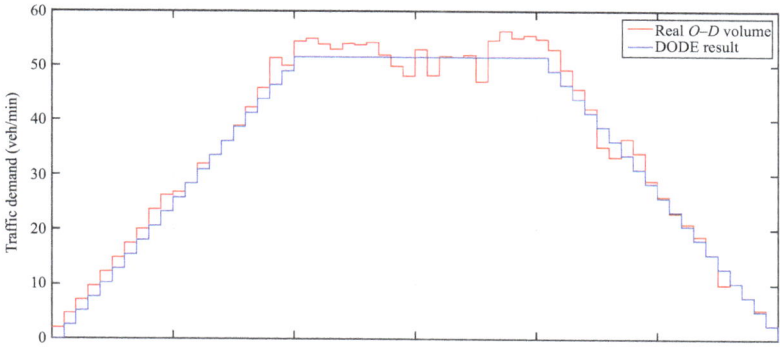

Figure 6.13   The O–D volume of O_1D_7

6.5   Conclusion

Dynamic traffic diversion is a critical problem in ITS. In this chapter, a novel traffic diversion model based on DODE is established by analyzing the existing diversion models. The traffic diversion model based on dynamic traffic demand estimation and prediction consists of the traffic diversion module and the DODE. Experiments are conducted to validate the efficiency of the proposed model. The proposed model is designed to have the following operational features: (1) carrying out the traffic status, including traffic flow, traffic speed and traffic density, so as to generate the calculation results of evaluation index about road network; (2) showing relationship between the change of the diversion rate and the change of the traffic flow in the on-ramps and exit ramps and (3) updating dynamic O–D demand to improve the efficiency and accuracy of the dynamic traffic diversion model.

References

[1]   Ran B and Boyce D E. Modeling dynamic transportation networks. *Transportation Science*, 1999, 33(4):431–433.
[2]   Allsop R E. Some possibilities for using traffic control to influence trip distribution and route choice. *Transportation and Traffic Theory, Proceedings*, 1974, 6:345–373.
[3]   McDonald M, Hounsell N B and Njoze S R. Strategies for route guidance systems taking account of driver response. *Vehicle Navigation & Information Systems Conference proceedings*. 1995: 328–333.
[4]   Cascetta E, Russo F, Viola F A and Vitetta A. A model of route perception in urban road networks. *Transportation Research Part B: Methodological*, 2002, 36(7):577–592.

[5] Abdel-Aly M A, Kitamura R and Jovani P P. Using stated preference data for studying the effect of advanced traffic information on drivers' route choice. *Transportation Research Part C: Emerging Technologies*, 1997, 5(1):39–50.

[6] Khattak A J and Palma A D. The impact of adverse weather conditions on the propensity to change travel decisions: a survey of Brussels commuters. *Transportation Research Part A: Policy and Practice*, 1997, 31(3):181–203.

[7] Peeta S, Ramos J and Pasupathy R. Content of variable message signs and on-line driver behavior. *Transportation Research Record Journal of the Transportation Research Board*, 2000, 1725:102–108.

[8] Xu R, Luo Q and Gao P. Study on passenger flow distribution model and algorithm of urban rail transit network based on multi-path. *Journal of the China Railway Society*, 2009, 02:114–118.

[9] Zhou B. Research on the solution algorithm of Logit stochastic user equilibrium model based on path. *Southeast University*, 2015.

[10] Yan Y. Research on passenger flow distribution of Urban Rail Transit Based on seamless transfer. *Southwest Jiaotong University*, 2015.

[11] Ling C. Research on the improved logit stochastic user equilibrium assignment model and algorithm. *Lanzhou Jiaotong University*, 2017.

[12] Bogers E A L and Van Zuylen H. The importance of reliability in route choices in freight transport for various actors on various levels. *Proceedings of the European Transport Conference*. 2004: 149–161.

[13] Peeta S and Yu J W. A hybrid model for driver route choice incorporating en-route attributes and real-time information effects. *Networks and Spatial Economics*, 2005, 5(1):21–40.

[14] Connors R D and Sumalee A. A network equilibrium model with travellers' perception of stochastic travel times. *Transportation Research Part B: Methodological*, 2009, 43(6):614–624.

[15] Rasouli S and Timmermans H. Applications of theories and models of choice and decision-making under conditions of uncertainty in travel behavior research. *Travel Behaviour & Society*, 2014, 1(3):79–90.

[16] Chen S. A decision-making method based on Dempster-Shafer theory and prospect theory. *Journal of Information & Computational Science*, 2014, 11(4):1263–1270.

[17] Gao S. Route choice in an uncertain environment: algorithms and behavioral studies. *General Information*, 2014.

[18] Hasuike T, Katagiri H, Tsubaki H and Tsuda H. Interactive approaches for sightseeing route planning under uncertain traffic and ambiguous tourist's satisfaction. *New Business Opportunities in the Growing E-Tourism Industry*. 2015: 75–96.

[19] Luca S D and Pace R D. Evaluation of risk perception in route choice experiments: an application of the Cumulative Prospect Theory. *International Conference on Intelligent Transportation Systems*. 2015: 309–315.

[20] Ridwan M. Fuzzy preference based traffic assignment problem. *Transportation Research Part C: Emerging Technologies*, 2004, 12:209–233.

[21] Chakroborthy P and Kikuchi S. Application of fuzzy set theory to the analysis of capacity and level of service of highways. *Proceedings of the First International Symposium on Uncertainty Modeling and Analysis*. 1990: 146–150.

[22] Lotan T and Koutsopoulos H N. Models route choice behavior in the presence of information using concepts from fuzzy set theory and approximate reasoning. *Transportation*, 1993, 20:129–155.

[23] Hounsell N and Chatterjee K. Variable message signs in London: evaluation in CLEOPATRA. *Road Transportation Information and Control*, 1998, 4:21–23.

[24] Hussein D. An agent-based approach to modeling driver route behavior under the influence of real-time information. *Transportation Research Part C: Emerging Technologies*, 2002, 10:331–349.

[25] Markku Ki and Heikki S. Effects of weather and weather forecasts on driver behavior. *Transportation Research Part F: Traffic Psychology and Behaviour*, 2007, 10:288–289.

[26] Lee C, Ran B, Yang F and Loh W Y. A hybrid tree approach to modeling alternate route choice behavior with online information. *Journal of Intelligent Transportation Systems*, 2010, 14(4):209–219.

[27] Mammar S, Messmer A, Jensen P, Papageorgiou M, Haj-Salem H and Jensen L. Automatic control of variable message signs in Aalborg. *Transportation Research*, 1996, 4C:131–150.

[28] Papageorgiou M. Dynamic modeling, assignment and route guidance in traffic networks. *Transportation Research Part B: Methodological*, 1990, 24(6):471–495.

[29] Lioogendoom S P and Bovy P H L. Optimal Routing Control Using Variable Message Signs. Estimators of travel times for road networks, new developments, evaluation results and application. *University Press*, 2000.

[30] Wang Y and Papageorgiou M. Feedback routing control strategy for freeway networks: a comparative study. *Proceeding of the 2nd International Conference on Traffic & Transportation Studies*. 2000: 642–649.

[31] Hawas Y and Mahmassani H S. A decentralized scheme for real-time route guidance in vehicular traffic networks. *Proceedings of 2nd World Congress on Intelligent Transport Systems*. 1995: 1956–1963.

[32] Pavlis Y and Papageorgiou M. Simple decentralized feedback strategies for route guidance in traffic networks. *Transportation Science*, 1999, 33:264–278.

[33] Morin J M. Aid-to-decision for variable message sign control in motorway networks during incident condition. *Proceedings of the 4th ASCE International Conference on Applications of Advanced Technologies in Transportation Engineering*. 1995: 378–382.

[34] Messmer A, Papageorgiou M and Mackenzie N. Automatic control of variable message signs in the interurban Scottish highway network. *Transportation Research Part C: Emerging Technologies*, 1998, 6:173–187.

[35] Wang Y, Papageorgiou M and Messmer A. A predictive feedback routing control strategy for freeway network traffic. *Proceedings of the American Control Conference*, 2002, 1856(5):3606–3611.

[36] Lafortune S, Sengupta R, Kaufman D E and Smith R L. Dynamic system-optimal traffic assignment using a state space model. *Transportation Research*, 1993, 27B:451–473.

[37] Iftar A. A decentralized routing controller for congested highways. *Proceedings of the IEEE Conference on Decision and Control*, 1995, 4:4089–4094.

[38] Messmer A and Papageorgiou M. Route diversion control in motorway networks via nonlinear optimization. *IEEE Transaction on Control System Technology*, 1995, 3(1):144–154.

[39] Wie B, Tobin R, Bernstein D and Friesz T. Comparison of system optimum and user equilibrium dynamic traffic assignment with schedule delays. *Transportation Research*, 1995, 3C:389–411.

[40] Mahmassani H S and Peeta S. Network performance under system optimal and user equilibrium dynamic assignments: implications for advanced traveler information systems. *Transportation Research Record*, 1993, 1408:83–93.

[41] Ben-Akiva M, Bierlaire M, Bottom J, Koutsopoulos H and Mishalani R. Development of a route guidance generation system for real-time application. *IFAC Symposium on Transportation Systems*. 1997: 433–439.

[42] Wisten M B and Smith M J. Distributed computation of dynamic traffic equilibria. *Transportation Research*, 1997, 5C:77–93.

[43] Wang Y, Messmer A and Papageorgiou M. Freeway network simulation and dynamic traffic assignment using METANET tools. *Transportation Research Record*, 2001, 1776:178–188.

[44] He Z, Guan W and Ma S. A traffic-condition-based route guidance strategy for a single destination road network. *Transportation Research Part C: Emerging Technologies*, 2013, 32(4):89–102.

[45] He Z, Chen B, Jia N, Guan W, Lin B and Wang B. Route guidance strategies revisited: comparison and evaluation in an asymmetric two-route traffic network. *International Journal of Modern Physics C*, 2014, 25(4):1450005.

[46] He Z, Zheng L, Guan W and Mao B. A self-regulation traffic-condition-based route guidance strategy with realistic considerations: overlapping routes, stochastic traffic and signalized intersections. *Journal of Intelligent Transportation Systems*, 2016, 20(6):545–558.

[47] Cascetta E and Nguyen S. A unified framework for estimating or updating origin-destination matrices from traffic counts. *Transportation Research Part B: Methodology*, 1988, 22:437–455.

[48] Cremer M and Keller H. Dynamic identification of O-D flows from traffic counts at complex intersections. *Proceedings of the 8th International Symposium on Transportation and Traffic Theory*. 1981: 121–142.

[49] Cremer M and Keller H. A systems dynamics approach to the estimation of entry and exit O-D flows. *Proceedings of the 9th International Symposium on Transportation and Traffic Theory*. 1984: 431–450.

[50] Cremer M and Keller H. A new class of dynamic methods for the identification of origin-destination flows. *Transportation Research Part B: Methodology*, 1987, 21(2):117–132.

[51] Nihan N L and Davis G A. Application of prediction-error minimization and maximum likelihood to estimate intersection O-D matrices from traffic counts. *Transportation Science*, 1989, 23(2):77–90.

[52] Sherali H D, Arora N and Hobeika A G. 1997. Parameter optimization methods for estimating dynamic origin-destination trip-tables. *Transportation Research Part B: Methodology*, 1997, 31(2):141–157.

[53] Jiao P, Lu H, Liu Y and Yang S. A study of models and algorithms of dynamic OD matrix estimation for intersection. *Proceedings of the Eastern Asia Society for Transportation Studies*, 2003, 4:885–897.

[54] Chang G and Wu J. Recursive estimation time-varying O-D flows from traffic counts in freeway corridors. *Transportation Research Part B: Methodology*, 1994, 28(2):141–160.

[55] Lin P and Chang G. A generalized model and solution algorithm for estimation of the dynamic freeway origin-destination matrix. *Transportation Research Part B: Methodology*, 2007, 41(5):554–572.

[56] Sherali H D and Park T Y. Estimation of dynamic origin-destination trip tables for a general network. *Transportation Research Part B: Methodology*, 2001, 35(3):217–235.

[57] Okutani I and Stephanedes Y J. Dynamic prediction of traffic volume through Kalman filtering theory. *Transportation Research Part B: Methodology*, 1984, 18(1):1–11.

[58] Bell M G H. The estimation of origin-destination matrices by constrained generalized least squares. *Transportation Research Part B: Methodology*, 1991, 25(1):13–22.

[59] Chang G and Tao X. An integrated for estimating time-varying network origin-destination distributions. *Transportation Research Part A: Policy and Practice*, 1999, 33(5):381–399.

[60] Tavana H. Internally-consistent estimation of dynamic network origin-destination flow from intelligent transportation systems data using Bi-level optimization. *The University of Texas*, 2001.

[61] Zhou X. Dynamic origin-destination demand estimation and prediction for off-line and on-line dynamic traffic assignment operation. *University of Maryland*, 2004.

[62] Bert M. Dynamic urban origin-destination matrix estimation methodology. École Polytechnique Fédérale de Lausanne, Lausanne, 2009.

[63] Allsop R E. Introduction to the theory of traffic flow. *Speed*, 1984, 50(02):69–75.

[64] Chowdhury D, Santen L and Schadschneider A. Statistical physics of vehicular traffic and some related systems. *Physics Reports*, 2000, 329(4):199–329.

[65] Gunay B. Car following theory with lateral discomfort. *Transportation Research Part B: Methodological*, 2007, 41(7):722–735.

Chapter 7
Game theoretic lane change strategy for cooperative vehicles under perfect information

Andres Ladino[1] and Meng Wang[2]

Lane change maneuvers are main causes of traffic turbulence at highway bottlenecks. We propose a dynamic game framework to derive the system optimum strategy for a network of cooperative vehicles interacting at a merging bottleneck. Cooperative vehicles on the highway mainline seek for optimal strategies (i.e., whether and when to perform courtesy lane change to facilitate the merging vehicle) to minimize their cost, while taking into account potential future interactions at the merging section while minimizing the distance traveled on the acceleration lane. An optimal strategy is found by minimizing the joint cost of all interacting vehicles while respecting behavioral and physical constraints. Numerical examples show the feasibility of the approach in capturing the nature of conflict and cooperation during the merging process, and demonstrate the benefits of sharing information and cooperative control for connected and automated vehicles.

7.1 Introduction

Social and economic development comes with an increased effect on traffic congestion and safety risks in terms of accidents. Planning, design, and deployment of such systems face new challenges every day [1]. In particular, when multiple connected vehicles interact and exchange information, the problem of decision-making under conflicting situations with multiple vehicles as players emerges, especially at network discontinuities such as highway on-ramps [2]. In the context of traffic flow theory, these discontinuities are often bottlenecks with characterized highway capacity. In order to optimize the utility of the road network at merges, vehicular flow control has been proposed on the infrastructure side via ramp metering and variable speed limits strategies [3].

From cooperative systems' perspective, quite some advancements have been achieved in designing platoon or vehicle longitudinal controllers to stabilize platoon and traffic via Cooperative Adaptive Cruise Control (CACC) [4–6]. Nonetheless, it

---
[1]Laboratoire d'Ingénierie Circulation et Transports, COSYS Department, Lyon, France
[2]Delft University of Technology, Department of Transport & Planning, Delft, The Netherlands

remains an open research question as to how to design the decision-making and (trajectory) planning systems for lane change maneuvers of Connected & Automated Vehicles (CAVs) such that a collision is avoided, safety is guaranteed, and traffic efficiency is maximized.

Several strategies were reported to deal with merging situation, most of which act on the longitudinal speed regulation. A set of scenarios was summarized in [4], describing vehicle interactions and active platoon policies at specific time instants. This finite state description did not formally address the theoretical properties of the solution. An optimal acceleration trajectory planning method for merging vehicles was proposed in [7], relying on a passing order decided by a higher decision layer. A specific trajectory design is proposed and fuzzy controllers are used as regulation strategies [8]. More recently, [9] proposed a bi-level control strategy to solve the merging process under mixed traffic conditions. However, it does not include the courtesy lane change of the mainline vehicles as the decision alternative. Reference [10] formulated a stochastic switched system model to analyze how platoon-induced congestion varies with the fraction of platooned vehicles at merge, yet the decisions on when and where to split the platoon is not addressed. For a more complete literature review on this topic, we refer the reader to [11].

The merge situation can be modeled as a negotiation process between vehicles on the main carriageway and vehicles on the on-ramp willing to join the highway (see Figure 7.1). A game theoretical framework was proposed in [12] where interacting CAVs predict and determine discrete desired lane sequences and continuous accelerations to minimize a cost function reflecting undesirable future situations.

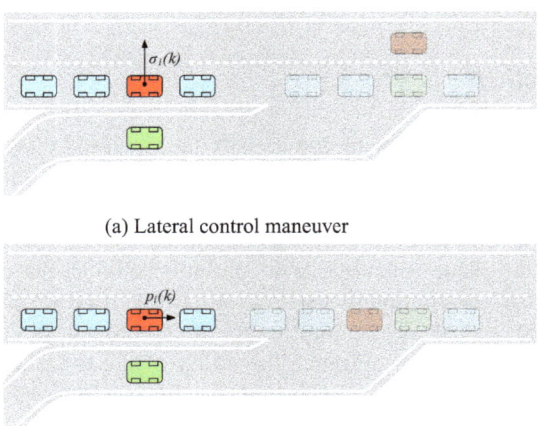

(a) Lateral control maneuver

(b) Longitudinal control maneuver

Figure 7.1 Control actions for cooperative lane change maneuvers. In this case, the red CAV illustrates two behaviors to open gaps for the inserting vehicle in green

The computational load of this approach makes the real-time application a daunting task. A similar approach was considered in [13], where the constraints of the changing lane are formulated as a Mixed Logical Dynamical (MLD) model and the final control problem is cast via Mixed Integer Linear Programming (MILP). The framework assumes noncooperative nature of automated vehicles. The majority of the aforementioned work did not consider the cooperative nature of interacting vehicles and some even needs another framework to put in the operational layer acceleration maneuvers. The lane change decisions are assumed to be selfish just for the benefit of a single agent, instead of cooperative to optimize the collective traffic performance.

This chapter puts forward a dynamic game framework to derive system optimum strategies for a network of cooperative vehicles interacting at a merging bottleneck. Cooperative vehicles traveling along the highway main lane seek to maximize an individual payoff by minimizing a running cost in a finite time horizon (i.e., whether and when to perform courtesy lane change to facilitate the merging vehicle). To minimize such cost, specific penalties are given for deviations from their desired driving conditions while taking into account the predicted action of merging vehicles. Merging vehicles minimize the distance traveled on the acceleration lane in addition to the same cost terms of the mainline vehicles and predict the reaction of mainline vehicles responding to their merging decisions (i.e., whether and when to merge in the prediction horizon). An optimum strategy is found by minimizing the joint cost of interacting vehicles while respecting behavioral and physical constraints. Properties of the games and existence of solutions will be provided in this work.

To solve the problem, a simplified discrete formulation of longitudinal vehicle dynamics is formulated. The longitudinal model is distributed, e.g., only interacting under predecessor-follower topology, and can be easily adapted to capture platooning systems dynamics. The full dynamic game is then cast as a set of subproblems regularly expressed as standard optimal control problems that can be solved by mixed-integer quadratic/linear programming. Several examples at simulation level show the feasibility of the approach in capturing the nature of cooperation.

The operational assumptions and problem setup are explained with more detail in Section 7.2. Then the model, including longitudinal and lateral dynamics, is explained in Section 7.3. The lane change decision action is cast as a dynamic game in Section 7.4. Numerical examples are shown in Section 7.5.

Notation

We denote, in general, single states with italic letters; x_i, hence, denotes the value of the variable x for a specific vehicle i, vectors collecting a set of variables are denoted as $\mathbf{x} \in \mathbb{R}^{|\mathscr{A}|}$. The symbol $\mathbb{R}^{|\mathscr{A}|}$ for a finite set $\mathscr{A}$ denotes the set of real vectors indexed by elements of $\mathscr{A}$ and $|\mathscr{A}|$ is the number of elements of the set $\mathscr{A}$. For $x \in \mathbb{R}^n$, $x > 0$ is meant component-wise. Matrices are denoted with capital letters $A = [a_{ij}]$ where elements are denoted as a_{ij}, $\rho(A)$ denotes an operation such as the spectral radius of the matrix. The identity matrix of suitable dimensions is denoted by $\mathbb{I}$.

7.2 Problem formulation

In this chapter, we consider the situation shown in Figure 7.1. Let $\mathscr{V} = \{1,\ldots,n\}$ be a group of CAV traveling along a road infrastructure composed by specific lanes labeled $\sigma = \{1,2,3\} \in \mathbb{N}$ from most right to most left. Let denote $\sigma_i(k)$ the lane occupied by vehicle i at a specific instant of time k. Two vehicles i, j traveling in different lanes $\sigma_i \neq \sigma_j$ are going to perform a merging negotiation at a current time k_0 in a time horizon of N steps.

Two types of vehicle maneuvers can be conceived as possible in this situation. First, as shown in Figure 7.1(a) the i-th vehicle in the platoon can modify its lateral position (in discrete lanes) to a new state $\sigma_i(k) = \sigma_i(k_0) + 1$, while other vehicles in the platoon will keep the same position $\sigma_{i^-}(k) = \sigma_{i^-}(k_0) \ \forall \ i^- \in \mathscr{I}\setminus i$. In this case, a *lateral* decision operates over the vehicle i. A second situation can be envisaged as shown in Figure 7.1(b), the decision is taken at the level of the longitudinal control where a vehicle i performs a maneuver to pass vehicle j or yields in courtesy to open a gap where the j vehicle will insert in front of vehicle i. Control maneuvers for this situation can be designed under knowledge of the state of the inserting vehicle j [9]. In this case, a *longitudinal* decision operates over vehicle i.

The decision-making and control system follows a hierarchical setting, where the decision-making module is placed on the top of a motion-control module [9]. This decision-making is based on a dynamic game framework [12]. It takes into account the current state information of the dynamic driving environment, which consists of surrounding cooperative/noncooperative vehicles. The state information can be estimated from measurements of on-board sensors or information transmitted via Vehicle-to-Vehicle (V2V) communication devices. The interacting vehicles negotiate and jointly decide whether and when to change lane to optimize a joint cost/payoff function, taking into account the dynamic process as a response to the lane change actions. The lane change time is transferred to the lower level as the command to start lane change execution process. We focus on the tactical decisions while omitting the details in the vehicle lateral dynamics. The control problem can be cast as follows: *Determine the lateral optimal control strategy such that a joint payoff/cost for vehicle i and j is maximized/minimized.*

7.3 Highway traffic system dynamics

For the setting under consideration, we consider two dimensions in the space of decision: longitudinal and lateral dynamics.

7.3.1 Longitudinal dynamics

The spacing with respect to the preceding vehicle and longitudinal position for vehicle i is cast into discrete formulation as:

$$s_i(k+1) = s_i(k) + (v_l(k) - v_i(k))\Delta t \\ p_i(k+1) = p_i(k) + v_i(k)\Delta t \tag{7.1}$$

where $k \in \mathbb{Z}^+$ denotes the discrete time index and Δt is the time step size. The collection $\mathbf{p}, \mathbf{s}, \mathbf{v} \in \mathbb{R}^n$ denote vehicle's position, the headway space, and the longitudinal speed, respectively. It is convenient to define error terms to design control law. Hence, let:

$$e^v_{0,i}(k) = v_{0,i} - v_i(k) \tag{7.2}$$

$$e^v_{l,i}(k) = v_l(k) - v_i(k) \tag{7.3}$$

where $v_{0,i}$ denotes the desired speed of vehicle i and the subscript $l \in \mathcal{V} \cup \{j\}$ denotes the index of the direct leader of vehicle i. A feedback-control law can be formulated as:

$$v_i(k+1) = k_0 e^v_{0,i}(k) + k_l e^v_{l,i}(k) \tag{7.4}$$

k_0, k_l are feedback gains for the errors to the desired speed and the predecessor speed, respectively.

Safety conditions are guaranteed by imposing constraints on these dynamics. Hence, the vehicle dynamics are subject to the following linear constraints:

$$a_{\min}\Delta t \leq v_i(k+1) - v_i(k) \leq a_{\max}\Delta t \tag{7.5}$$

$$v_{\min} \leq v_i(k) \leq v_{\max} \tag{7.6}$$

$$s_i(k) \geq v_i(k) t_{\min} + s_0 \tag{7.7}$$

where $t_{\min}$ denotes the minimum time gap between two vehicles on the same lane. s_0 denotes the minimum spacing between two vehicles. Constraint (7.7) states that any leader-follower space headway should keep some safe distance at any time instant k. $a_{\min}, a_{\max}, v_{\min}, v_{\max}$ represent boundaries in acceleration and speed correspondingly.

We remark that the simplified vehicle dynamic model with control input is flexible in terms of the error term definition. We choose the current form to capture the heterogeneous choice of desired speed by system users, while acknowledging that this is not the unique model for CAV platoons. If we use the gap error:

$$e^s_i(k+1) = s_i(k) - v_i(k)t_d - s_0 \\ v_i(k+1) = k_s e^s_i(k) + k_l e^v_{l,i}(k)$$

where t_d denotes the desired time gap of ACC/CACC systems and k_s denotes the feedback gain. The model can describe CACC platoon dynamics with proper tuning of feedback gains [14].

7.3.2 Lateral dynamics

We use the discrete lane change decision δ as the control decision variable, $\delta_i \in \mathscr{D} := \{-1, 0, 1\}$, where $\{-1, 0, 1\} := \{$change right, no lane change, change left$\}$. In the chapter, we assume only one lane change during the prediction horizon, but the framework is general to include multiple lane changes in the horizon [12]. This single switch aims to reduce the computational burden of the approach.

We use the travel lane of vehicle i, $\sigma_i(k)$ as the discrete state variable at time k. The dynamics of the lateral behavior is determined by:

$$\sigma_i(k+1) = \sigma_i(k) + \delta_i(k) \tag{7.8}$$

We assume lane change can take place as long as the gap is sufficiently large according to (7.7). We introduce the general possible maneuvers for all vehicles traveling along highways; nonetheless, it is worth mentioning that the space of decision for the lateral maneuvers can be constrained for each vehicle depending on the trip lane or specific infrastructure policies.

7.3.3 Lane change and dynamic communication topology

The leader-follower pair is dynamic as a result of lane changes for the group of n CAVs. Let a graph $\mathscr{G} = \{\mathscr{V}, \mathscr{E}\}$; $\mathscr{V}$ represents the set nodes consisting in all CAVs within the network and $\mathscr{E} = \{\mathscr{V} \times \mathscr{V}\}$ represents the set of edges representing a relationship between leaders and followers. Then $\mathscr{E} = \{\varepsilon_{il} = 1\}$ if vehicle l is the leader of vehicle i at specific sample time k, 0 otherwise. The adjacency matrix of $\mathscr{G}$ is concentrated in the squared matrix $A_g = [\varepsilon_{ij}]$. In general, due to the lane change model (7.8), the set $\mathscr{E}$ is dynamic in time.

7.3.4 Closed-loop dynamics

Let's suppose a uniform formation where the desired speeds for all vehicles are the same and constant $v_{0,i} = \bar{v}_0$. For system (7.1), in combination with (7.4), it is possible to write the closed-loop system as:

$$\begin{aligned} s_i(k+1) &= s_i(k) + (v_l(k) - v_i(k))\Delta t \\ v_i(k+1) &= k_0(\bar{v}_0 - v_i(k)) + k_l(v_l(k) - v_i(k)) \end{aligned} \tag{7.9}$$

Gathering all individual systems i into an algebraic equation, it can be expressed as:

$$\begin{bmatrix} \mathbf{s}(k+1) \\ \mathbf{v}(k+1) \end{bmatrix} = \underbrace{\begin{pmatrix} \mathbb{I} & (A_g - \mathbb{I})T \\ \mathbb{O} & K_l(A_g - \mathbb{I}) - K_0 \end{pmatrix}}_{A} \begin{pmatrix} \mathbf{s}(k) \\ \mathbf{v}(k) \end{pmatrix} + \begin{pmatrix} \mathbb{O} \\ K_0 \bar{v}_0 \end{pmatrix} \tag{7.10}$$

where K_0, K_l, T are diagonal matrices in $\mathbb{R}^{n \times n}$ with corresponding elements $k_0, k_l, \Delta t$ in their diagonal. $\mathbb{I}, \mathbb{O}$ are the identity and the zero matrices of corresponding dimensions. $\bar{v}_0 \in \mathbb{R}^n$ is the constant vector containing on each element $\bar{v}_0$. A_g is the adjacency matrix of the network topology (see Section 7.3.2).

Game theoretic lane change strategy for cooperative vehicles 153

It can be shown that if the spectral radius $\rho(\bar{A}) \leq 1, \rho(A) := \{\max|\lambda| : \lambda = \text{eig}(A)\}$, then the system (7.10) is stable.

7.4 Game theoretic formulation of the lane change decision problem

In this section, we propose the dynamic game formulation for the lane change control maneuver.

7.4.1 Dynamic lane change game formulation

A vehicle traveling along a specific lane can establish a lateral decision denoted as a lane change strategy within a finite future time horizon N.

Definition 7.1 (Lane change strategy). A vehicle lane change strategy from lane $\sigma_\ell \to \sigma_{\ell^+}$ is defined as the sequence:

$$\begin{aligned}
&\xi_\delta = \{\sigma(k_0), \sigma(k_0+1), \ldots, \sigma(k_0+N-1)\} \\
&\sigma(k^*) = \sigma_\ell \\
&\sigma(k^*+1) = \sigma_{\ell^+} \\
&\sigma(k+1) = \sigma(k) + \delta(k), \\
&\sum_{k=0}^{N-1} |\delta(k)| = 1
\end{aligned} \quad (7.11)$$

ξ_δ represents the sequence associated to a particular lateral control $\delta(k)$ which induces the choice lane changing maneuver at k^* in the horizon N.

Consider the case of Figure 7.2 where cooperative vehicles work together to find the best strategy that maximizes the utility of the system as a whole.

The objective of the dynamic game is to create a decision block that considers the tradeoff between two possible cases. In the first situation, vehicle i performs a lane change maneuver to create the necessary gap for insertion as shown in Figure 7.1(a). In

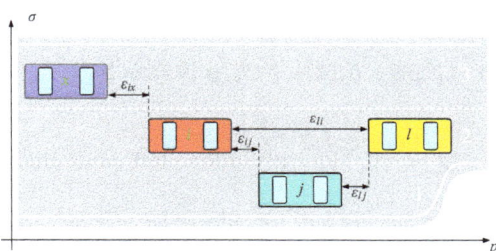

Figure 7.2 Lane change dynamic game. The controlled CAV in red optimizes the decision making between yielding at the merging time and changing lane

the second situation, vehicle j should wait for the mainline vehicle to yield the necessary gap to so that the merging maneuver is performed without violating constraints. It is worth remarking that both vehicles i, j will play a game in a finite horizon time that lead to a decision on lane changing for vehicle j. The cost for each vehicle is measured by undesirable situations:

$$\begin{aligned} L_i(\mathbf{p}(k), \mathbf{v}(k), \delta_i(k)) = & \beta_1 |e_{0,i}^v(k)| + \beta_2 |e_{l,i}^v(k)| \\ & + \beta_3 |v_i(k+1) - v_i(k)| \\ & + \beta_4 |\sigma_i(k) - \sigma_i^*| \\ & + \beta_5 |\delta_i(k)| \\ & - \min\{0, \beta_6(p_j(k) - p_{j,end})\} \end{aligned} \quad (7.12)$$

where $\beta_g, g \in \{1, 2, 3, 4, 5, 6\}$ are the weights on different cost terms. $p_{j,end}$ denotes the position of the end of a mandatory lane change section for vehicle j. The running cost function can be interpreted as follows:

- The first term encourages the vehicle to travel at its desired speed;
- The second term encourages consensus on speed for each leader-follower pair;
- The third term favors smooth speed change and hence discourage sharp acceleration and deceleration;
- The fourth term penalizes deviation from desired lane σ_i^* and the fifth term penalizes lane changes;
- The last terms penalize potential failure for mandatory lane change. It favors early mandatory lane changes and increases when the distance to the end of the merging lane p_{end} is decreasing.

The optimal control problem can be cast as an optimization of the running cost L_i for each one of the players while other players have already decided. A dynamic game can be integrated within an optimal control problem where each one of the players fixes a specific strategy in particular for the lane change by targeting the specific value σ_i^*. Notice that each player i has a finite number of strategies to choose by selecting specific δ_i. In particular, when playing the game in between vehicle i and vehicle j, it is possible to write the following finite horizon problem:

$$\begin{aligned} \min_{\delta_i(\cdot) \in \mathscr{D}} \quad & \sum_{g=i,j} \sum_{k=0}^{N-1} L_g(\mathbf{p}(k), \delta_g(k)) + \Phi_g(\mathbf{p}(N)), \delta(N)) \\ \text{s.t.} \quad & (7.1), (7.5), (7.7), (7.6), (7.8) \\ & \delta_i(k) \in \mathscr{D} = \{0, 1\}, \text{ only allow left lane changes} \\ & \sum_{k=0}^{N-1} \delta_i(k) \leq 1, \text{ only allow one lane change} \end{aligned} \quad (7.13)$$

The objective of the former optimal control problem is to promote the minimization of the individual costs. This is formulated as an optimization problem, where one seeks the optimal lane change decision trajectories for each vehicle i in a prediction horizon N

to maximize the payoff function of the whole group. In fact each one of the player should maximize a payoff given by:

$$J_i(\mathbf{p}(k), \mathbf{v}(k), \delta_i(k)) = -\sum_{k=0}^{N-1} L_i(\mathbf{p}(k), \mathbf{v}(k), \delta_i(k)) \quad (7.14)$$

The dynamic game entails prediction of the payoff over a time horizon with N steps: $[0, N]$. We consider N to be sufficiently large and, therefore, set the terminal cost $\Phi = 0$. The player i will select a strategy among a finite set $\mathscr{D}$ of strategies.

7.4.2 Existence of equilibrium

We first introduce general conditions for the existence of equilibrium in a two-player-game setup. Let consider the vehicle i and all the possible sets of finite strategies $\mathscr{A} = \{a_1, a_2, \ldots, a_r\}$ to be chosen for the lateral decisions. Let $\mathscr{B} = \{b_1, b_2, \ldots, b_q\}$ the possible decisions for the j vehicle traveling in the on-ramp lane. It is worth to remark that vehicles i, j have at most $|\mathscr{A}^i| = |\mathscr{B}^i| = 2^{N-1}$ possibilities to change lane during a future finite horizon.

Theorem 7.1 *(Existence of Nash equilibrium [15]). Assume that the sets of strategies $\mathscr{A}, \mathscr{B}$ are compact, convex subsets of $\mathbb{R}^n$. Let the payoff functions $\phi^{\mathscr{A}}, \phi^{\mathscr{B}}$ be continuous. If $a \mapsto \phi^{\mathscr{A}}(a, b)$ results in a concave function of a, $\forall~b \in \mathscr{B}$ and $b \mapsto \phi^{\mathscr{B}}(a, b)$ is a concave function of b, $\forall~a \in \mathscr{A}$, then the game is called noncooperative and it admits a Nash equilibrium.*

Proof To proof the existence of the Nash equilibrium, it is important to proof the compactness and convexity of the best replies. The existence of the equilibrium is then supported by Kakutani's fixed point theorem that shows that each best response correspondence has a fixed point. Consider the best response maps $r^{\mathscr{A}}(b), r^{\mathscr{B}}(a)$ defined as:

$$\begin{aligned} r^{\mathscr{A}}(b) &= \{a \in \mathscr{A}, \phi^{\mathscr{A}}(a, b) = \max_\theta~\phi^{\mathscr{A}}(\theta, b)\} \\ r^{\mathscr{B}}(a) &= \{b \in \mathscr{B}, \phi^{\mathscr{B}}(a, b) = \max_\theta~\phi^{\mathscr{A}}(a, \theta)\} \end{aligned} \quad (7.15)$$

The continuity of $\phi^{\mathscr{B}}$ causes the map $\phi^{\mathscr{B}}(a, b)$ to be continuous and establishes a closed set $\mathscr{G}(b) = \{(a, b) | b \in r^{\mathscr{B}}(a)\}$ where $\mathscr{G}(b) \subset \mathscr{B}$, correspondingly $\mathscr{G}(a) \subset \mathscr{A}$.

Since $\mathscr{B}$ is convex $b_1\theta + (1 - \theta)b_2$ is and admissible convex combination for $\mathscr{B}$. In this case, $\theta b_1 + (1 - \theta)b_2 \in r^{\mathscr{B}}(a)$ since $\phi^{\mathscr{B}}(a, \theta b_1 + (1 - \theta)b_2) \geq \theta \phi^{\mathscr{B}}(a, b_1) + (1 - \theta)\phi^{\mathscr{B}}(a, b_2)$ due to the concavity property of $\phi^{\mathscr{B}}$. The same analysis can be conducted for $\phi^{\mathscr{A}}(b)$. This result leads to the fact that any combination of responses (a, b) in fact belongs to the map of combinations $m(a, b) \mapsto r^{\mathscr{A}}(b) \times r^{\mathscr{B}}(a)$. Given that the sets $\mathscr{A}, \mathscr{B}$ are nonempty and convex the map $m(a, b)$ is not empty, a fixed value can be found according to Kakutani's theorem.

Lemma 7.1 *(Existence of Nash equilibrium at fixed time). If $\sigma_i(k) \in \mathscr{D}$, then sets of strategies A, B for two players admit a Nash equilibrium at time k.*

Proof Given that $\sigma = |3|$ and according to Definition 7.1, the number of strategies is countable. By extension of Theorem 7.1, it is straight forward to admit a Nash equilibrium at sample time k.

Definition 7.2 (Payoff function). Let be $J_i^A(\mathbf{p}(k), \mathbf{v}(k), a_\delta, b_\delta)$ the function defining the payoff after a player decides among the sets of strategies A as:

$$J_i^A(\mathbf{p}(k), \mathbf{v}(k), a_\delta, b_\delta) = \psi_i(p(N)) - \sum_{k=0}^{N-1} L_i(\mathbf{p}(k), a_\delta, b_\delta) \qquad (7.16)$$

In Definition 7.2, L_i is defined as the running cost while the ψ_i is called the final cost.

Assumption 7.1 (Available game information). Dynamic (7.1) is well known for each one of the participants of the games.
The same as $J_i^A(\mathbf{p}(k), \mathbf{v}(k), a_\delta, b_\delta), J_j^B(\mathbf{p}(k), \mathbf{v}(k), a_\delta, b_\delta)$ and the sample time k is considered synchronous in vehicles i, j.

7.4.3 Properties of the lane change dynamic game

Consider the full dynamics expressed in (7.10) jointly with (7.8) and enclosed in the form $\mathbf{x}(k+1) = f(\mathbf{x}(k), \delta(k)) = A\mathbf{x}(k) + M\delta(k)$. $\mathbf{x}^T = \begin{pmatrix} \mathbf{p}^T & \mathbf{v}^T & \sigma^T \end{pmatrix}$. In a particular case where two players are defining a game, it is possible to define split dynamics and running costs as:

$$\begin{aligned} \mathbf{x}(k+1) &= A\mathbf{x}(k) + M_1\delta_1(k) + M_2\delta_2(k) \\ L_i(k) &= L_{i1}(\mathbf{x}(k), \delta_1(k)) + L_{i2}(\mathbf{x}(k), \delta_2(k)) \end{aligned} \qquad (7.17)$$

Remark 7.1 (Finding equilibrium via Pontryagin Maximum Principle (PMP)). Let consider the system (7.17) with associated running cost (7.17). Let $\mathbf{x}^*(\cdot), \delta_1^*(\cdot), \delta_2^*(\cdot)$ be, respectively, the trajectory and open-loop controls of two players in a Nash equilibrium. By definition, these two controls provide corresponding solutions to the associated optimal control problems for each player. Applying the PMP, the following are necessary conditions for the Nash equilibrium [16]:

$$\begin{aligned} \mathbf{x}(k+1) &= \mathbf{x}(k) + M_1\delta_1^\#(k) + M_2\delta_2^\#(k) \\ \lambda_1(k) &= \bar{A}\lambda_1(k+1) + \nabla_\mathbf{x} L_{11}(\mathbf{x}(k), \delta_1^\#(k)) \\ \lambda_2(k) &= \bar{A}\lambda_2(k+1) + \nabla_\mathbf{x} L_{22}(\mathbf{x}(k), \delta_2^\#(k)) \end{aligned} \qquad (7.18)$$

where

$$\delta_1^\sharp = \underset{\delta_1 \in \mathscr{D}}{\arg\max}\, \lambda_1 M_1 \omega L_{11}(t, \mathbf{x}, \delta_1) \tag{7.19a}$$

$$\delta_2^\sharp = \underset{\delta_2 \in \mathscr{D}}{\arg\max}\, \lambda_2 M_2 \omega L_{22}(t, \mathbf{x}, \delta_2) \tag{7.19b}$$

Let define the Hamiltonian for the control problem (7.13) based on (7.17):

$$H(\mathbf{x}(k), \delta_1(k), \delta_2(k)) = \sum_{i \in \mathscr{I}} (L_{i1}(\mathbf{x}(k), \delta_1) L_{i2}(x(k), \delta_2(k))) - \lambda_i(\mathbf{x}(k) + M_1 \delta_1(k) + M_2 \delta_2(k)) \tag{7.20}$$

By considering the costate condition for $\lambda \in \mathbb{R}^n$ from the PMP [16]:

$$\lambda_i(k) = \frac{\partial H}{\partial x_i} = \left(\frac{\partial H}{\partial x_i}\right)^T \lambda_i(k+1) + \frac{\partial L(k)}{\partial x_i} \tag{7.21}$$

with the final condition $x(T) = 0$, then it is possible to obtain the conditions in (7.18). The optimal condition is derived from the fact that for a fixed lateral control $\bar{\delta}_2(\cdot)$, the optimal $\delta_1(\cdot)$ can be found via:

$$\delta_1^*(\cdot) = \underset{\delta_1 \in \mathscr{D}}{\arg\min}\, H(\mathbf{x}(k), \delta_1(k), \bar{\delta}_2(k)) \tag{7.22}$$

which can be transformed into a maximization problem where the player is maximizing the payoff function similar to (7.16), leading to

$$\delta_1^*(\cdot) = \underset{\delta_1 \in \mathscr{D}}{\arg\max}\, -H(\mathbf{x}(k), \delta_1(k), \bar{\delta}_2(k)) \tag{7.23}$$

The stationary condition is necessary for optimality; then by introducing (7.20) into (7.23), we obtain:

$$0 = -\frac{\partial L_{11}(\mathbf{x}(k), \delta_1(k)) + L_{12}(\mathbf{x}(k), \bar{\delta}_2(k))}{\partial \delta_1} + \lambda_1 \frac{\partial (x(k) + M_1 \delta_1(k) + M_2 \bar{\delta}_2(k))}{\partial \delta_1} \tag{7.24}$$

leading to (7.19a). In the same way, equation (7.19b) can be obtained when the first player fixes its own strategy to a value $\bar{\delta}_1 = \delta_1^*$. The Nash equilibrium is obtained when the payoff for player 1 is maximized (7.23) with the best response of player 2 and vice versa [15]. In other words, no player can increase his payoff by single-mindedly changing his strategy, as long as the other player sticks to the equilibrium strategy.

In general, the game presented here is a nonzero-sum game and players, in fact, cooperate toward the common objective, given by the successful lane change. On the other hand, the scalability of this approach may suffer with long time horizons. In this case, we propose an heuristic way to solve this algorithm [17].

7.5 Numerical examples

7.5.1 Experimental setting

To test the working of the dynamic game framework, we conducted numerical examples. The scenario is set up as in Figure 7.3. We simulate three vehicles, with Vehicle 2 and Vehicle 3 interacting with each other in the merging section. The initial conditions are: $p_1(0) = 0m, p_2(0) = -50m$ $v_{l1}(0) = v_{l2}(0) = v_1(1) = v_2(0) = 30m/s, \sigma_1(0) = \sigma_2(0) = 2, \sigma_3(0) = 1, \sigma_2^* = \sigma_3^* = 2$ (The desired lanes for both Vehicle 2 and Vehicle 3 are Lane 2, the right lane on the main freeway). $v_3(1) = 25m/s, p_3(1) = -45m$ for Scenario 1 and $p_3(1) = -35m$ for Scenario 2. $v_0 = 30m/s, v_{min} = 0m/s, v_{max} = 35m/s, a_{min} = -5m/s^2, a_{max} = 2m/s^2$ $\beta_1 = 0.2, \beta_2 = 0.2, \beta_3 = 0.5, \beta_4 = 5, \beta_5 = 5, \beta_6 = 0.05, t_{min} = 0.5s$.

7.5.2 Scenario 1: delayed merge

In Scenario 1, Vehicle 3 is 5 m in front of Vehicle 2 but with a slower speed. The resulting cost of all vehicles and the cost of Vehicle 3 and Vehicle 2 are shown in Figure 7.4. The best situation for Vehicle 3 is that Vehicle 2 performs courtesy lane change from Lane 2 to Lane 3, so that Vehicle 3 has a conflict-free merge. In this case, Vehicle 3 can change lane immediately to minimize its own cost. However, this leads to deviation from the desired lane of Vehicle 2, leading to a cost function value of 50. The overall cost is not the optimum for the whole vehicle group.

From the collective system perspective, the best strategy is that Vehicle 2 stays in the same lane and passes Vehicle 3. Vehicle 3 waits for Vehicle 2 to pass until sufficient safety gap is developed in front and changes lane at $k = 7$ s.

Interestingly, if following a first-in-first-out strategy that is widely used in cooperative merging systems [11], it leads to the feasible strategy that is best for Vehicle 3 but not the best for the collective vehicle group.

Figure 7.5 shows the system optimal solution, where the error on desired speed e_0^v, speed error to predecessor e_l^v, vehicle speed, and lane sequence are depicted. Note that the change of increasing rate in speed for Vehicle 3 is due to the fact that before the lane change, Vehicle 3 has no leader and it only accelerates toward the desired speed. When it changes lane, both the error on desired speed and speed

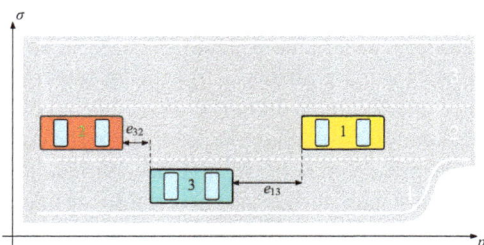

Figure 7.3 Scenarios

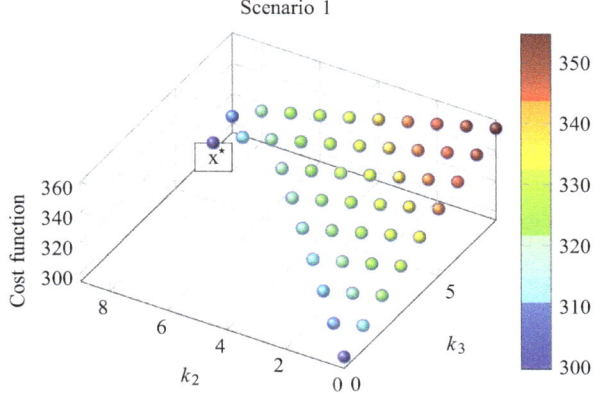

Figure 7.4   Overall cost for Scenario 1. k_i represents the time to change lane of vehicle i

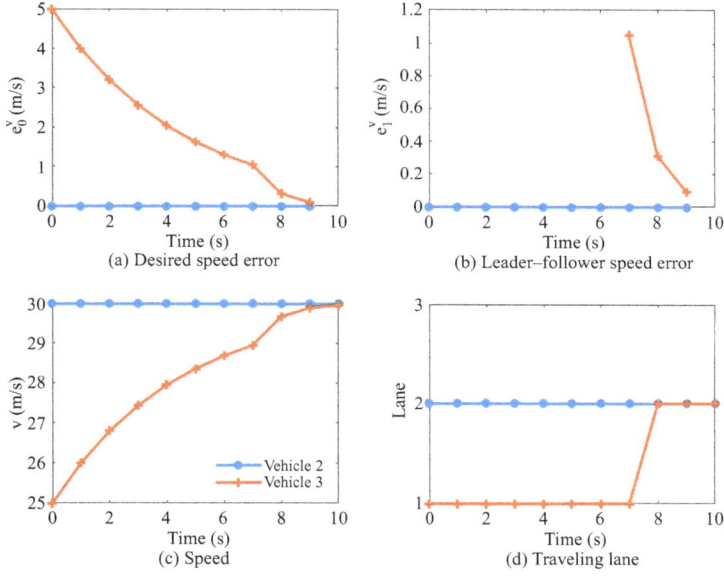

Figure 7.5   Delayed merge

error to predecessor demand it to accelerate, resulting in an increase in speed change rate.

7.5.3   Scenario 2: courtesy lane change

In Scenario 2, Vehicle 3 is 15 m ahead of Vehicle 2. The resulting cost of all vehicles and the cost of Vehicles 2 and 3 are shown in Figure 7.6. In this case, it is not safe for Vehicle

160 *Traffic information and control*

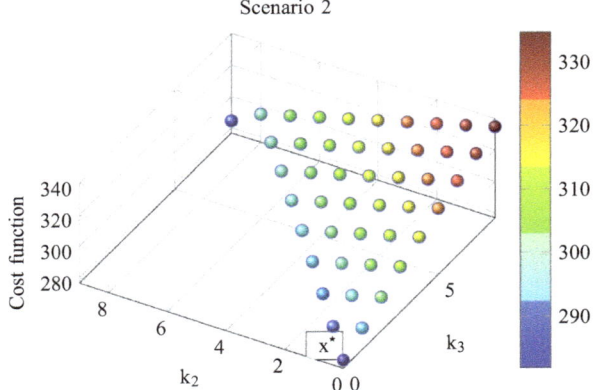

Figure 7.6 *Overall cost for Scenario 2. Blank areas are infeasible strategies, i.e., due to violation of safety constraints*

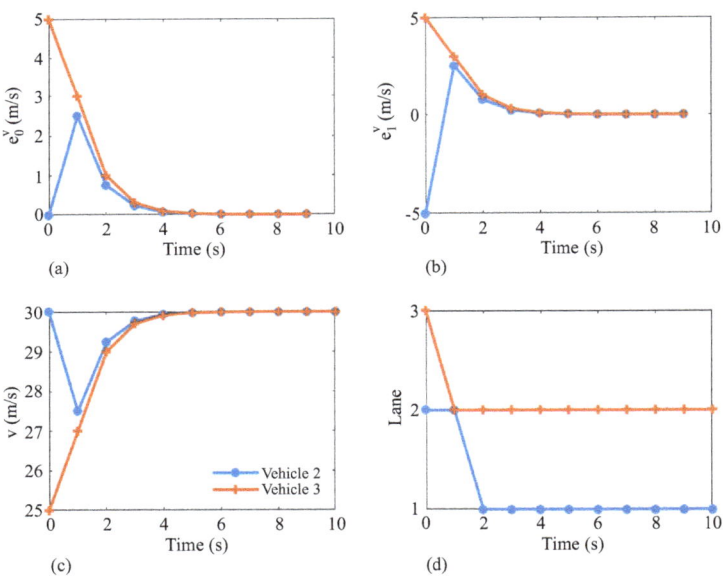

Figure 7.7 *Courtesy lane change*

3 anymore if Vehicle 2 still tries to pass Vehicle 3. Instead, the best strategy from the whole vehicle network perspective is that Vehicle 2 changes to left immediately and Vehicle 3 changes left immediately. This will minimize the lane preference cost and the cost of reaching the end of the merging section for Vehicle 3, at the expenses of lane preference cost for Vehicle 2. Figure 7.7 shows the system optimal solution.

This illustrates the very nature of cooperation: *individual agents may have to compromise their own benefits for the best performance of the whole system.*

7.6 Conclusion

In this chapter, we proposed a dynamic game formulation for cooperative lane change maneuvers of automated vehicles at highway merges. Simplified vehicle longitudinal and lateral dynamics models are used to predict the system process under different lane change strategies. The framework captures the competitive and cooperative nature of the interactions between the merging vehicle and the mainline vehicle, and renders the design tractable to a range of mathematical tools related to optimal control and integer programming. The discrete dynamic model with control input substantially reduces the computational load for the dynamic merging game compared to previous work. Numerical examples demonstrate the potential of the approach in generating system optimum strategies as opposed to existing noncooperative merging algorithms and the benefits of sharing information and joint optimization under a common goal.

Future research is directed to the scalability analysis of the proposed framework and efficient solution algorithms to a large network of cooperative vehicles and the assessment of effects of this framework on traffic operations.

References

[1] Hanappe F, Hudson A, Pelloux P, *et al.* Impacts and potential benefits of autonomous vehicles: From an international context to Grand Paris. Apur; 2018.

[2] Rios-Torres J, and Malikopoulos AA. Automated and cooperative vehicle merging at highway on-ramps. IEEE Transactions on Intelligent Transportation Systems. 2017;18(4):780–789.

[3] Papageorgiou M, Kiakaki C, Dinopoulou V, *et al.* Review of road traffic control strategies. Proceedings of the IEEE. 2003;91(12):2043–2067.

[4] Nowakowski C, Thompson D, Shladover SE, *et al.* Operational concepts for truck maneuvers with cooperative adaptive cruise control. Transportation Research Record: Journal of the Transportation Research Board. 2016;2559:57–64.

[5] Zhang Y, Wang M, Hu J, *et al.* Semi-constant spacing policy for leader-predecessor-follower platoon control via delayed measurements synchronization. In: 21st IFAC World Congress; 2020.

[6] Zhang Y, Bai Y, Hu J, *et al.* Control design, stability analysis, and traffic flow implications for cooperative adaptive cruise control systems with compensation of communication delay. Transportation Research Record. 2020.

[7] Ntousakis IA, Nikolos IK, and Papageorgiou M. Optimal vehicle trajectory planning in the context of cooperative merging on highways. Transportation Research Part C: Emerging Technologies. 2016;71:464–488.

[8] Milanés V, Godoy J, Villagra J, *et al.* Automated on-ramp merging system for congested traffic situations. IEEE Transactions on Intelligent Transportation Systems. 2011;12(2):500–508.

[9] Duret A, Wang M, and Ladino A. A hierarchical approach for splitting truck platoons near network discontinuities. Transportation Research Part B: Methodological. 2019 4.

[10] Jin L, Čičič M, Amin S, *et al.* Modeling the impact of vehicle platooning on highway congestion. In: Proceedings of the 21st International Conference on Hybrid Systems: Computation and Control (part of CPS Week) - HSCC '18. ACM Press; 2018. pp. 237–246.

[11] Rios-Torres J, and Malikopoulos AA. A survey on the coordination of connected and automated vehicles at intersections and merging at highway on-ramps. IEEE Transactions on Intelligent Transportation Systems. 2017;18(5):1066–1077.

[12] Wang M, Hoogendoorn SP, Daamen W, *et al.* Game theoretic approach for predictive lane-changing and car-following control. Transportation Research Part C: Emerging Technologies. 2015;58:73–92.

[13] Fabiani F, and Grammatico S. A mixed-logical-dynamical model for automated driving on highways. In: 2018 IEEE Conference on Decision and Control (CDC). IEEE; 2018. pp. 1011–1015.

[14] Xiao L, Wang M, Schakel W, *et al.* Unravelling effects of cooperative adaptive cruise control deactivation on traffic flow characteristics at merging bottlenecks. Transportation Research Part C: Emerging Technologies. 2018;96:380–397.

[15] Bressan A. Non cooperative differential games. A tutorial. Penn State University; 2010.

[16] Lewis FL, Vrabie D, and Vassilis LS. Optimal control. 3rd ed. John Wiley & Sons; 2012.

[17] Ladino A, and Wang M. A dynamic game formulation for cooperative lane change strategies at highway merges. In: 21st IFAC World Congress; 2020.

Chapter 8
Cooperative driving and a lane change-free road transportation system

Zhengbing He[1]

8.1 Introduction

Over the past 10 years, Connected and Automated Vehicle (CAV) technology has attracted substantial attention from industry to academia. This technology is absolutely among the hottest topics and research fields. Recently, Automatic Cruise Control (ACC) has been widely deployed in newly produced high-end vehicles and Cooperative Automatic Cruise Control (CACC) is expected to be practically applied within a few years. Even fully automated vehicles, which completely abandon human drivers, have been tested on roads (Figure 8.1). Going forward, it may not be surprising to observe a vehicle with no driver operating beside us.

Connected and automated vehicles

Taking advantage of vehicle-to-vehicle and vehicle-to-infrastructure technologies, the so-called *connected vehicles* connect and share (status) information with each other when moving on roads, allowing a variety of applications that are able to enhance traffic safety and efficiency. An *automated vehicle*, the definition of which is very close to that of a self-driving vehicle or an autonomous vehicle, is capable of sensing its environment and moving safely with little or no human input. The two terminologies are usually combined together and called a *connected and automated vehicle*, indicating various advanced vehicles supported with advanced communication and information technology

In 2016, the Society of Automotive Engineers (SAE) identified six levels of driving automation, ranging from 'no automation' to 'full automation', as follows [1].

- *Level-0, No Automation.* The full-time performance by the human driver of all aspects of the dynamic driving task, even when enhanced by warning or intervention systems.

---
[1]Beijing Key Laboratory of Traffic Engineering, College of Metropolitan Transportation, Beijing University of Technology, Beijing, China

Figure 8.1   A Waymo self-driving car

- *Level-1, Driver Assistance.* The driving mode-specific execution by a driver-assistance system of either steering or acceleration/deceleration using information about the driving environment and with the expectation that the human driver performs all remaining aspects of the dynamic driving task.
- *Level-2, Partial Automation.* The driving mode-specific execution by one or more driver-assistance systems of both steering and acceleration/deceleration using information about the driving environment and with the expectation that the human driver performs all remaining aspects of the dynamic driving task.
- *Level-3, Conditional Automation.* The driving mode-specific performance by an automated driving system of all aspects of the dynamic driving task with the expectation that the human driver will respond appropriately to a request to intervene.
- *Level-4, High Automation.* The driving mode-specific performance by an automated driving system of all aspects of the dynamic driving task, even if a human driver does not respond appropriately to a request to intervene.
- *Level-5, Full Automation.* The full-time performance by an automated driving system of all aspects of the dynamic driving task under all roadway and environmental conditions that can be managed by a human driver.

As indicated by the SAE, human drivers still monitor the driving environment in *Levels 0–2*, while automated driving systems will take over the driving environment in *Levels 3–5*. Considering the remarkable impact of Connected and Automated Vehicles (CAVs), people have begun to rethink the forms and features of future traffic mixing human-driven vehicles and machine-driven vehicles [2–8].

Meanwhile, the prospects of the future road transportation systems in *Levels 3–5* also attract substantial interests from researchers and engineers, although they will require much time to be fully realized *Levels 3–5*. Among those prospects, cooperative driving based on automated driving systems at intersections/on-ramps is a cutting-edge topic (Figure 8.2). With the aid of cooperative driving strategies, traffic will show better continuity rather than platooned flow segmented by signal controls; thus, traffic efficiency and safety will be greatly improved, and traffic emissions will be reduced accordingly. Most of the existing studies regarding cooperative driving at intersections/on-ramps aim to design various optimization algorithms to improve the traffic efficiency of passing through intersections and on-ramps.

Cooperative driving strategies at intersections/on-ramps

Before entering an intersection/freeway, CAVs that are about to cross/merge at a road-intersection/on-ramp coordinate in a centralized or decentralized fashion. Detailed crossing/merging plans will be assigned or generated through negotiation with every single vehicle, making a vehicle cross/merge into the intersection/freeway just after the conflicting vehicle leaves so that the traffic efficiency and safety at intersections/freeways will be guaranteed to be as maximized as possible. Traffic signals may not be needed anymore and, thus, this is also called *signal-free (or autonomous) intersection/on-ramp management.*

To overcome the challenges of current road topology and fully utilize the advantages of CAV, some pioneer studies attempted to propose *Star Wars*-like future transportation systems. For example, He *et al.* [9] proposed a signal-free autonomous intersection with all-direction turn lanes, at which vehicles are allowed to load on any downstream outgoing road from any upstream incoming lanes, i.e., the turning direction associated with incoming lanes is removed at the proposed intersection. Stevanovic and Mitrovic [10] proposed a flexible arterial utilization simulation model, in which directional driving paths are altered between neighbouring lanes to align vehicles for decreased conflict for left and right turns at intersections (Figure 8.3).

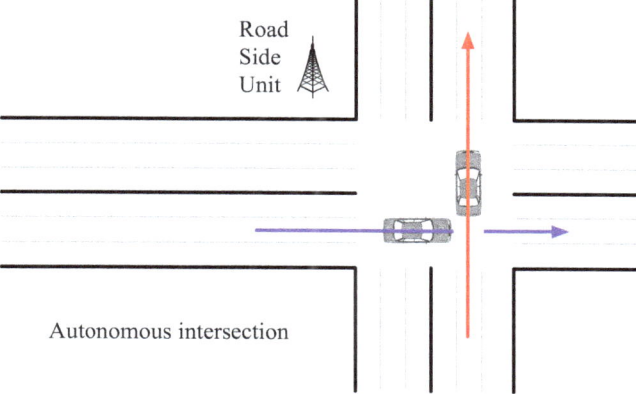

Figure 8.2 An illustration of cooperative driving at an intersection, where a vehicle is just passing by the conflicting vehicle at the intersection

166 *Traffic information and control*

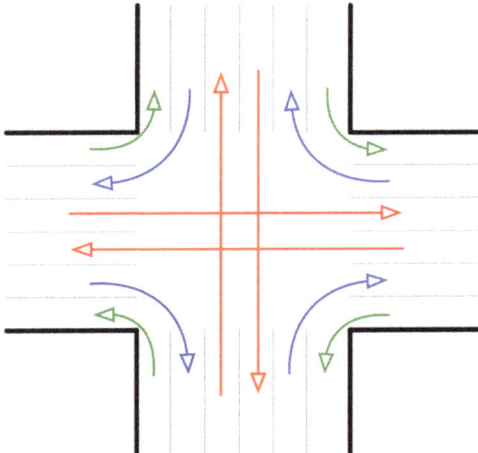

Figure 8.3 An illustration of the intersection with flexible arterial utilization simulation modelling [10]. It can be observed that moving directions on lanes are quite flexible at the intersection

The remainder of this chapter is organized as follows. Sections 8.2 and 8.3 deliberately introduce some representative cooperative driving strategies at intersections and on-ramps, respectively. Section 8.4 proposes a novel future lane change-free road transportation system. Section 8.5 makes conclusions and introduces future directions.

8.2 Cooperative driving strategies at intersections

This section deliberately introduces some representative cooperative driving strategies at intersections, including the safety driving pattern-based strategy [11], the reservation-based strategy [12], and trajectory optimization-based strategy [13]. For systematic reviews of this active field, one may refer to [14–16].

8.2.1 *Safety driving pattern-based strategy*

Li and Wang [11] are one of the first researchers to propose cooperative driving at intersections. Many concepts in the seminal work, such as the coordination area, vehicle group, intersection zone partition and trajectory planning, have inspired the follow-up studies.

To achieve cooperative crossings without the control of traffic lights, this chapter formulated the cooperative intersection driving as follows. A virtual circle centred at the intersection point is defined as the area of coordinating vehicles (Figure 8.4). Within the area, lane changes are not allowed and intervehicle communication is employed to make sure that vehicles in the coordination area have the status information of all other vehicles. Then, vehicles self-organize into small groups that are constructed by randomly assigning several temporary dominant

Cooperative driving and a lane change-free road transportation system 167

vehicles and letting them group with their neighbouring vehicles (Figure 8.4). A zone-blocking strategy is employed to ensure collision-free intersections, i.e., dividing the crossing zone into subzones and allowing only one vehicle to enter a zone at any time.

The cooperative driving schedule and cooperative trajectory planning are further proposed to solve the above-formulated problem. The cooperative driving schedule indicates the sequence of vehicles entering an intersection. To this end, a tree generation algorithm in which each node represents a particular driving sequence is developed to obtain the optimal cooperative driving schedule, and a safety pair labelling algorithm is incorporated to take safety into account.

After acquiring the sequence of vehicles entering an intersection, the cooperative trajectory planning is conducted to guarantee the proper vehicle movements on intersection approaches. The virtual vehicle mapping techniques proposed by Uno *et al.* [17] and Sakaguchi *et al.* [18] (which will be introduced in Section 8.3.1 in detail) are modified for the cooperative driving problem. The basic logic of the virtual vehicle mapping technique is to virtually map a vehicle in a lane onto an object lane; then, the interested vehicle in the object lane can be controlled with respect to the virtual vehicle to guarantee safety.

8.2.2 Reservation-based strategy

Dresner and Stone [12] proposed another influential seminal study on cooperative driving at intersections. The proposed concept of 'reservation' is widely adopted currently.

In the study, the multiagent framework is applied to assign vehicles to pass through intersections efficiently; i.e., *driver agents* control the vehicles and an

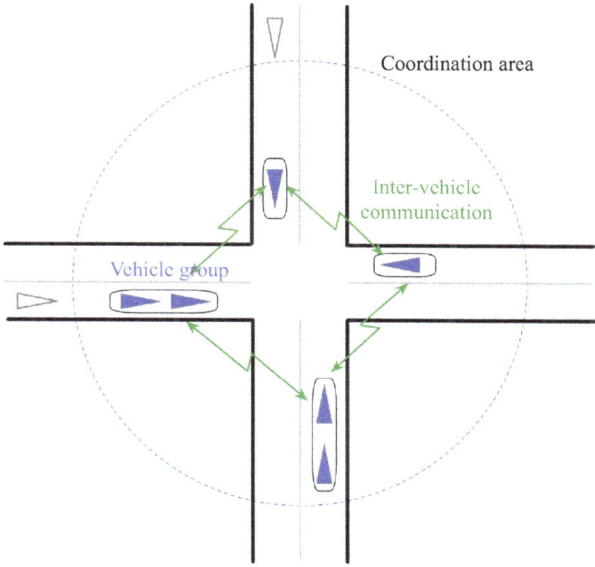

Figure 8.4 Problem formulation of the safety driving pattern-based strategy [11]

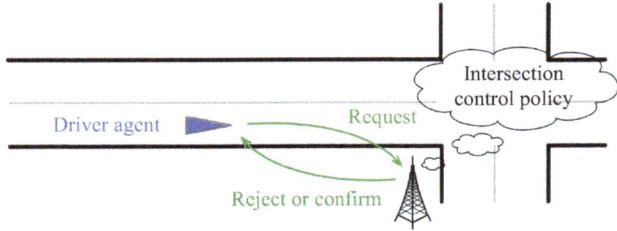

Figure 8.5 The framework of the multiagent approach and the reservation-based strategy [12]

intersection agent is placed at each intersection (Figure 8.5). The driver agents 'call ahead' and attempt to reserve a space-time block for passing through the intersection; then, the intersection agent decides to grant or reject the requested reservations according to an intersection-control policy. This is similar to making a reservation for a hotel: the potential guest specifies when he/she will arrive, how much space is required, and how long the stay will be; the hotel determines whether to grant the reservation.

More specifically, the proposed reservation-based strategy follows the First-Come First-Served (FCFS) rule; i.e., the first vehicle that requests to pass through the intersection will be served first. To properly determine whether granting reservations is possible, the intersection is divided into an $n \times n$ grid of *reservation tiles*, where n is the granularity of the policy (Figure 8.6). Vehicles and reservations are then represented by using the discretized grids. After a reservation request, the intersection agent runs an *internal simulation* to obtain the trajectory of the vehicle across the intersection on the basis of current traffic conditions. At each simulation step, if the requesting vehicle occupies a reservation tile that is already reserved by another vehicle, the request will be rejected; otherwise, the requested reservation tiles will be reserved.

This simulation-based reservation strategy is logically straightforward, although it is not computationally tractable and the computational burden is relatively high, particularly when the number of the vehicles considered is large.

Later, the strategy for an isolated intersection was extended to a network with multiple interconnected intersections and different navigation policies with dynamic planned paths were examined [19]. Because human-controlled vehicles cannot be replaced in a short period, the problem of simultaneously managing human- and machine-controlled vehicles was investigated [20] in the framework of this reservation-based strategy. A mixed reality platform was also developed on this basis, allowing an autonomous vehicle to interact with multiple virtual vehicles in a computer simulation at a real-world intersection [21].

8.2.3 Trajectory optimization-based strategy

By considering the spatiotemporal trajectories of vehicles that are approaching and crossing an intersection, Lee and Park [13] proposed a trajectory

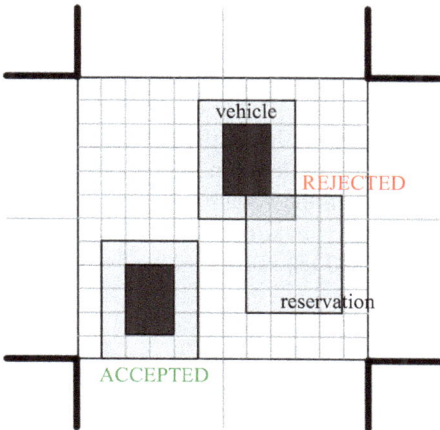

Figure 8.6 Discretized intersection and reservation tiles

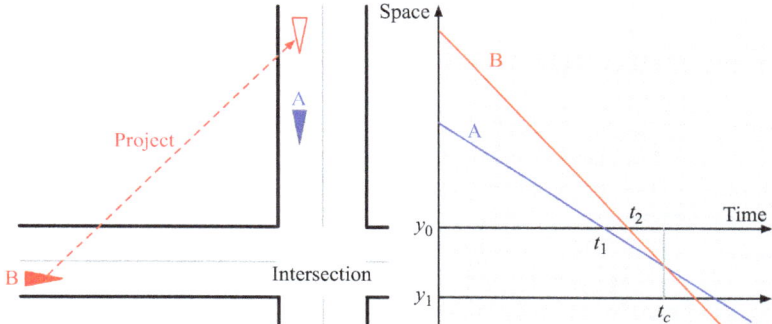

*Figure 8.7 Time-space diagram of trajectory conflict that occurs at an intersection [13]. A and B indicate two trajectories of vehicles approaching and entering an intersection (**B** is the trajectory that is projected to the lane Vehicle A is on), respectively; y_0 and y_1 are the intersection edges where a vehicle enters and leaves the intersection, respectively; t_1 and t_2 are the times when Vehicles A and B enter the intersection; t_c is the potential conflict time at the intersection*

optimization-based strategy for cooperative driving at intersections. Compared with previous studies, the proposed strategy is able to globally optimize the entry of the vehicles approaching an intersection so as to improve the system efficiency.

As shown in Figure 8.7, the idea of the trajectory optimization-based strategy is to avoid the occurrence of trajectory overlaps at an intersection (in Figure 8.7, A and B represent potential conflicts at time t_c).

A nonlinear optimization problem is then formulated by considering vehicle acceleration/deceleration and safety. The object function is to minimize the total length of overlapped trajectories, where the length of a trajectory overlap is defined as the length of a time-space trajectory segment of a vehicle that potentially appears inside an intersection together with another vehicle trajectory segment. Given vehicle speed and acceleration/deceleration, the length of an overlap can be written as an integral equation based on the equations of motion.

The constraints include (1) maximum acceleration or deceleration, (2) maximum and minimum speeds and (3) minimum headways between two consecutive vehicles on the same lane.

To solve the nonlinear programming problem, heuristic algorithms, i.e., the active set method based on sequential quadratic programming and the interior point method based on the calculation of the Karush-Khun-Tucker conditions [22], are employed.

Later, Lee et al. [23] expanded the trajectory optimization-based strategy and implemented it in a corridor consisting of multiple intersections. Furthermore, the simulation test based on VISSIM demonstrated that the proposed strategy can significantly improve the traffic efficiency and safety and the air quality.

8.3 Cooperative driving strategies at on-ramps

In general, cooperative on-ramp driving is a simplified case of cooperative intersection driving because, in essence, both merging from on-ramps and passing through intersections are vehicle-conflicting, while the on-ramp merging areas are simpler than intersections in terms of topology. Therefore, it may not be difficult to transfer the strategies for intersections to those for on-ramp merging through simple revisions. To present more diverse strategies, we attempt to introduce two distinguishing cooperative on-ramp driving strategies, namely, the virtual vehicle mapping strategy [17,18] and the slot-based strategy [24,25].

8.3.1 Virtual vehicle mapping strategy

Uno et al. [17] and Sakaguchi et al. [18] proposed a virtual vehicle mapping strategy for on-ramp merging. The strategy generates a virtual vehicle by mapping a vehicle on one lane onto another lane in order to enable smooth longitudinal control between a vehicle on a freeway and the one at an on-ramp. The strategy is straightforward and safety-guaranteed, although it is obviously not optimal.

More specifically, three cases are classified to generate a virtual vehicle; see Figure 8.8.

- Case 1: When Vehicle B is estimated to be positioned following Vehicle A after merging, Vehicle B generates *Virtual* Vehicle A by mapping Vehicle A onto the lane where Vehicle B is.
- Case 2: When Vehicle B is estimated to be positioned in front of Vehicle A after merging, it is mapped onto the lane where Vehicle A is as *Virtual* Vehicle B.

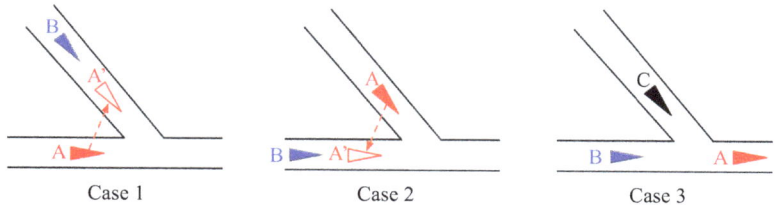

Figure 8.8 Classification of merging and generation of a virtual vehicle at an on-ramp [17]

- Case 3: There are three vehicles, Vehicles A, B and C, in this case. In terms of the interaction between Vehicle A and Vehicle C, the merging of Vehicle C after Vehicle A is the same as in Case 1. In terms of the interaction between Vehicle A and Vehicle B, the longitudinal control will eventually create the headway 2L.

After the generation, both real and virtual vehicles are equally considered in a car-following process, resulting in proper gaps (governed by the safety constraint in car-following models) in advance and smooth lateral merging.

8.3.2 Slot-based strategy

In contrast to all above-introduced strategies, Marinescu *et al.* [24,25] introduced a conveyor-like highway system, called a slot-based system. The system treats vehicles as the products on a conveyor belt and gives drivers little freedom of choice. Although the technical feasibility of the system is still unknown, the idea is interesting and enlightening.

In the slot-based system, when passing a potential bottleneck, e.g., ramps, vehicles are compelled to strictly follow virtual slots, whose behaviour is (periodically) predefined by the slot-based traffic management system. The slot S is defined as $S\{z, p, t, b, o\}$, where z is the size of the slot, p is the position of the slot at time t, b is a predefined sequence of acceleration, deceleration and lane changing manoeuvres, and o indicates the status of the slot, i.e., free (empty) or occupied by a vehicle.

Some basic problems regarding the slot-based system were discussed. One is the slot-changing problem, i.e., the determination of a new empty slot for a vehicle to move into. Two approaches are introduced: (1) the hierarchical approach, in which the traffic management system makes a decision for all vehicles and (2) the distributed approach, in which vehicles make decisions themselves through intervehicle communication and cooperation. Another problem is the optimized lane assignment problem, i.e., optimally assigning the vehicles on the outer lane to the empty slots on the inner lane in order to allow more space (slots) for the merging vehicles.

For on-ramp merging, a vehicle-to-slot strategy was proposed [24,25], which is separated into two phases: slot selection and moving into the slot. For the slot selection, a hybrid method of hierarchical and distributed approaches was proposed. When moving into a slot, a cloned slot was first generated on ramps (Figure 8.9).

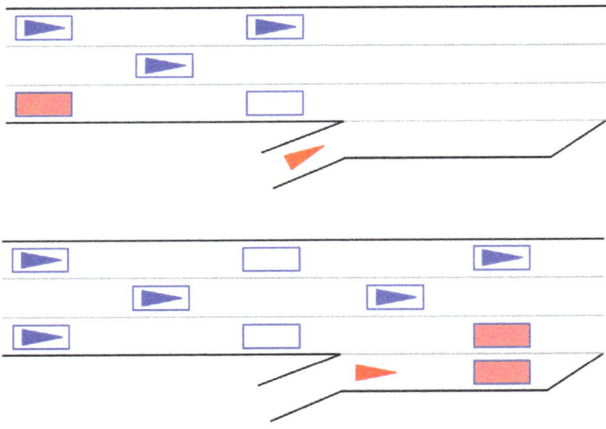

Figure 8.9 *On-ramp merging in the slot-based highway system [24]. The blue boxes are virtual slots predefined by the traffic management system. The blue box with/without a vehicle is an occupied/free slot. The blue box filled with red on the highway is the slot that the red vehicle would like to move into. The blue box filled with red on the ramp is the cloned slot that helps the red vehicle to move into the target slot*

8.4 Lane change-free road transportation system

In this section, we propose a novel cooperative driving road transportation system, called the *lane change-free road transportation system*. The distinguishing feature of this road system is that vehicles are allowed to turn for any direction on any incoming lane of an intersection, so that, with the guidance of a cooperative driving strategy, vehicles can load on any downstream outgoing road of the intersection from any lane of an upstream incoming road of the intersection. For a freeway within the system, vehicles cooperatively and intensively merge/diverge at the merging/diverging area based on multivehicle coordination, making lane changes unnecessary when entering/leaving a freeway. Therefore, when driving in such a transportation system, vehicles can reach their trip destinations without making any on-road lane changes. Road transportation is expected to achieve high traffic efficiency and enhanced driving safety after erasing the lane-changing manoeuvre that is considered one of the most challenging automated vehicle technologies on roads.

Besides conventional roads and ramps, the system consists of a cooperative driving strategy that coordinates vehicles' movements and three key components:

- Conflict Avoidance-Based Cooperative Driving (CACD) strategy,
- Cooperative Intersection with All-direction turn lanes (CI-A),
- Cooperative Merging Area (CMA), and
- Cooperative Diverging Area (CDA).

Cooperative driving and a lane change-free road transportation system 173

For simplicity, we collectively refer to CI-A, CMA and CDA as IMD areas.

Lane change-free road transportation system

The lane change-free road transportation system consists of *cooperative intersections with all-direction turn lanes, cooperative merging areas, cooperative diverging areas* and *a conflict avoidance-based cooperative driving strategy*. Taking advantage of communication and CAV technology, vehicles that move within the road system can reach their trip destinations with no need for making any lane changes. It is expected that automated vehicle technology could be greatly simplified and that driving safety and efficiency could be improved when lane-changing processes, i.e., the lane-changing module of automated vehicle technology, are removed from roads.

8.4.1 Lane change-free road transportation system: an illustration

This subsection introduces an illustration of the proposed lane change-free road transportation system in the scenario that vehicles can reach their trip destinations without making any on-road lane changes.

When driving in the road network shown in Figure 8.10, a vehicle enters the network from the rightmost lane of a road. If the network is a conventional road network, the vehicle (shown as the dashed arrow lines in Figure 8.10(a)) will make two lane changes before entering Intersection A and turning left. Subsequently, it enters Intersection B through a movement and makes one lane change before turning left to enter Intersection C. At the rightmost lane of the road where the trip destination is located, the trip is completed and the vehicle leaves the network.

If driving in a lane change-free road network, the vehicle (shown as the arrow lines in Figure 8.10(a)) maintains its through movement after entering the network from the rightmost lane. Then, it enters Intersection A from the rightmost lane and makes a left turn, which is allowed by the CI-A. Similarly, the vehicle enters Intersections B and C from the lane in which it is currently located since vehicle turning of any direction is allowed at the CI-A. It is obvious that no lane change occurs on the roads in the proposed lane change-free road system, meaning that only simple car-following technology is needed if driving in the system and that the system is obviously much simpler and should be safer than systems involving lane changes.

For merging/diverging (Figures 8.10(b) and (c)), the proposed CACD strategy coordinates vehicles passing through those potential conflicts and the merging/diverging vehicles can enter/leave the freeway along those virtual paths. The strategy centrally controls merging/diverging with schedules, in contrast to conventional self-organized individual-based merging/diverging.

In short, as we illustrated, vehicles can reach their trip destinations without making any lane changes if driving in the lane-change road system with the CACD strategy and the IMD areas.

174 *Traffic information and control*

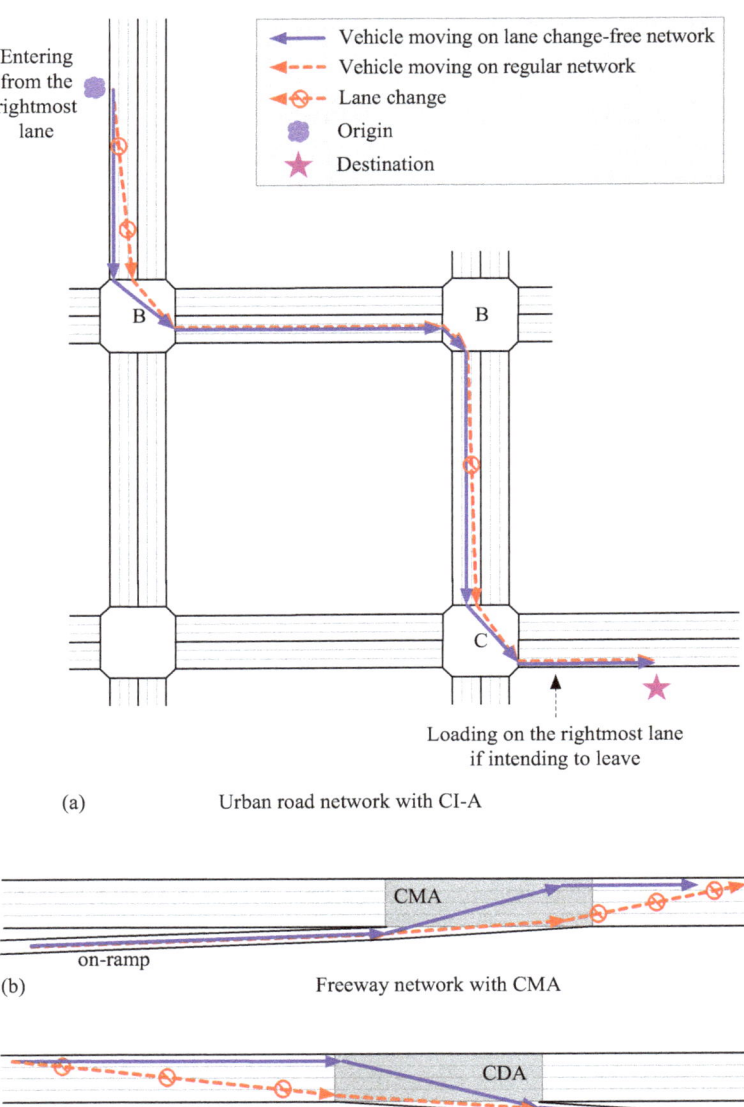

Figure 8.10 *Illustration of the lane change-free road network*

8.4.2 *System design*
8.4.2.1 **Overall approaching process**
The overall process by which a vehicle approaches an IMD area consists of the following three steps (Figure 8.11).

Cooperative driving and a lane change-free road transportation system 175

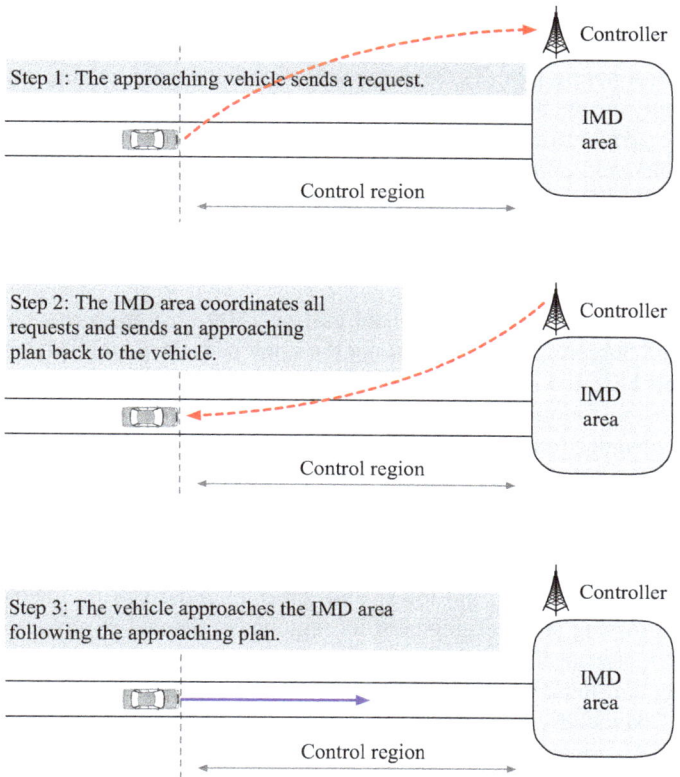

Figure 8.11 Overall process of a vehicle approaching an IMD area

Step 1: The approaching vehicle sends a request. When a vehicle (denoted by i) moves into the control region (within the control region, vehicles move following the approaching plan given by the IMD controller), it sends a request to the IMD controller for its passing.

Step 2: The IMD controller coordinates all requests and sends an approaching plan back to the vehicle. After receiving the request, the IMD controller coordinates with other already-received requests considering collision avoidance around conflicts inside the IMD area and, then, by employing the CACD strategy, calculates the time (denoted by $t_{i,\text{arrive}}^{\text{allow}}$) that allows vehicle i to enter the IMD area.

Based on $t_{i,\text{arrive}}^{\text{allow}}$ and the current state of vehicle i, the IMD controller proposes an approaching plan (i.e., the time-space trajectory) for vehicle i and sends the plan back to vehicle i. This represents a trajectory planning problem, given a vehicle's speed, position and time at the origin and destination in a time-space plane. In the existing literature, one can easily find various trajectory planning algorithms, such as [26–31]. He *et al.* [9] also provided a simplified trajectory planning algorithm that was used to test the proposed strategy.

Step 3: The vehicle approaches the IMD area following the approaching plan. Vehicle i receives the plan and approaches the IMD area by following it.

It is worth noting that (1) the communication between a vehicle and the IMD controller only occurs once at the moment that the vehicle enters the control region; (2) once the arrival time and the trajectory of an entering vehicle are settled, they cannot be changed and (3) when coordinating requests, the arrival time and the requests of all other vehicles within the control region have been fixed.

8.4.2.2 Conflict avoidance-based cooperative driving strategy

Conflict region

Assume that all vehicles that enter and exit an IMD area from the same pair of incoming and outgoing lanes move along the same path inside the IMD area, called the 'channel' here and denoted by c. The conflicts inside an IMD area can then be divided into the following two types (Figure 8.12). *Crossing conflict*: the point where two channels cross each other inside an IMD area; *Merging conflict*: the point where two or more channels merge to a downstream outgoing lane on the edge of an IMD area.

If the length of a vehicle is taken into account, a conflict can be extended to a conflict region along the conflicting channels. Specifically, as denoted by $b_{c,\text{cross}}^{\text{lower}}$, $b_{c,\text{cross}}^{\text{upper}}$, $b_{c,\text{merge}}^{\text{lower}}$ and $b_{c,\text{merge}}^{\text{upper}}$, the lower and upper boundaries (i.e., distance to the conflict) of crossing and merging conflict regions on channel c, respectively; refer to Figure 8.12 for an illustration, where $c = \{A, B\}$ for two conflicting channels. For crossing conflicts, when the vehicle on Channel B is passing lower boundary $b_{B,\text{cross}}^{\text{lower}}$ on Channel B, the vehicle on Channel A will be unable to pass lower boundary $b_{A,\text{cross}}^{\text{lower}}$ on Channel A until the front bumper of the vehicle on Channel B exits upper boundary $b_{B,\text{cross}}^{\text{upper}}$. For merging conflicts, when the vehicle on Channel B is passing lower boundary $b_{c,\text{merge}}^{\text{lower}}$ on Channel B, the vehicle on Channel A cannot pass lower boundary $b_{A,\text{merge}}^{\text{lower}}$ on Channel A. Upper boundary $b_{c,\text{merge}}^{\text{upper}}$ extends to the outgoing lane, and its location might depend on car-following behaviour and vehicle speed.

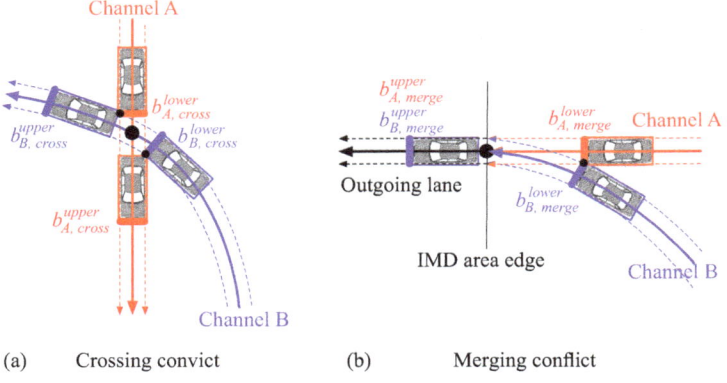

(a)   Crossing convict   (b)   Merging conflict

Figure 8.12 Conflicts within an IMD area

It is worth noting that, if one intends to make the system tolerate more errors, enlarging the conflict regions by adding an extra region (denoted by Δb) could be helpful. Then, the intersection makes coordination based on the conflict regions (a hard constraint) plus Δb (a soft constraint), and the time window assigned to a vehicle to pass through an intersection is enlarged: thus, more errors could be tolerated. However, the traffic efficiency could be compromised.

Coordination

Let all vehicles move inside an IMD area at the same constant speed (denoted by v_{max}^{IMD}), which can be achieved by modifying the vehicles' initial speed of entering the intersection. Since there are many conflicts inside an IMD area, letting vehicles move at a constant speed will greatly reduce the complexity of the coordination. The reliability of the whole system could be increased due to the fact that it is easier to maintain constant speed than to keep changing speeds within a limited area.

Then, *Step 2* in Section 8.4.2.1 can be unfolded as follows:

Step 2.1: Estimate the arrival time of the vehicle under normal circumstances. The step estimates the arrival time (denoted by $t_{i,\text{arrive}}^{\text{normal}}$) of the approaching vehicle (i.e., vehicle i) by only considering its car-following behaviour. Since the leading vehicle has sent a request and its approach plan has been generated by the intersection, the arrival time of this vehicle could be accordingly calculated based on car-following models.

Step 2.2: Calculate the arrival time allowed by the IMD area. This step checks whether or not vehicle i will conflict with other approaching vehicles around conflicts, given time $t_{i,\text{arrive}}^{\text{normal}}$ when vehicle i will arrive at the IMD area; if conflicts will occur, a proper arrival time for vehicle i to avoid the conflict, i.e., $t_{i,\text{arrive}}^{\text{allow}}$, $t_{i,\text{arrive}}^{\text{allow}} \geq t_{i,\text{arrive}}^{\text{normal}}$, will be calculated according to the following time-space optimization.

Imagine a time-space plane where the space is along a channel and there are several conflict regions; see Figure 8.13. Along the conflict region-time band, there have been reservations made for other vehicles that have entered the control region (squiggle rectangles in Figure 8.13). Let times $t_{c,k}^{j,\text{start}}$ and $t_{c,k}^{j,\text{end}}$ be the starting and ending times of the reservation, respectively, where $k = 1, 2, \ldots, K_c$ indicates a conflict and K_c is the total number of the conflicts on channel c; $j = 1, 2, \ldots, J_{c,k}$ indicates a reservation and $J_{c,k}$ is the total number of the reservations for conflict k on channel c.

For a newly incoming vehicle i, the earliest time of arrival at the IMD area is known (denoted by $t_{i,\text{arrive}}^{\text{normal}}$), i.e., the arrival time under the normal car-following condition. The objective of the coordination is to find a trajectory for vehicle i, whose corresponding occupation on all time-space conflict regions will not overlay with any existing region and whose starting time (i.e., $t_{i,\text{arrive}}^{\text{allow}}$) has minimum delay with reference to $t_{i,\text{arrive}}^{\text{normal}}$. Minimum delay ensures that the vehicles first arriving at the intersection are first served or considered.

Note that, in finding the optimal trajectory, avoiding overlap with the existing reservations is the only requirement. Thus, deceleration and acceleration could be considered, meaning that, if necessary, one could relax the previously assumed constraint that all vehicles have to move at constant speed inside an IMD area.

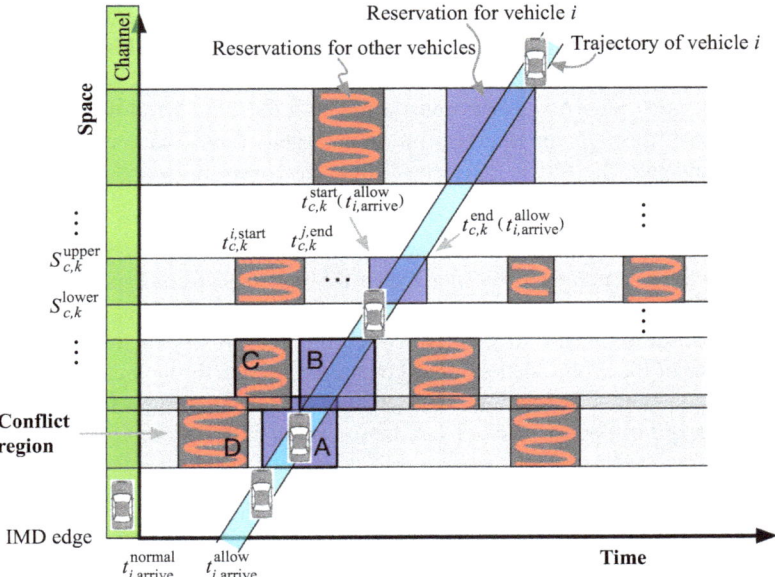

Figure 8.13 A time-space diagram for illustrating the coordination among different incoming vehicles

The optimization models can be written as follows.
Objective:

$$\text{minimize} \quad t_{i,\text{arrive}}^{\text{allow}} \tag{8.1}$$

Subject to:

$$t_{c,k}^{j,\text{end}} \leq t_{c,k}^{\text{start}}(t_{i,\text{arrive}}^{\text{allow}}) \tag{8.2}$$

$$t_{c,k}^{\text{end}}(t_{i,\text{arrive}}^{\text{allow}}) \leq t_{c,k}^{j+1,\text{start}} \tag{8.3}$$

$$t_{i,\text{arrive}}^{\text{normal}} \leq t_{i,\text{arrive}}^{\text{allow}} \tag{8.4}$$

$$k = 1, 2, \ldots, K_c \tag{8.5}$$

$$j = 1, 2, \ldots, J_{c,k} \tag{8.6}$$

where $t_{c,k}^{\text{start}}(t_{i,\text{arrive}}^{\text{allow}})$ and $t_{c,k}^{\text{end}}(t_{i,\text{arrive}}^{\text{allow}})$ are the starting and ending times that vehicle i would pass conflict k on channel c, respectively, given time $t_{i,\text{arrive}}^{\text{allow}}$ when vehicle i enters the intersection. The formulas can be written as follows:

$$t_{c,k}^{\text{start}}(t_{i,\text{arrive}}^{\text{allow}}) = t_{i,\text{arrive}}^{\text{allow}} + \frac{S_{c,k}^{\text{lower}}}{v_{\text{max}}^{\text{int}}} \tag{8.7}$$

$$t_{c,k}^{\text{end}}(t_{i,\text{arrive}}^{\text{allow}}) = t_{i,\text{arrive}}^{\text{allow}} + \frac{S_{c,k}^{\text{upper}} + L}{v_{\text{max}}^{\text{int}}} \tag{8.8}$$

where L is the length of vehicles, which is assumed to be identical in the paper; $S_{c,k}^{\text{lower}}$ and $S_{c,k}^{\text{upper}}$ are the distances from the starting point of the channel to the lower and upper boundaries of conflict k on channel c, respectively.

To solve the optimization models, we discretize $t_{i,\text{arrive}}^{\text{allow}}$ and enumerate all possible values starting from $t_{i,\text{arrive}}^{\text{normal}}$. It will be quite efficient to obtain the optimal value of $t_{i,\text{arrive}}^{\text{allow}}$, since the number of reservations, i.e., $J_{c,k}$, is usually limited.

After obtaining the optimal value of $t_{i,\text{arrive}}^{\text{allow}}$, we calculate and save the corresponding $t_{c,k}^{\text{start}}(t_{i,\text{arrive}}^{\text{allow}})$ and $t_{c,k}^{\text{end}}(t_{i,\text{arrive}}^{\text{allow}})$ for each conflict k to declare occupation.

8.4.3 Simulation test

He *et al.* [9] tested the CACD strategy in two CI-A scenarios with a three-lane four-arm isolated intersection: one is to validate the collision-free design and the other is to test the efficiency. In summary, the proposed CI-A can guarantee traffic safety and its efficiency outperforms the cooperative intersection with specific-direction turn lanes and the conventional fixed-timing signal-controlled intersection. Interested readers may refer to [9]; we do not repeat it here.

In addition, animations for the CI-A, CMA and CDA are developed to better visualize the proposed lane change-free road transportation system. Figure 8.14 presents snapshots of the animations and one can refer to the webpage (https://www.zhengbinghe.com/lcfree.html) for the complete animations.

8.5 Conclusion and future direction

This chapter first reviewed recent advancements in cooperative driving at intersections and on-ramps and comprehensively introduced three cooperative driving strategies at intersections, namely, the safety driving pattern-based strategy, the reservation-based strategy and the trajectory optimization-based strategy, as well as two cooperative driving strategies at on-ramps, namely, the virtual vehicle mapping strategy and the lot-based strategy.

Then, a novel concept, i.e., the lane change-free road transportation system, was newly proposed for the future world. This future road transportation system consists of a CI-A, a CMA, a CDA and a CACD strategy, and vehicles that move in such a road system are able to reach their destinations without making any on-road lane changes. It is expected that, through this system with no lane changes, the CAV technology could be greatly simplified and driving safety, as well as efficiency, could be improved.

Future research directions for cooperative driving and the lane change-free system include:

- *Efficiency optimization considering heterogeneous lane demands.* Most of the existing strategies are proposed based on the FCFS rule. Although this can guarantee the equity of incoming vehicles, it does not ensure the optimal users or optimal system. Particularly, when traffic demands for turning directions are

180 Traffic information and control

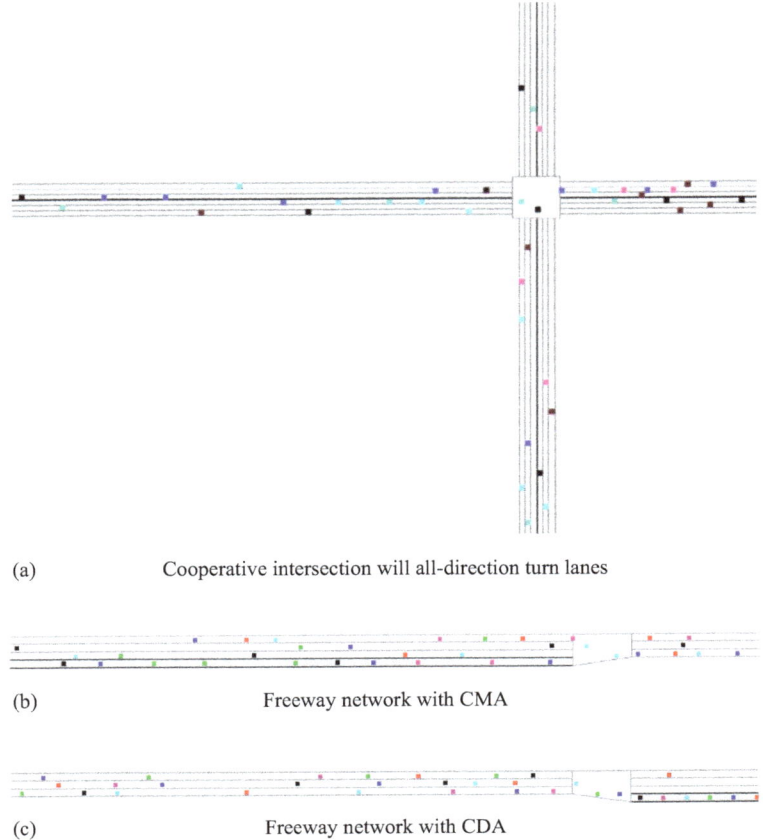

(a) Cooperative intersection will all-direction turn lanes

(b) Freeway network with CMA

(c) Freeway network with CDA

Figure 8.14 Snapshots for the animations of the lane change-free road transportation system

largely different, an algorithm that can maximize the system performance is necessitated.
- Network optimization. Most of the existing literature focuses on an isolated intersection or ramp. Obviously, the coordination of multiple cooperative intersection/ramps is of interest and importance. It is not difficult to imagine that there may be two types of coordination: (1) *centralized strategy*: vehicles are arranged before leaving their origins or vehicles obtain their schedules of passing every intersection and ramp in advance of leaving their origins and (2) *decentralized strategy*: instead of pre-trip scheduling, vehicles enter each intersection/ramp by following the guidance of that intersection/ramp, while the guidance is under the coordination of a system perspective.
- Incorporation of various scenarios and factors. Existing studies have considered many scenarios and factors, such as emergency vehicles, trucks and

pedestrians [11,12]. However, they are still far from sufficient due to the complexity of the real world. Therefore, various scenarios and factors must still be carefully considered. In particular, it may take a long time to evolve from human-driven vehicles to complete high automation. In between, there will be a long period during which human- and machine-driven vehicles will mix on roads. Therefore, efficient incorporation of human-driven vehicles is a challenging but meaningful task.

- *Step-by-step materialization.* Currently, most of the strategies are only tested in computer simulations. Obviously, they are still far from practical applications with real vehicles. However, it is not realistic to directly transfer those strategies with high-automation assumptions to the current CAV methods that are still in testing. Therefore, a step-by-step materialization is needed. More specifically, one may first transfer the cooperative driving strategy to smart vehicle models (i.e., the toy-like small vehicles in laboratories) with real communication devices and then to real vehicles in limited numbers and in simple scenarios.

Acknowledgements

This research is supported by the National Key Research and Development Program of China (2018YFB1600500).

References

[1] SAE International. Automated driving levels of driving automation are defined in new SAE International standard J3016. http://wwwsaeorg/misc/pdfs/automated_drivingpdf. 2016.

[2] Wang M, Treiber M, Daamen W, *et al.* Modelling supported driving as an optimal control cycle: Framework and model characteristics. Transportation Research Part C: Emerging Technologies. 2013;36:547–563.

[3] He Z, Zheng L, Song L, *et al.* A jam-absorption driving strategy for mitigating traffic oscillations. IEEE Transactions on Intelligent Transportation Systems. 2017;18(4):802–813.

[4] Zheng L, He Z, and He T. An anisotropic continuum model and its calibration with an improved monkey algorithm. Transportmetrica A: Transport Science. 2017.

[5] Wang M. Infrastructure assisted adaptive driving to stabilise heterogeneous vehicle strings. Transportation Research Part C: Emerging Technologies. 2018;91:276–295.

[6] Zheng L, Zhu C, He Z, *et al.* Safety rule-based cellular automaton modeling and simulation under V2V environment; 2018.

[7] Xie DF, Zhao XM, and He Z. Heterogeneous traffic mixing regular and connected vehicles: Modeling and stabilization. IEEE Transactions on Intelligent Transportation Systems. 2019;20(6):2060–2071.

[8] Wang H, Qin Y, Wang W, et al. Stability of CACC-manual heterogeneous vehicular flow with partial CACC performance degrading. Transportmetrica B. 2019;7(1):788–813.
[9] He Z, Zheng L, Lu L, et al. Erasing lane changes from roads: A design of future road intersections. IEEE Transactions on Intelligent Vehicles. 2018;3(2):173–184.
[10] Stevanovic A, and Mitrovic N. Traffic microsimulation for flexible utilization of urban roadways. Transportation Research Record. 2019;1–13.
[11] Li L, and Wang FY. Cooperative driving at blind crossings using inter-vehicle communication. IEEE Transactions on Vehicular Technology. 2006;55(6):1712–1724.
[12] Dresner K, and Stone P. A multiagent approach to autonomous intersection management. Journal of Artificial Intelligence Research. 2008;31:591–656.
[13] Lee J, and Park B. Vehicle intersection control algorithm under the connected vehicles environment. IEEE Transactions on Intelligent Transportation Systems. 2012;13(1):81–90.
[14] Chen L, and Englund C. Cooperative intersection management: A survey. IEEE Transactions on Intelligent Transportation Systems. 2016;17(2):570–586.
[15] Rios-Torres J, and Malikopoulos AA. Automated and cooperative vehicle merging at highway on-ramps. IEEE Transactions on Intelligent Transportation Systems. 2017;18(4):780–789.
[16] Namazi E, Li J, and Lu C. Intelligent intersection management systems considering autonomous vehicles: A systematic literature review. IEEE Access. 2019;7:91946–91965.
[17] Uno A, Sakaguchi T, and Tsugawa S. A merging control algorithm based on inter-vehicle communication. IEEE Conference on Intelligent Transportation Systems, Proceedings, ITSC. 1999;783–787.
[18] Sakaguchi T, Uno A, Kato S, et al. Cooperative driving of automated vehicles with inter-vehicle communications. IEEE Intelligent Vehicles Symposium, Proceedings. 2000;(Mi):516–521.
[19] Hausknecht M, Au TC, and Stone P. Autonomous intersection management: Multi-intersection optimization. IEEE International Conference on Intelligent Robots and Systems. 2011;4581–4586.
[20] Au TC, Zhang S, and Stone P. Autonomous intersection management for semi-autonomous vehicles. Handbook of Transportation. 2015;88–104.
[21] Quinlan M, Au TC, Zhu J, et al. Bringing simulation to life: A mixed reality autonomous intersection. IEEE/RSJ 2010 International Conference on Intelligent Robots and Systems. 2010;55(6):6083–6088.
[22] Nocedal J, and Wright S. Numerical optimization. New York: Springer-Verlag; 2000.
[23] Lee J, Park B, Malakorn K, et al. Sustainability assessments of cooperative vehicle intersection control at an urban corridor. Transportation Research Part C: Emerging Technologies. 2013;32:193–206.
[24] Marinescu D, Čurn J, Bouroche M, et al. An active approach to guaranteed arrival times based on traffic shaping. In: 13th International IEEE

Conference on Intelligent Transportation Systems (ITSC). Madeira Island, Portugal; 2010:1711–1717.

[25] Marinescu D, Čurn J, Bouroche M, et al. On-ramp traffic merging using cooperative intelligent vehicles: A slot-based approach. IEEE Conference on Intelligent Transportation Systems. 2012;900–906.

[26] Wu X, He X, Yu G, et al. Energy-optimal speed control for electric vehicles on signalized arterials. IEEE Transactions on Intelligent Transportation Systems. 2015;16(5):2786–2796.

[27] Katrakazas C, Quddus M, Chen WH, et al. Real-time motion planning methods for autonomous on-road driving: State-of-the-art and future research directions. Transportation Research Part C: Emerging Technologies. 2015;60:416–442.

[28] Paden B, Cap M, Yong SZ, et al. A survey of motion planning and control techniques for self-driving urban vehicles. IEEE Transactions on Intelligent Vehicles. 2016;1(1):33–55.

[29] Zhou F, Li X, and Ma J. Parsimonious shooting heuristic for trajectory design of connected automated traffic part I: Theoretical analysis with generalized time geography. Transportation Research Part B: Methodological. 2017;95:394–420.

[30] Ma J, Li X, Zhou F, et al. Parsimonious shooting heuristic for trajectory design of connected automated traffic part II: Computational issues and optimization. Transportation Research Part B: Methodological. 2017;95:421–441.

[31] Jiang H, Hu J, An S, et al. Eco approaching at an isolated signalized intersection under partially connected and automated vehicles environment. Transportation Research Part C: Emerging Technologies. 2017;79:290–307.

Part II

Modern traffic signal control

Chapter 9
Urban traffic control systems: architecture, methods and development

Fusheng Zhang[1] and Lu Wei[1]

Traffic control system, as one of the core applications in an intelligent transportation system, plays a key role in alleviating traffic congestion. This chapter presents Split Cycle Offset Optimization Technology (SCOOT) and Sydney Coordinated Adaptive Traffic System (SCATS), two representative and widely used among the world control systems, given their basic architecture, principles and main functions. Besides, a comparison of data usage and optimization methods between them is provided. A discussion and analysis on the future development of urban traffic control system are finally proposed under connected vehicles or other future technologies environments.

9.1 Introduction

9.1.1 Brief description

The urban traffic control system is a key part of the intelligent transportation system, which can effectively avoid traffic conflict, reduce emissions and improve the efficiency of the urban road network. The development of urban traffic control system is driven by control, communication and detection technologies. Since the first interconnected traffic signal control system that, including six intersections, was introduce in Salt Lake City in 1917, various traffic control systems were developed and implemented around the world to solve area traffic congestion. Two representatives which also with highest market shares are Split Cycle Offset Optimization Technology (SCOOT) [1] and Sydney Coordinated Adaptive Traffic System (SCATS) [2]. The brief history of the urban traffic control system in countries and regions around the world is provided as follows.

9.1.1.1 Europe
The development and application of Intelligent Transportation System (ITS) in Europe is closely linked to the EU's process of transport integration. In 1969, the

---

[1]Beijing Key Laboratory of Urban Road Traffic Intelligent Control Technology, North China University of Technology, Beijing, China

European Commission proposed the development of electronic technology for traffic control among member states.

In general, the development of the EU ITS is divided into three stages:

1. From 1970 to 1986, it was the initial stage of development of European ITS. With the launch of a European project on electronic aids for traffic on major roads (EUCO-COST 30 plan), the application of information technology in transportation has become a hotspot, and relevant institutions in various EU countries have carried out research, development and application of related technologies.
2. From 1985 to 2003, it was the stage of comprehensive research, popularization and application of European ITS. This stage of work involves the design of the ITS standard framework, as well as advanced travel information systems, vehicle control systems, commercial vehicle management systems and electronic toll systems.
3. From 2003 to the present, it is the integration application stage of EU ITS and the research and development stage of key areas. With the introduction of the e-Safety concept and the EU's 'Trans-European Transport Network (TEN-T)', 'Telematics Plan', 'EasyWay Programme' and other related plans to promote the integration of the EU, research and development related to intelligent transportation in Europe is gradually turning to integration, standardization and coordinated development. At the end of 2016, Europe adopted the 'European Cooperative Intelligent Transportation System Strategy', and its goal is to deploy large-scale cooperative intelligent transportation systems on the roads of EU countries by 2019 to achieve 'intelligent communication' between different vehicles.

In the area of traffic control, the representative traffic signal control systems in Europe are TRANSYT and SCOOT which developed by the British Transport Research Laboratory (TRL) Institute. TRANSYT [3] was developed by D. I. Robertson and others of the British Road Research Institute in 1966, and it is currently widely used in the design, simulation and optimization of coordinated timing of road network signals. SCOOT was developed by the British Transport Research Institute in 1973, successfully developed in 1975 and officially put into use in 1979. It has been used by more than 170 cities around the world. The largest application scale is to control about 2000 intersections in London, UK.

As we known, London and Siemens signed a contract on next-generation traffic-management systems in 2018. Currently, a large number of SCOOT systems are used in London. In the future, Siemens will develop a new Real-Time Optimization (RTO) system. A major feature is the comprehensive consideration of various types of traffic detection information, including information about the vehicle network and future autonomous vehicles; another feature is the comprehensive consideration of all traffic modes on the road, based on which to optimize the transportation system.

At the opening ceremony of Gulf Traffic in Dubai in 2018, Siemens announced an innovation project 'Flow AI' for signal optimization. Simulations at an intersection in German showed that it can reduce waiting time by 47% (from 35 s to 18 s).

9.1.1.2 The United States

In the United States, RHODES [4] (Real Time Hierarchical Optimized Distributed Effective System) and DYNASMART [5] (DYnamic Network Assignment Simulation. Model for Advanced Road Telematics) are the frameworks of the Advanced Traffic-Management System (ATMS). Under the guidance, intending to improve the level of transportation services, integrated control systems for travel and traffic-management services, such as traffic signal control, traveller information, traffic demand, event management and emissions detection, are integrated.

The RHODES system relies on the adaptive traffic signal control system to expand to the combination with buses, to realize the priority of bus signals and the release of bus information. The DYNASMART system is based on the information collected in multiple ways, analyses the traffic status in real-time, designs offline and evaluates the operation effect of real-time signal control, trunk route guidance and other traffic-management strategies. It provides Public Travel Information System traffic status to determine the optimal strategy for reducing congestion.

9.1.1.3 Australia

Australia is one of the world's earliest countries that engaged in intelligent traffic control technology research. The famous SCATS is used in almost all cities in Australia. Currently, Shanghai, Tianjin, Hangzhou and other cities also use this system.

The advantage of the traffic control system (SCATS) is its ability to automatically adapt to changes in traffic conditions. It obtains road traffic information in real time through intersection traffic detectors, optimizes control schemes and automatically adapts to changes in traffic flow.

The SCATS system was developed by the RTA in the 1970s. The SCATS regional traffic control system has played an important role in road traffic management in cities such as Sydney, Canberra and Melbourne. At present, more than 80 cities worldwide have introduced this system. The Traffic Management Center (TMC) reflects the level of traffic management and traffic informatization in Australia. There are many methods and experiences that we can learn from.

Sydney TMC mainly uses the SCATS regional traffic adaptive coordination system. It is an advanced system for controlling the Sydney traffic information network. It controls more than 2,200 intersections and 3,000 traffic signals on the main roads in and around Sydney, monitoring 3,600-square-kilometres coverage area. There are more than 3 million cars in Sydney, with an average of more than 2 million trips per day. Under the control of the SCATS system, it is rare to see queues or blockages at intersections and road sections. Generally, cars only need to wait for a signal cycle to pass through an intersection. During the rush hour of commuting, there are special systems for adjustment of the main traffic routes in and out of the city. For example, the 'Electronic Road Direction Adjustment System (TELCS)', which is used in the Sydney Harbour Bridge, can automatically adjust the road direction. It is automatically adjusted according to the traffic conditions. The SCATS system can reduce traffic stoppages, delays and fuel consumption on a large scale.

190   Traffic information and control

Infrastructure has an 'absolutely critical' significance to Australia's successful development in the twenty-first century. The federal government has been committed to building a good infrastructure to grasp the opportunities brought by the rise of the Asian middle class. In 2016, the Australian Infrastructure Plan was proposed by the Australian government. The plan sets out the infrastructure challenges and opportunities Australia faces over the next 15 years and the solutions required to drive productivity growth, maintain and enhance our standard of living, and ensures our cities and regions remain world class.

9.1.2 Classification

There are various classifications of urban traffic control. The main classification criteria include control scope, control methods and some special control modes.

Despite more and more traffic data are provided, there are still many issues that need to be considered when optimizing traffic control. Different strategies and methods will produce different results. We classify traffic signal control system technologies from two perspectives: optimal control frequency and operation mode, as shown in Table 9.1.

From the frequency perspective, traffic signal control systems can be classified into two categories: plan-level optimization and real-time-level optimization. Plan-level optimization refers to generating a signal timing plan based on statistical data or choosing the plan from the alternative plans which are most suitable for the current traffic situation. The second category, real-time level optimization, refers to the system optimization that signal timing parameters is executed immediately by controllers based on second-by-second vehicles or pedestrians detection.

From the operation mode perspective, there are three operation modes when the optimum signal timing plan or group of parameters is acquired. The first one is local operation. Signal timing plans need to be download to local controller via communication network based on a certain protocol. Then, the local controller runs or switches the plan based on the current traffic status. The second one is the optimum plan running in the central computers. The centre system only interacts with local controllers with control commands and response words, not the whole signal timing plan. As we have known, if a completely different plan replaces the existing one, the local controller usually needs a transition, which usually causes a disturbance to the current traffic situation to a certain extent. The benefit of the second one is avoiding the disturbance caused by the plan transition. The third one is operating hierarchically and combining the system and local controller effectively. The centre

Table 9.1   Classification of optimization type

Frequency	Timing plan	Auto plan selection
	Real time	Auto plan generation
		Second-by-second optimization
Operation	Local controller	
	Centre system	

Urban traffic control systems: architecture, methods and development 191

system responds to traffic demand detection in real time and provides strategical decisions while the local controllers make tactical decisions and corrections based on the specific demands.

9.1.3 Level of traffic control system

Different operation mode and various demand for traffic data will produce different levels of traffic signal control systems. Four levels are concluded based on optimization frequency and functions of control, as shown in Table 9.2.

9.2 SCOOT

9.2.1 Overview

SCOOT, as a dynamic, real-time, on-line adaptive traffic control system, was first introduced in 1983 and had been continuously developed to meet the needs of traffic managers from around the world. The most recent version is SCOOT MMX, which implement new features directly aimed at the multi-modal nature of today's traffic signal installations and the need to optimize for all users. Being used in over 350 towns and cities globally, SCOOT has been an effective and efficient solution for controlling traffic on signalized road networks around the world.

SCOOT measures traffic demand on all approaches to intersections in the urban network and optimizes the signal timings at each intersection to minimize delays and stops. Timing changes are small, to avoid major disruption to traffic flows, and frequent, to allow rapid response to changing traffic conditions. Therefore, SCOOT is suitable for implementation on all types of urban roads, grids, corridors and arteries.

The benefits of SCOOT can be concluded as follows:

- SCOOT, which is an acronym for the Split Cycle and Offset Optimization Technique, is a dynamic, on-line, real-time method of signal control.

Table 9.2 Levels of traffic signal control systems

Levels Functions	Level 0	Level 1	Level 2	Level 3
Comm	N	Y	Y	Y
Detection	N	N	Y	Y
Real-time data	N	N	N	Y
Monitoring	N	Y	Y	Y
Clock sync	N	Y	Y	Y
Timing	N	Y	Y	Y
Central control	N	Y	Y	Y
Timing optimization	N	N	Y	Y
Real-time optimization	N	N	N	Y

- Optimization of the network is achieved using small, regular changes designed to avoid major disturbance of traffic flow.
- Cycle Split and Offset are all optimized.
- Performance indexes (stops and delays) are used to measure effectiveness similar to TRANSYT.
- Update your signal timing plans every 3 s.
- SCOOT makes an ideal complement to standard Time of Day and Traffic Responsive-based traffic control.
- The Stratos Outstation has a future-proof platform – easily updatable to support new SCOOT functionality.
- The Stratos Outstation is capable of full implementation of UG405.
- The Stratos Outstation also has a wide range of National Environment Management Authority (NEMA) controller and detection technology compatibility.

9.2.2 Basic principles

SCOOT is developed based on TRANSYT, and its models and optimization principles are similar to TRANSYT. The difference is that SCOOT is an online system. Vehicle arrival information collected by detectors installed at the upstream of each link is processed online, and then a signal timing plan is generated to realize the adjust the split, cycle and offset in real time.

SCOOT models traffic detected on-street to continuously adapt three key traffic control parameters – the amount of green for each approach (Split), the time between adjacent signals (Offset) and the time allowed for all approaches to a signalized intersection (Cycle time). These three optimizers are used to continuously adapt these parameters for all intersections in the SCOOT-controlled area, minimizing wasted green time at intersections and reducing stops and delays by synchronization of adjacent traffic signal installations. As a result, the signal timings evolve with the changing traffic situation without any of the traditional disruptions caused by changing fixed time plans on other urban traffic control system.

SCOOT uses detectors at the upstream end of links to measure demand and Cyclic Flow Profiles (CFP) in real time. The upstream detection also allows any congestion on the link to be monitored and the possible exit blocking effects of this congestion on upstream links. The SCOOT model predicts downstream arrival patterns using a calibrated link travel speed and travel time. The saturation flow rate for each signal stop line is validated when the system is implemented. This allows the growth and clearance of queues to be estimated accurately. The on-line traffic model is used in real-time by the signal optimizer. By the combination of relatively small changes to traffic signal timings, SCOOT can respond to both short-term local peaks in traffic demand, as well as following trends over time and, thus, maintain constant coordination of the signal network.

In addition to the optimization from the basic SCOOT model, the operation has considerable flexibility to override values and set parameters for different regions and different times. These may include gating strategies to protect an area from excessive levels of traffic, bus priorities, etc. In addition to network management, SCOOT has a substantial database facility for storing, manipulating and presenting traffic data including flows, journey times and queues.

9.2.3 System architecture

The basic architecture of SCOOT is shown in Figure 9.1, SCOOT can implement a centralized strategic traffic policy, responding to fluctuations in demand in real time. The centralized system also allows system-wide strategies to be employed which are as follows:

- Peak hour routes,
- Keep emergency and evacuation routes clear,
- Traffic metering (gating) on the outskirts of congested areas and
- Central bus priority.

Traditionally, SCOOT has been using dedicated multi-drop transmission lines to outstations. SCOOT requires second-by-second communication between the central computer and outstations. Typically, six-to-eight intersections can be served by 1,200 baud rate. Recently, the PC version of SCOOT was enhanced to enable the use of modern communication technologies used by ITS solutions. This approach absorbs inconsistencies and delays in data delivery and has less impact on the system. This new approach reduces dependency on traditional leased-line communication techniques and opens up the potential to use a wide range of modern communication technologies previously unavailable to SCOOT systems.

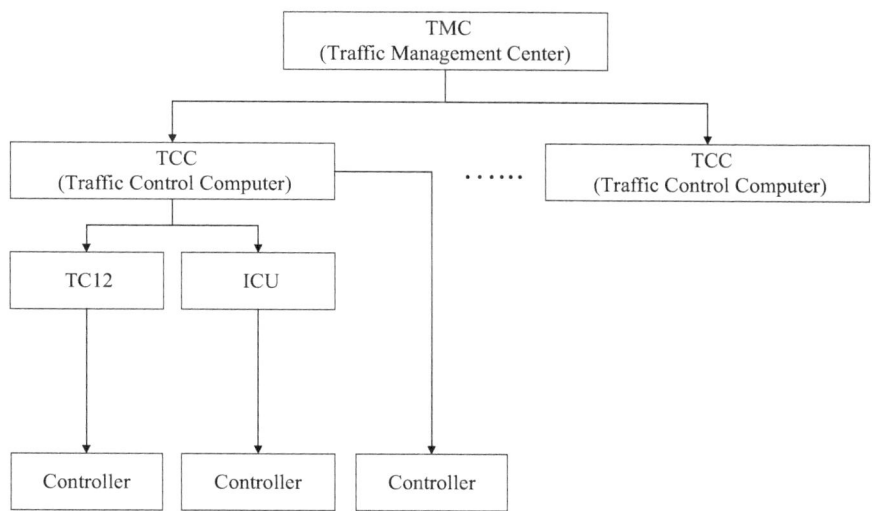

Figure 9.1 Architecture of SCOOT

9.2.4 Optimization process

9.2.4.1 Demand detection

Based on traffic data collected from presence detectors on each approach of the intersections, SCOOT calculates optimum signal timing parameters to response to the fluctuation of traffic on-street.

Detector data collected in 1/4-s intervals are processed and used to establish the Flow Profiles, one for each link, as shown in Figure 9.2. The traffic predicted to be crossing the downstream stop-line in each interval is stored in these profiles.

The CFP in SCOOT is similar to TRANSYT. The only difference is that the CFP in SCOOT is generated in real time. Based on the real-time CFP and platoon-dispersion model, SCOOT predicts the arrival information at the stop-line. The unit of the y-axis of CFP is LPU (Link Profile Unit), which is a mixed measurement unit for traffic flow and occupancy. LPU represents the time needed for corresponding traffic flows. While the same car runs fast or slow, the LPU is different. The same as traffic flow upstream; the saturation flow at stop-line is also measured by LPU.

9.2.4.2 Queue prediction

The principle of queue prediction process is shown in Figure 9.3; the upper right is the real-time vehicle arrival profile at the detector section, while the lower right is queue profile at the stop-line. Figure 9.3 is updated second by second. When the traffic light is red, the arrival vehicles will be added to the tail of the queue. As the light turns green, the queue will be discharged with the saturation flow. However,

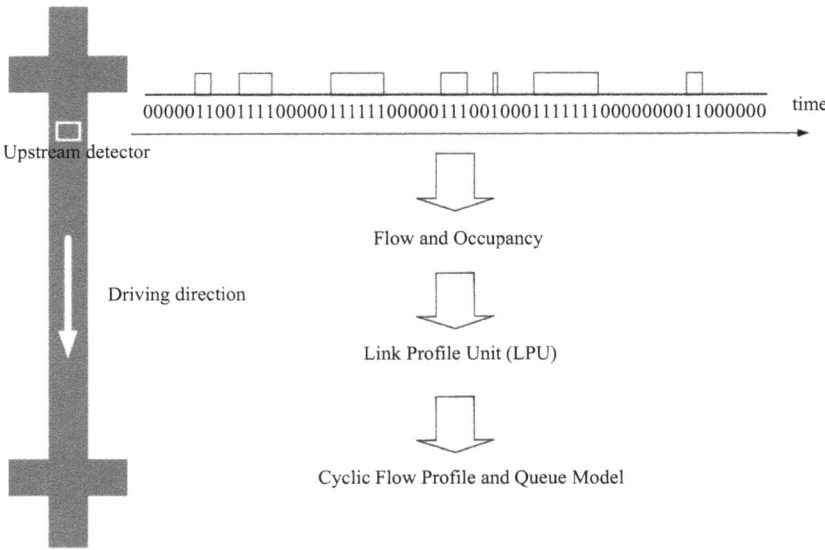

Figure 9.2 Data collection of SCOOT

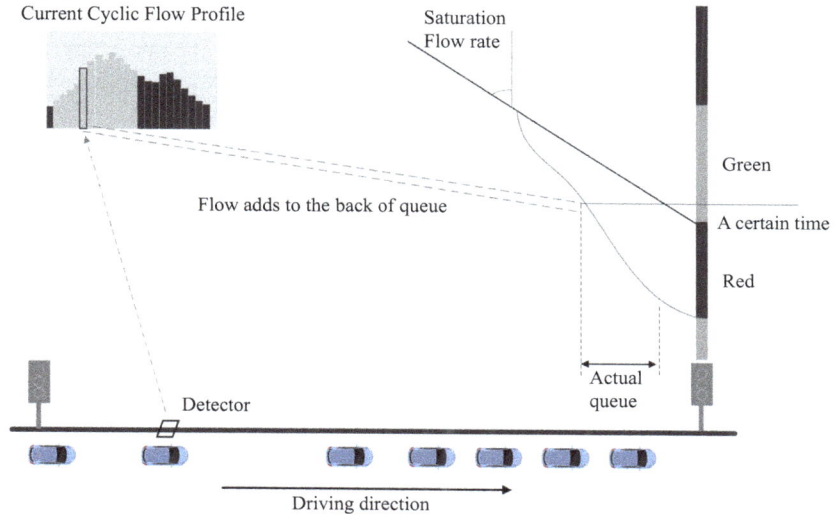

Figure 9.3 Queue model of SCOOT

due to difficulty of estimating vehicle speed and platoon dispersion accurately, the parameter validation is required before commissions.

9.2.4.3 Congestion prediction

Congestion is directly measured from the detector. The detector is placed beyond the normal end of the queue in the street where rarely covered by stationary vehicles. In other words, if the detector is always covered by stopped vehicles, the congestion may have been occurs. If any detector shows standing traffic for the whole of an interval, this is recorded. The number of intervals of congestion in any cycle is also recorded.

As described in the SCOOT USER GUIDE [6], the percentage of congestion is calculated from:

$$\frac{No.\ of\ congested\ intervals \times 4 \times 100}{Cycle\ time\ in\ seconds}$$

This percentage of congestion is available to view and more importantly for the optimizers to be considered.

9.2.4.4 Performance prediction

The same as TRANSYT, SCOOT takes the weighted sum of delays and stops as the comprehensive indicator of traffic situations, named PI (Performance Index). Sometimes, congestion factor is also used as a PI. SCOOT predicts the PI of each signal timing plans based on its queue model.

9.2.4.5 Signal optimization

SCOOT provides three optimizers (Table 9.3) to generate the most appropriate signal timing parameters. The three optimizers are split optimizer, cycle optimizer and offset optimizer.

Split optimizer

The split optimizer runs for a node at an optimum point before each stage change. It considers the effect of advancing, retarding or holding the stage change and the effect this has on the degrees of saturation of the links. The degree of saturation is defined as 'the ratio of the average flow to the maximum flow that can pass a stop-line'. In SCOOT terms, this becomes 'the ratio of the demand flow to the maximum possible discharge flow'; it is the ratio of the demand to the discharge rate (Saturation Occupancy) multiplied by the duration of the effective green time. The split optimizer will try to minimize the maximum degree of saturation on links approaching the node.

If congestion is present on a link, this must be taken into account by the split optimizer. To enable this, the proportion of the previous cycle, that was congested, is included with the degree of saturation used by the optimizer when making its decision. The congested link tend to obtain more green time whatever the degree of saturation is shown in the model. The degree of congested link is determined by the link's congestion importance factor.

Cycle optimizer

The cycle optimizer operates on a region of nodes that can sensibly be expected to have progression between them. The engineer chooses this grouping.

The cycle optimizer usually runs every 5 min for each region, although this can be varied by the user in version 4 and later. There is a provision within SCOOT for the optimizer to run twice as often if a trend of rising or falling flows has been established. At this time, it will work out the degree of saturation level (user-configurable but typically 90%). The Minimum Practical Cycle Time (MPCY) for the node is increased by a small fixed-step. If all are below the ideal saturation level, the MPCY of the node is reduced by a small fixed-step. The optimizer assesses all the nodes' MPCYs and considers all cycle times from MPCY of the critical node up to the maximum permitted region cycle time before deciding to increase or decrease the region cycle time.

Table 9.3 Optimizers of SCOOT

Optimizer	Frequency	Max change	Min change
Split	Every stage	4	1
Cycle	Every 2.5 or 5 min	16/8/4	16/8/4
Offset	Once per cycle	4	4

Offset optimizer

Once per cycle, the offset optimizer uses the flow profile to predict the stops and delays throughout the cycle for all the links upstream and downstream of a particular node. This allows the best overall offset for the node to be established and the 'time now' for the node to be adjusted to move towards this best point. Cyclic profile over the detector, the profile of the flow arriving at the stop-line after platoon dispersion and the journey time from the loop to the stop-line have been taken into account. If the stage start time was advanced a few seconds, the few vehicles that start queuing up at the end of the red time would not have to wait at all.

The optimizer predicts the stops and delays for each link and sums up those predictions together for the node. The choice is then made for a move towards the offset giving minimum delay and stops.

The congestion on a link is also used in the offset optimizer so that a congested link can be given priority over links without congestion.

9.2.5 Additional features

In various past versions, SCOOT has developed a series of additional features and modules to alleviate traffic congestion, including communication, gating control, bus priority, emission estimation, etc.

9.2.5.1 Gating

The objective of 'gating' is to control the inflow of traffic into sensitive areas to prevent the formation of long queues or congestion. The gating logic restricts traffic-entering areas susceptible to congestion and redistributes queuing up vehicles to other roads more suitable for storing traffic. Alternatively, gating may be used to favour one route or local area at the expense of another.

To implement gating, SCOOT must be able to take 'action at a distance' by modifying the traffic signal settings at junctions which may be far removed from the problem area. In some situations, such as where major problems in one region may cause congestion to spread rapidly to adjoining areas, the use of gating could provide overall benefits to the network.

It is emphasized that gating can be an uncompromising method of control and care must be taken when tuning the system to provide an adequate level of restraint, but without excessively delaying the traffic which is being gated.

9.2.5.2 Bus priority

The logic of bus priority is contained within the SCOOT kernel which has been modified by the TRL. The bus priority optimizer provide bus priority types including extending a current green signal (an extension) or causing succeeding stage to occur early (a recall). The decision of bus priority type is decided with reference to the degree of saturation or spare capacity of the road network. Buses are modeled by SCOOT as queuing with other vehicles. This allows buses to be given priority even though other vehicles may delay them. The effect of bus lanes also be modeled.

9.3 SCATS

9.3.1 Overview

SCATS, developed by the Roads and Traffic Authority (RTA) in Australia, uses a real-time traffic adaptive approach to urban traffic control by measuring current traffic conditions and then adjusting cycle time, split and offset. The real-time response to changing traffic conditions ensures the most appropriate traffic signal phasing to safely direct traffic through intersections. Using in-road and/or above-road sensors and by sending instructions to traffic signal controllers to manage signal timings, SCATS automatically maximizes traffic throughput, minimizes delays and minimizes the number of stops.

Rather than changing individual intersections in isolation, SCATS manages groups of intersection called 'subsystems', the basic unit of the system. Each subsystem will consist of several intersections, usually between one and ten. One of those intersections is designated as the controlling or 'critical' intersection. SCATS adapts and coordinates the intersections within each subsystem and can coordinate adjacent subsystems. This coordination aims to divide the traffic on major roads into 'platoons' (groups of vehicles) and to allow just enough time for each platoon of vehicles to progress through the system while allowing the green time required for competing flows. This maximizes the network capacity for the benefit of all users.

9.3.2 Basic principles

SCATS is known for its effective feedback control that continuously and autonomously self-calibrates efficient traffic conditions and measures road utilization at each detector. SCATS tracks critical traffic demand to adjust the effective road capacity with cycle time changes and optimizes phase (or stage) times to fit the varying demands of competing movements. SCATS dynamically balances the local site optimization and inter-site coordination to capture the efficiency benefits of platoon progression.

The same as SCOOT, SCATS also manages three main parameters to achieve traffic adaptive and coordination control: cycle time, phase split and offset. In SCATS parlance, decision-making occurs at two levels: strategic and tactical.

At the strategic level, SCATS provides network-level coordination and optimal control. Strategic control is managed by the regional computers. Using flow and occupancy data collected from vehicle detectors, the regional computer determines on an area basis the optimum cycle length, phase splits and offsets to suit the prevailing traffic conditions.

Strategic control bases its adjustments on a traffic demand measurement known as 'Degree of Saturation' (DS). However, in this context, DS represents how effectively the road is being used. Using the in-ground loop detectors at the critical intersections, the local controller collects flow and occupancy data during the green phase. The data are sent to the regional computer, which calculates the degree of saturation. Values of DS greater than unity (insufficient green time to satisfy

Urban traffic control systems: architecture, methods and development 199

demand) will occur in congested conditions and SCATS will quickly respond to such an over-saturated situation.

At the tactical level, SCATS responds in real time to significant changes in the traffic state to reduce inefficiencies through terminating underutilized movements and to capture efficiencies by reallocating time to competing movements.

SCATS strategic decision-making considers the interrelatedness of the traffic controllers that are networked by the characteristics of the road network. Tactical control is undertaken by the local controllers and meets the cyclic variation in demand at each intersection. Tactical control primarily allows for green phases to be terminated early when the demand is low, and for phases to be omitted entirely from the sequence if there is no demand. The local controller bases its tactical decisions on information from vehicle detectors at the intersection. It should be emphasized, however, that the degree to which tactical control can modify the signal operation always remains entirely under the control of the regional computer.

Finally, there is a tight integration between the SCATS controllers, SCATS region computers and SCATS communication network. The result is a 'SCATS platform' that provides the catalyst for hierarchical control to function effectively and robustly.

9.3.3 System architecture

SCATS has been designed in a modular configuration to suit the varying needs of small, medium and large cities. In its simplest form, a single regional computer can control up to 250 intersections. Expansion of the system is achieved by installing additional regional computers on a Transmission Control Protocol/Internet Protocol (TCP/IP) network. SCATS can also internally manage several instances of the regional traffic control software on one physical computer. This provides flexibility in hardware configuration and for simulation use. All systems have a Central Management Computer (CMC) to manage global data, access control, graphics data and data backup. A typical SCATS system is shown in Figure 9.4 as follows.

1. **Central management computer**
   The CMC is also a personal computer operating under the Windows operating system. Communications with regional computers and workstations are via TCP/IP. In addition to monitoring of the working status of each equipment in SCATS, the CMC also stores the various data and signal parameters. Traffic engineers can expand new functions by utilizing these storing data.
2. **Regional computer**
   The regional traffic control function utilizes standard personal computers operating under the Windows operating system. A range of intersection communication methods are provided and include network (TCP/IP), serial, dial-out and dial-in. In SCATS, the regional computer can be installed according to the specific situation. Each regional computer can control 250 traffic signal controllers. Key functions of a regional computer are analysing traffic flow data and providing control decisions for the local area. Besides, data collected by the regional computer will also be sent to the CMC.

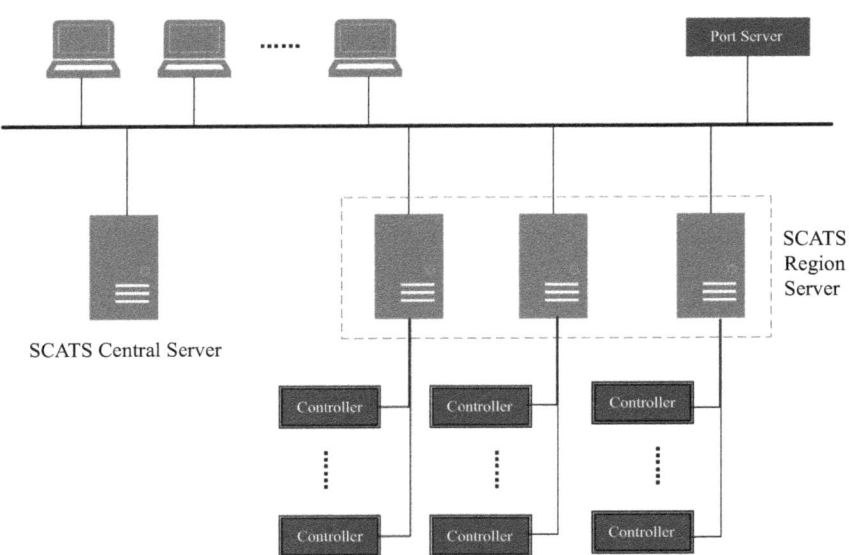

Figure 9.4   Architecture of SCATS

3. **Traffic signal controller**
   In SCATS, each intersection will be installed a traffic signal controller. The core functions of the traffic signal controller are to collect and analyse real-time traffic data provided by each detector, as well as to share the data with the regional computer through the communication network for adjusting the signal timing plan. Also, control commands that from the regional computer will also be executed by the local controller.

9.3.4   Optimization process

9.3.4.1   Demand detection

SCATS collection real-time traffic data from detectors installed at the stop-line of each approach of intersections. Based on the real-time traffic data, SCATS calculates DS, which represents the ratio of effective green and the duration of the whole green time, as follows:

$$DS = \frac{g'}{g}$$

where g' is the effective green time and g is green time that allocated by the current signal timing plan. SCATS considers that traditional saturation is usually related to the size of the vehicle and headway. For mixed traffic flows, the arrival of vehicles at the section of the detector is random. Therefore, a parameter that is not directly related to vehicle length is required to reflect real-time traffic situations.

To avoid the effect of vehicle type and length, SCATS introduces a virtual measurement, comprehensive flow, to indicate the flow of mixed vehicles at stopline. The comprehensive flow can be expressed as:

$$q' = \frac{DS \times g \times S}{3,600}$$

where q' is comprehensive flow and S is the maximum flow rate.

9.3.4.2 Cycle determination

Cycle length determination is performed for each subsystem of one or more intersections. The detector with the largest DS value that measured in the past cycle will be selected to determine cycle length time in the next.

The cycle length in SCATS is computed differently for low and high traffic demands. For each subsystem, four restrict parameters are determined, they are:

(a) the minimum cycle C_{min},
(b) the maximum cycle C_{max},
(c) the median of cycle that could provide continuous flow in both directions C_s and
(d) the cycle time that a little greater than C_s, as C_x.

The value lower than C_x or C_{min} is only selected when the arrival rate detected at the key position is lower than the pre-set threshold.

To maintain the continuity of the traffic signal at adjacent intersections, the cycle time is adjusted step by step in a small rate. Compared to the past cycle, the variation of a new cycle is limited in 6 s.

Cycle length is increased or decreased to maintain the DS at around 0.9 on the lane with the greatest saturation. Cycle time can range between 20 and 240 s, but a lower limit for cycle time (usually 30–40 s) and an upper limit (usually 100–150 s) are specified by the user. Cycle time can vary by up to 21 s, but this upper limit is resisted unless a strong trend is recognized.

9.3.4.3 Split determination

Phase splits are specified as a percentage of the cycle time and are varied by a small amount each cycle in such a way as to maintain equal degrees of saturation on competing approaches. Split in SCATS equalizes saturation of an intersection by minimizing the maximum DS on all participating approaches.

The minimum split which can be allocated to a phase is either a user-definable minimum or, more usually, a value determined from the local controller's minimum phase length. The current cycle length and the minimum requirements of the other phases limit the maximum split that can be allocated to a phase. Fixed time phases can have their phase time specified in seconds.

9.3.4.4 Offset determination

Optimal offsets between adjacent intersections or subsystems are calculated and selected. Subsystems carrying lower flows may not receive good coordination if the cycle time is inappropriate. However, when traffic conditions permit the use of a cycle time that can provide good offsets over several subsystems, the system tends to maintain this cycle time even though a smaller cycle time would provide sufficient capacity. SCATS does this because optimal offsets on the heavy flow links minimize the total number of stops in the system, thus reducing fuel consumption and increasing the capacity of the network.

9.4 Summaries and limitation analysis

SCOOT and SCATS are the most widely used traffic control systems around the world. Although they have provided better efficiency than fixed or other traditional traffic control methods – however, as they are developed in the 1970s and 1980s – limitations also exist and need to be improved.

First, SCATS and SCOOT are all responsiveness control systems. The adjustments of the cycle, split and offset cannot be implemented instantaneously or frequently. Rather, transitions may take some time. In general, it is not feasible to implement many different coordination patterns with a short period of time. Therefore, the current research trends focus on proactive control in a predictive way.

Second, saturated or over-saturated traffic situations have appeared in many metropolises or even some medium-sized cities. When approaches of intersections have achieved over-saturated status, it will often lead to the detection or model invalid. Such as queue model in SCOOT. So, how to improve these systems to deal with over-saturated traffic situations is also a hotspot research topic.

Third, with the rapid development of vehicle detection and wireless communication technologies, many new spatial-temporal traffic data will be available for urban traffic management and control. More rich information of vehicles, including real-time position, speed, acceleration and so on, can be collected easily and economically. This will greatly improve the capability of perceptron for urban traffic control systems. As we have known, the detector at the stop-line in SCATS can only acquire vehicle occupancies in a point, the platoon arrival and progression cannot be acquired. SCOOT detectors are placed upstream of approaches to the intersection, so the discharge rate at the stop-line and travel time between the upstream detector and stop-line can only be estimated or validated by traffic engineers frequently. These are not only costly but also reduce the accuracy of control models or strategies.

9.5 Future analysis of urban traffic control system

The evolution of urban traffic control system has always been driven by the technologies of vehicle sensing, network communication, computer and intelligent control. Comprehensive surveys were provided in [7,8]. The environments of traffic signal control has been greatly improved by these new technologies. The changes mainly include three aspects: the changes with input traffic data, the changes with control objects and changes with computational power. Therefore, the expectation of the existing traffic control system can be concluded as follows:

1. The existing model of the traffic control system does not have self-learning ability. Therefore, the relevant parameters need to be calibrated by human base on experiences or a lot of extensive field investigation, such as SCOOT validation.
2. With the expansion of urban traffic network, large-scale traffic-centralized control is a great challenge.
3. The existing traffic control methods mostly simplify the control constraints to establish the precise mathematical model, but these methods are different from the actual traffic flow conditions and the control effect is poor.
4. The existing traffic control system optimizes signal timing parameters mostly based on fix-point traffic detection. However, the fix-point detector has lower space coverage and less useful traffic flow information. How to use new traffic information provided by Vehicle to Everything (V2X) or connected vehicles to improve the efficiency of the traffic control system need to be considered.
5. Regional road network lacks timely response of the actual traffic fluctuation so that it is difficult to achieve real-time control.

9.5.1 Changes in system environments
9.5.1.1 Traffic data

Traffic data are the bases of urban traffic optimal control. Traditional traffic signal control system, such as SCOOT and SCATS, uses fix-point detection data to trace the variation of on-street traffic flow. However, the fix-point detection can only acquire vehicle presence information at a point; other important vehicle information, such as vehicle id and vehicle speed, cannot be directly obtained. The limitation of space coverage on the urban road also leads to low accuracy on detecting traffic flow parameters and low efficiency of the traffic control system.

The cooperative vehicle-infrastructure system, connected and autonomous vehicle technology provides the urban traffic control system with a new spatial-temporal data environment. They enable vehicles to communicate with each other and also roadside infrastructure or central system. The advent of these technologies provides an opportunity to significantly enhance the traffic control system in maximizing the throughput and minimizing delay and stops. The differences between traditional fix point detection and new spatial-temporal data environment can be concluded as follows:

1. The space coverage is wider and more evenly cover the whole urban road network. However, the penetration rate in the short future is less.
2. Providing richer traffic and vehicle information, such as vehicle position, vehicle speed and route information, trajectory data and so on.
3. By using data-fusion technology to incorporate multi-source heterogeneous traffic data, the perception and model expression abilities of the traffic control system can be significantly enhanced.

9.5.1.2 Traffic control objects and variables

The development of new technologies has also caused the control object and variables of urban traffic control systems.

Traditional traffic control optimizes signal timing plan and adapts green signal duration time to control the vehicles running on the urban roads based on artificial experience or real-time vehicle detection data. The control methods are passive response controls. However, with the help of connected vehicles, reverse control of vehicle such as speed guidance and so on will be realized. Actually, some researchers and systems have applied this to real-world traffic control.

9.5.1.3 The demand of computational power

With the advent of big traffic data era, the demand for computational power, whether central or local, is increasing. The architecture of distributed edge computing has become the hotspot of current research on the urban traffic control system. Edge computing, as a distributed computing architecture, move the data, services and functions from the central node to the edge.

In traditional traffic control systems, data are stored and processed at the centre of a network. This architecture has some problems including higher network transmission demand, higher storage demand and lower security. Now, with the development of the local intelligent device, the computation, storage and transmission abilities of the local controller are improved significantly. Significant processing can be completed at smaller local sites which are effectively at the edge of the network. Edge computing eliminates the need for data to be sent across long routes to data centres and clouds. Ultimately, it is shifting processing power out of the data centre and closer to end-users.

In the urban traffic control system, edge computing shifts the data processing and information services that executed by a central system originally to each local controller which installed at intersections. The benefits can be summarized as follows:

1. Compared to central computing, edge computing integrates data collection, processing, and execution of control decisions, avoiding the delay caused by data upload and download. Therefore, the speed of response will be significantly improved.
2. Decentralized traffic signal control system takes each intersection of the urban road network as an intelligent node which processes local information and provides local control strategies. Simultaneously, a powerful communication network enables data sharing and interaction among each intelligent node. A hierarchical calculation for different control objectives is provided to achieve distributed collaborative optimization.

9.5.2 Standardization

The traditional traffic control field does not have a perfect and consistent standard system, which has caused the chaotic state of the hardware platform to a certain

Table 9.4 Comparison of existing and future control system

Existing systems	Future systems
Localized automation	Swarm intelligence
Adaptive central control	Cooperative adaption
Hierarchy tree structure	Meshwork structure
Layered control	Coordinated control
Single physical centre	Virtual centre
Predefined parameters	Self-learning parameters
Backup plan for failure	Crossed controlling for failure
Predefined rules set	Evolving rules set

extent. How to make full use of the existing achievements of IT technology, unify basic agreements, standardize data definition, description and operation methods, and reserve space for customization and expansion for urban traffic signal control system is still a big challenge in most countries.

On the other hand, the traffic control language also needs to be further standardized, including the definitions of traffic control objects, properties, operations, basic rules and so on. All of these will provide traffic signal control standard basic support on operable objects and methods.

9.5.3 Summary

With the development of edge computing, internet of things, connected vehicles and big traffic data, future urban traffic signal control will be improved in several aspects, as follows (Table 9.4).

Future urban traffic signal control systems will be a cooperative and network system, which contains abilities of detection data sharing, real-time optimization, synchronized control, comprehensive-monitoring, fault-tolerant and coordination evaluation. With these new features and functions, urban signal control systems will be intelligent and efficient.

References

[1] R. Bretherton, "SCOOT urban traffic control system – philosophy and evaluation" in Control Computers Communications in Transportation, Amsterdam, The Netherlands: Elsevier, 1990.
[2] P. Lowrie, "SCATS – A Traffic Responsive Method of Controlling Urban Traffic" in Sales information brochure, Australia, Sydney: Roads & Traffic Authority, 1990.
[3] D. I. Robertson, "TRANSYT – A traffic network study tool", IVth Int. Symp. Theory of Traffic Flow, 1968.
[4] P. M. Mirchandani, and F.-Y. Wang, "Rhodes to Intelligent Transportation Systems", IEEE Intelligent Systems, vol. 20, no. 1, pp. 10–15, 2005.

[5] H. Mahmassani, T. Hu, and R. Jayakrishnan, "Dynamic traffic assignment and simulation for advanced network informatics (Dynasmart)", Proceedings of the 2nd international CAPRI seminar on Urban Traffic Networks..
[6] TRL Limited, "System Handbook for an STC UTC System."
[7] Y. Wang, X. Yang, H Liang, and Y. Liu, "A Review of the Self-Adaptive Traffic Signal Control System Based on Future Traffic Environment", Journal of Advanced Transportation, vol. 2018, 2018.
[8] H. Wei, G. Zhang, V. Gayah, and Z. Li. *"A survey on traffic signal control methods."* arXiv preprint arXiv:1904.08117 (2019).

Chapter 10
Algorithms and models for signal coordination

Hao Wang[1] and Changze Li[1]

This chapter briefs about the history of the development of signal coordination and the achievements so far in this field. The review on the field is focused on a methodology called MAXBAND and its various derivatives.

10.1 Introduction

Prior to the illustration of signal coordination, traffic control at a single intersection needs to be explained because it serves as the basis of the field. Single intersection traffic control first appeared as a method with fixed cycle when Clayton [1] proposed a traffic signal timing optimization model, which was founded on green time and average delay of vehicles. The model is considered, however, unpractical for its hypothesis largely deviates from reality. By relaxing constraints, researchers have been improving the Clayton model and put forward their own models [2], amid which stands the Webster model [3] widely employed today.

On account of the limitation of method with fixed cycle when reacting to the stochastic feature of traffic flow, actuated control method emerged initially in America with its flexibility and efficiency. Miller [4] proposed an actuated signal control method, on the basis of which Vincent [5] developed a system with a microprocessor. Meanwhile, the study on actuated control also sparked new ideas concerning its counterpart in the field of fixed method. In 1981, Akcelik [6] defined a compensational coefficient and proposed a model targeted at optimizing both delay and stop times.

With traffic demand increasing, neither fixed nor actuated method performed satisfyingly and, hence, the appearance of self-adaptive method. Through self-adaptation, condition at an intersection can be monitored by measuring various parameters, such as traffic volume, queuing length and delay, and adjusted by comparing those parameters to their ideal value. Some representative methods are Optimization Policies for Adaptive Control (OPAC) [7] and Controlled Optimization of Phases (COP) [8], which were created on the foundation of the self-adaptive model promoted by Robertson [9]. Notably, this model also serves as

[1]School of Transportation, Southeast University, Nanjing, China

the basis of the Split Cycle and Offset Optimization Technique (SCOOT), a traffic control system [10] dealing with network control. Steven [11] made a comparison among the typical model, providing an instruction to engineers as to how to choose a model properly.

Because of the rapid development of computer science, many optimizing algorithms have been introduced into the field of traffic control. Fuzzy control was for the first time used in signal optimization by Pappis [12]. Favilla [13] first designed a fuzzy controller with a self-adaptative strategy, which updated the function according to the dynamics of the traffic situation. Kaedi [14] put forward an optimization method based on neural network, which worked in two stages. The first was about the prediction of traffic flows in different phases and the second about the optimization of cycle and green time using the previous prediction. Teodorovic [15] realized the optimization of signal sequence by genetic algorithm. Moreover, Pham [16] applied a probabilistic fuzzy logic method to the signal control of an isolated intersection.

As mentioned at the beginning, traffic control at a single intersection laid the groundwork for signal coordination, which, in accordance with range, can be categorized into arterial and network coordination.

Arterial coordination made its debut in Salt Lake City, America in 1917, where an artery of six intersections was coordinated artificially. Then, an arterial signal coordination of 12 intersections was achieved in Houston, 1922, and gradually developed into a stepping control system. Signal coordination was designed in the first place to accomplish one-way traffic control but it certainly did not meet the demand from busy arteries with huge traffic volume. Therefore, the typical two-way signal coordination appeared.

In 1964, Morgan and Little [17] made discussion and analysis of different object function. On the basis of their previous work, Little went on and created a model for the maximization of green-wave bandwidth based on mixed integer linear programming [18], which is able to describe complex traffic scenarios as well as find solutions efficiently. Thereafter, Little also presented the relevant software acronymized MAXBAND [19], enabling the quick solution of parameters such as common cycle, average velocity and offset, as well as the adjustment of weight factor in correspondence with traffic demand. Built on MAXBAND, Gartner proposed MULTIBAND [20–23], which relaxes the constraints on bandwidths at different intersections and, hence, acquires a solution with variable bandwidths. To address left-turning traffic flow, Messer [24] designed PASSER II, a system capable of maximizing bandwidth and phase sequence, proffering solutions that take into consideration left-turning volume. With the model mentioned above came the software with the same name [25], applying solely to arterial coordination. But it was extended lately to the application of network coordination by Chaudhary [26]. Sripathi [27], Papola [28], and Pillai [29] also contributed greatly to the field.

Apart from aiming at the optimization of bandwidth of green-wave, models for minimizing delay were also invented. One salient result came from Hillier [30], who managed to create a method concentrating on the association between total delay and offset, which is abstracted from the analysis of queue dissipation between adjoining intersections. Lieberman [31] constructed a timing optimization model Signal Optimisation II (SIGOP II) by utilizing dynamic programming. Last but not least, the widespread control system, Traffic Network Study Tool of version 7F (TRANSTY 7-F) [32], and Synchro [33] are also devoted to the minimization of delay time. Liu [34] designed a signal-coordination model in which users could choose to minimize delay or maximize capacity according to the traffic situation. However, Yang [35] had verified by testing on real arteries that methods aimed at maximizing bandwidth were usually better than those minimizing trip delay were.

In regards to the arterial signal coordination under oversaturation, Chang [36] proposed a strategy, which could maximize capacity by suppressing the input of upstream flow. Hu [37] created a coordination model targeted in the maximization of dissipation in oversaturated links, on the basis of Wu's [38] method for the quantified definition of traffic saturation. Apart from this, Hu [39] put forward a model for the optimization of offsets at adjoining intersections.

As single intersection control is embracing intellectual algorithms, so does arterial coordination. Wu [40] combined decoupling and genetic algorithms and applied them in a model, which optimizes the total delay time, arriving at better results than those from traditional models. Chen [41] integrated heuristic and particle swarm algorithms and developed a model, which ameliorates traffic delay by modifying offsets. Li [42] applied fuzzy control to cope with the uncertainty and randomness of traffic flow, and made compensation and modification for the signal coordination.

Another part of signal coordination lies in network coordination. It focuses on the whole grid network and is advancing swiftly in recent years. As a matter of fact, research on network coordination is mainly split into two parts: one focused on the network-coordination models and the other the decomposition of a network into minimal ones. For the former, researchers have built many traffic systems and theoretical models. In 1968, TRRL in British released Traffic Network Study Tool (TRANSYT), a system which evaluates the total travel time and works offline. However, versions of the software in its early years suffered from complexity which increases considerably when the scale of network reaches a certain level and, thus, were updated and upgraded for multiple times [43,44]. Genetic algorithm was embedded in TRANSYT-7F V8.1 and its later versions [45,46]. MAXBAND [47] and MULTIBAND [22] were also integrated into the modeling and optimization of network coordination, resulting in MAXBAND-86 and MULTIBAND-96, respectively. Additionally, MULTIBAND-96 can also be used in conjunction with delay-based models such as TRANSYT to determine optimal phase sequences [48].

To cope with the tardiness of TRANSYT caused by working offline, SCOOT [49] was launched in British 1980, a system capable of collecting real-time data and generating relative optimizations. A self-adaptive mode was imbedded in SCOOT to ensure its stability but it also suffered from complexity issue as TRANSYT did. Wolshon [50] suggested that SCOOT acted greatly when traffic volume is moderate while performed poorly dealing with oversaturated traffic. In Sydney 1982, Sydney Coordinated Adaptive Traffic System (SCATS) [51] was activated to achieve network signal coordination. Control plans were stored in the database of the system and could be arranged into a particular combination according to demand if called upon.

Even though SCATS had no built-in models, its three-layer structure assured the efficiency when it ran on massive networks. However, it cannot handle oversaturation properly. In the early years of twenty-first century, China witnessed the installation of Hicon [52], a control system revolving around a coordination model featuring self-adaptation and possessing interfaces for further extension.

Then there are theoretical models for network traffic flow and coordination. With regard to the former, Gazis [53] created a simplified traffic flow model to solve oversaturation, which accidentally resulted in turning discrete signals into continuous ones, beaconing a new way for the research. One of the predecessors, Diakaki [54], joined the path by proposing TUC (Traffic-responsive Urban Control), an active coordination strategy which applies to heavily loaded traffic networks, and succeeded in practice. TUC's deficiency of not supporting self-adaptation was then remedied by Kouvelas [55], who combined TUC with Webster timing model and acquired a conditional mixed coordination strategy. Plus, Aboudolas [56] proposed a coordination model for massive traffic grids. As for the latter, it has assimilated many intellectual algorithms. Nakamiti [57] designed an artificial intelligence algorithm for network signal control. Yu [58] devised a self-adaptive model based on the Markov process in which he built in a random signal control mechanism. Dotoli [59] came up with a network-coordination model and it could cope with traffic scenarios under different demand. McKenney [60] designed a distributed self-adaptive system where nodes could communicate mutually and exchange information and statistics. With that being said, the system featured great robustness and stability. Dahal [61] created a system which made decisions in view of the prediction of the condition of a block and its adjoining blocks. Eltantawy [62] injected reinforcement learning into self-adaptive network control and successfully simulated traffic networks under different circumstances. Zhang [63] employed cellular automata and rough fuzzy neural network jointly in the fixed control of traffic system, obtaining a model capable of pattern recognition.

The decomposition of a network actually sees the network coordination from a different angle. Subnetworks are the basic unit of coordination and whether they are properly chosen at the core of the smooth operation of network control strategy. The decomposition is usually carried out either statically or dynamically and the differences between the two ways can be told literally. Studies in this field are more or less concentrated on how to measure the extent to which intersections should be combined or correlated and how the decomposition is administered. In 1971, Wlinchus [64] for the first time proposed the concept of subnetworks within a big network and suggested that they should be separated by the dynamical variation on traffic flow. In 1973, Yagoda [65] made an analysis on intersection correlation. His work manifested that the intersection correlation is positively connected to traffic volume and negatively to the length of the road. In 1976, Pinnell [66] *et al.* offered their consideration over how the upstream volume, road length, and arrival distribution would affect the correlation. Moore [67] *et al.* analyzed not only the correlation between two intersections, but also all of them with an area and gave his dividing method of clustering. Ma [68] *et al.* developed the fuzzy version of dynamical network decomposition. Based on the path, their work gave consideration on intersection distance, the uncertainty of vehicle arrival, parade length, traffic volume, and cycle duration. In 2011, Bie [69] *et al.* provided a calculation method for intersection correlation on the basis of cycle correlation and parade length correlation. In perspective, most of the work in this area is fixated on the measurement of similarities between intersections and the topology of a network is scarcely mentioned.

Due to the characteristics of oversaturated traffic network, the control methods for systems under normal condition do not apply and, thus, methodology should be developed to deal with such occasion specifically. As for oversaturated networks, Nagase [70] managed the transformation of traffic jam into a dynamic optimizing problem and solved it by linear programming. Lo [71] described network coordination as a mixed integer programming and used genetic algorithm to solve it. Putha [72] put forward an ant colony-optimizing algorithm for the timing parameters in Girianna's [73] model and made a comparison between the results from his method and from genetic algorithm.

10.2 Basic MAXBAND approach

In 1964, John T. Morgan and John D.C. Little published a thesis [17], in which they gave preliminary discussion on synchronization of traffic signals and a prototype for the later MAXBAND. In the paper, they first introduced the concept of bandwidth. Then, they focused on two problems: (1) Given an arbitrary number of signals along a street, a common signal period, the green and red times for each signal, and specified vehicle speeds in each direction between adjacent signals, synchronize the signals to produce bandwidths that are equal in each direction and as large as possible. (2) Adjust the synchronization to increase one bandwidth to some specified, feasible value and maintain the other as large as is then possible. The model proposed then was still at its early stage since many factors appeared as parameters instead of decision variables. Despite its deficiency, the model laid the fundamental base for the field.

In 1965, John D.C. Little took a step further and proposed a model for the synchronization of traffic signals by mixed integer linear programming [18]. This time they concentrated instead on problems where they set up some prerequisites and tried to achieve certain goals. The prerequisites are: (1) an arbitrary number of signals, (2) the red-green split at each signal, (3) upper and lower limits on signal period, (4) upper and lower limits on speed between adjacent signals, and (5) limits on change in speed. The goals are to find: (1) common signal period, (2) speeds between signals, and (3) the relative phasing of the signals, in order to maximize the sum of the bandwidths for the two directions. Founded in the preceding work, the model in 1965 was deemed more fledged compared to its predecessor. It takes more decision variables, gives more margins and, thus, is more practical.

In 1981, John D. C. Little and Mark D. Kelson released MAXBAND [19], a versatile program for setting signals on arteries and triangular networks, as a refinement of the previous work. As a matter of fact, MAXBAND is the name of the computer program. But the methodology behind the program is often addressed as MAXBAND out of conciseness. MAXBAND offers rigorous method to concurrently generate the offsets between adjacent signals, optimize the prevailing speed at each link, and determine the proper left-turn phases. The key variables of the model are shown in Figure 10.1 and its primary formulations are quoted below:

$$MaxB = (b + k\overline{b}) \tag{10.1}$$

s.t.

$$(1-\rho)\overline{b} \geq (1-\rho)\rho b \tag{10.2}$$

$$1/C_{max} \leq z \leq 1/C_{min} \tag{10.3}$$

$$w_i + b \leq 1 - r_i, i = 1,\ldots,n \tag{10.4}$$

$$\overline{w}_i + \overline{b} \leq 1 - \overline{r}_i, i = 1,\ldots,n \tag{10.5}$$

$$(w_i + \overline{w}_i) - (w_{i+1} + \overline{w}_{i+1}) + (t_i + \overline{t}_i) + \delta_i L_i - \overline{\delta}_i \overline{L}_i + m_i = \\ (r_{i+1} - r_i) + (\tau_{i+1} + \overline{\tau}_i) + \delta_{i+1} L_{i+1} - \overline{\delta}_{i+1}\overline{L}_{i+1}, i = 1,\ldots,n-1 \tag{10.6}$$

$$(d_i/f_i)z \leq t \leq (d_i/e_i)z, i = 1,\ldots,n-1 \tag{10.7}$$

$$(\overline{d}_i/\overline{f}_i)z \leq \overline{t}_i \leq (\overline{d}_i/\overline{e}_i)z, i = 1,\ldots,n-1 \tag{10.8}$$

$$(d_i/h_i)z \leq (d_i/d_{i+1})t_{i+1} - t_i \leq (d_i/g_i)z, i = 1,\ldots,n-2 \tag{10.9}$$

$$(\overline{d}_i/\overline{h}_i)z \leq (\overline{d}_i/\overline{d}_{i+1})\overline{t}_{i+1} - \overline{t}_i \leq (\overline{d}_i/\overline{g}_i)z, i = 1,\ldots,n-2 \tag{10.10}$$

$$b, \overline{b}, z, w_i, \overline{w}_i, t_i, \overline{t}_i \geq 0, i = 1,\ldots,n \tag{10.11}$$

$$m_i, \text{integer}; \delta_i, \overline{\delta}_i, \text{binary integers}, i = 1,\ldots,n \tag{10.12}$$

Parameters included in above formulations are: $A, B, C(\overline{C})$ are the reference points against which inbound and outbound green signal are calculated; , $D(\overline{D})$ are the reference points of outbound (inbound) green-wave; $\theta(\overline{\theta})$ are the difference of

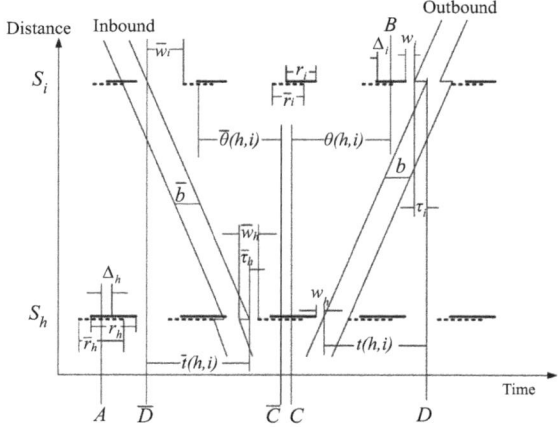

Figure 10.1 Space-time diagram for MAXBAND [19]

signal offsets at different intersection calculated based on outbound (inbound); ρ, the preference parameter; C_{max} and C_{min}, the boundaries of the cycle length; r_i, the common red time at signal i; $L_i(\overline{L}_i)$, the time allocated to the left-turn movements; $\tau_i(\overline{\tau}_i)$, the queue clearance time; $e_i, f_i(\overline{e}_i, \overline{f}_i)$, the lower and upper limits for the outbound (inbound) speeds; $g_i, h_i(\overline{g}_i, \overline{h}_i)$, the lower and upper limits for the outbound (inbound) speed change.

Decision variables include bandwidth $b(\overline{b})$, cycle length $(1/z)$, time between the start of a green phase and the boundary of its green band $w_i(\overline{w}_i)$, prevailing speed $t_i(\overline{t}_i)$, and integer variables $\delta_i, \overline{\delta}_i, m_i$. Particularly, different values of the binary variables, δ_i and $\overline{\delta}_i$, can result in four possible phase designs to accommodate major arterial flows: outbound left leads and inbound left lags; outbound left lags and inbound left leads; outbound (or inbound) left leads; and outbound (or inbound) left lags. All the possible phase patterns are shown in Figure 10.2 and their corresponding δ_i and $\overline{\delta}_i$ in Table 10.1.

The objective function (10.1) for MAXBAND is to maximize the weighted sum of the two-way bandwidths. Constraint (10.2) allocates the progression preference to either the inbound or outbound direction. Constraint (10.3) limits the upper and lower bounds of the selected cycle length. The directional interference constraints in (10.4) and (10.5) can ensure the green bandwidth to be within the available green time. The loop integer constraint in (10.6) is specified to guarantee that the signals will not cause traffic flows to stop in the green bands. The variation of travel times (a proxy of speed) is constrained by (10.7–10.10).

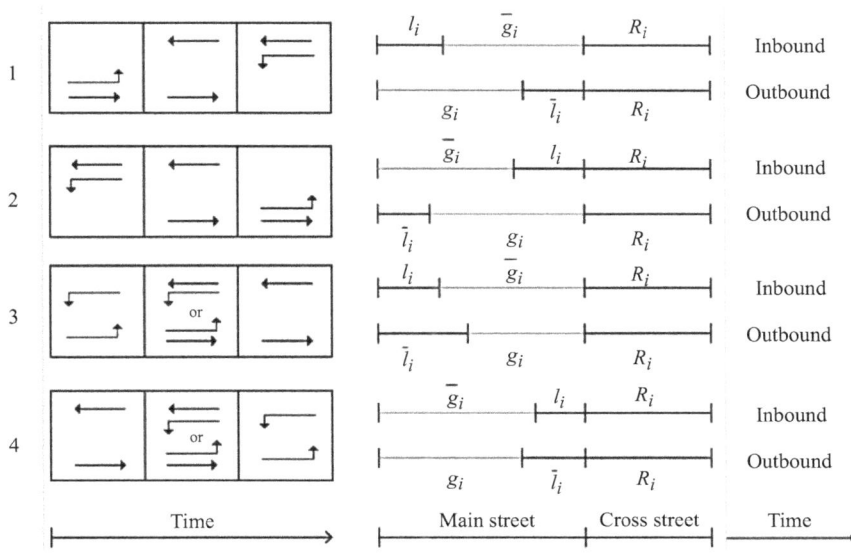

Figure 10.2 Phase patterns

214 Traffic information and control

Table 10.1 Values of δ_i and $\bar{\delta}_i$ of different patterns

Pattern	δ_i	$\bar{\delta}_i$
1	0	1
2	1	0
3	0	0
4	1	1

As shown above, MAXBAND eventually took form and was able to model complicated details in varied traffic scenarios. Since then, it has received increasing attention as the solution of mixed-integer problems became more efficient with the help of computers.

10.3 Extended MAXBAND approach

MAXBAND provides an effective way to synchronize traffic signals and gain bandwidth under variant conditions. With that being said, it still can be improved for flexibility and versatility, in regard to which many scholars have made their contributions. Exhibited below are three salient branches.

10.3.1 Variable bandwidth method

To address MAXBAND's incapacity for consideration over flows on different links, Stamatiadis C. and Gartner N. H. presented MULTIBAND [20–23]. Precisely, the program's name is MULTIBAND, after which the model is usually named. The core idea of MULTIBAND lies in modifying some constraints of MAXBAND so that the solution space is enlarged, thus resulting in more flexible resolutions. Time-space diagram is given in Figure 10.3 and key formulations of MULTIBAND are shown below:

$$MaxB = \frac{1}{n-1}\sum_{i=1}^{n-1}(a_i b_i + \bar{a}_i \bar{b}_i) \tag{10.13}$$

s.t.

$$a_i = \left(\frac{\gamma_i}{\psi_i}\right)^{\varsigma} \tag{10.14}$$

$$\bar{a}_i = \left(\frac{\bar{\gamma}_i}{\bar{\psi}_i}\right)^{\varsigma} \tag{10.15}$$

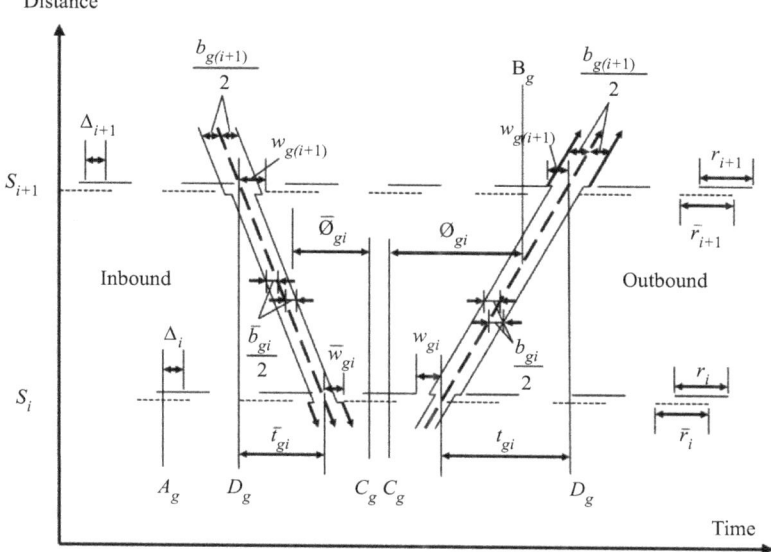

Figure 10.3 Time-space diagram for MULTIBAND [20]

$$(1 - k_i)\overline{b}_i \geq (1 - k_i)k_i b_i \tag{10.16}$$

$$\frac{1}{2}b_i \leq w_i \leq 1 - r_i - \frac{1}{2}b_i, i = 1,\ldots,n \tag{10.17}$$

$$\frac{1}{2}b_i \leq w_{i+1} + \tau_{i+1} \leq 1 - r_{i+1} - \frac{1}{2}b_i, i = 1,\ldots,n \tag{10.18}$$

$$\frac{1}{2}\overline{b}_i \leq \overline{w}_i - \overline{\tau}_i \leq 1 - \overline{r}_i - \frac{1}{2}\overline{b}_i, i = 1,\ldots,n \tag{10.19}$$

$$\frac{1}{2}\overline{b}_i \leq \overline{w}_{i+1} \leq 1 - \overline{r}_{i+1} - \frac{1}{2}\overline{b}_i, i = 1,\ldots,n \tag{10.20}$$

$$b_i, \overline{b}_i, z, w_i, \overline{w}_i, t_i, \overline{t}_i \geq 0, i = 1,\ldots,n \tag{10.21}$$

Compared to MAXBAND, new parameters included in above formulations are: a_i and $\overline{a}_i$, the link-specific weights in the two directions which could be calculated by following equations; k_i, the target ratio of inbound to outbound bandwidth on section i (taken as the ratio of the corresponding volumes in each direction); $\Upsilon_i(\overline{\Upsilon}_i)$, directional volume on section i, outbound (inbound), either total volume or through (platoon) volume can be used; $\psi_i(\overline{\psi}_i)$, saturation flow on section i, outbound (inbound); ς = exponential power, the values = 0, 1, 2, 4 were used. New decision variables include $b_i(\overline{b}_i)$, the outbound (inbound) bandwidth between signals Si and $Si+1$. $b_i, \overline{b}_i$

were defined and weighted with respect to its contribution to the overall objective function.

Equations (10.13–10.21), together with constraints (10.3), (10.6–10.10) and (10.12), form the MULTIBAND model. MULTIBAND inherited from MAXBAND parts concerning travel time and relationship between adjoining intersections while it variated on objective function and set up directional interference constrains (10.15–10.18). Note in Figure 10.3 the redefinition of the time reference points to the centerline of the bands (or, the progression line), which is indicated by dashed lines, rather than the edges. And the ratio constraint (10.14) was also changed to reflect the multiband situation. The most important change occurred in the objective function (10.13). Since the bands are link-specific, they can be weighted in disaggregation to reflect desirable traffic objectives for each link. Thus, they obtained a method sensitive to varying traffic conditions and can tailor the progression scheme to the different possible traffic flow patterns. And then, the authors proved that MULTIBAND outperformed MAXBAND by carrying out statistical analysis.

10.3.2 Multimode band method

Despite the ability to facilitate traffic progression along the artery, the typical bandwidth optimization approaches fail to consider the characteristics of different systems under the same circumstance. For instance, bus-related studies have demonstrated that the characteristics of bus systems differ from the characteristics of general vehicles. Specifically, the factors of bus systems, such as the bus speed and the dwell time, are characterized in a different way. Therefore, in order to achieve better performance when the model deals with scenarios where there are more than one traffic components, researchers have made efforts and found new approaches. And, a model is called "multimode" when it takes into consideration two or more different traffic systems at the same time.

In 2015, Dai and Wang [74] published their work, revolving around a model considering characteristics of both general vehicles and buses simultaneously. The bus systems in their paper have exclusive bus lanes; thus, all buses can travel along the bus lanes with less interference from general vehicles. Instead of maximizing bandwidths, their model pursues the minimum total bus travel time for both directions. The formulations of the model, along with the time-space diagram (Figure 10.4), are given below:

$$MinT = (\sum_{i=1}^{n-1} t_{bi} + k_b \sum_{i=1}^{n-1} \bar{t}_{bi}) \tag{10.22}$$

s.t.

$$(1 - k_b)k_b \sum_{i=1}^{n-1} t_{bi} \geq (1 - k_b) \sum_{i=1}^{n-1} \bar{t}_{bi} \tag{10.23}$$

Algorithms and models for signal coordination 217

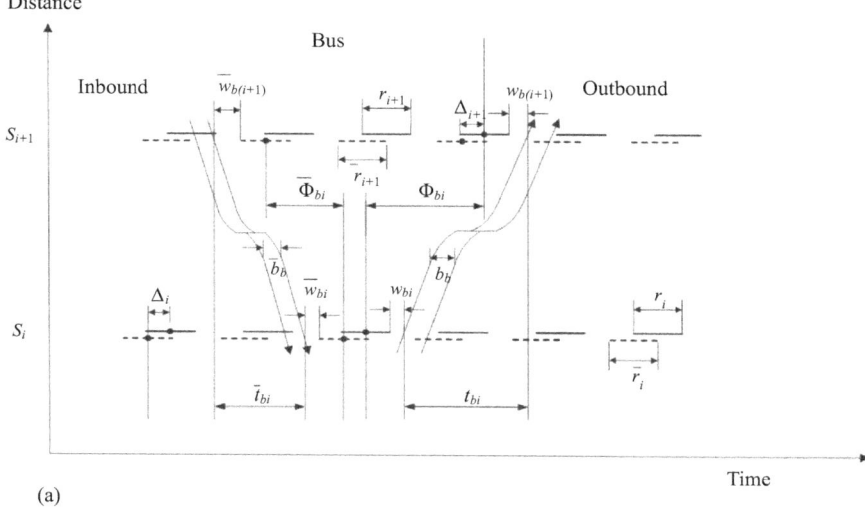

(a)

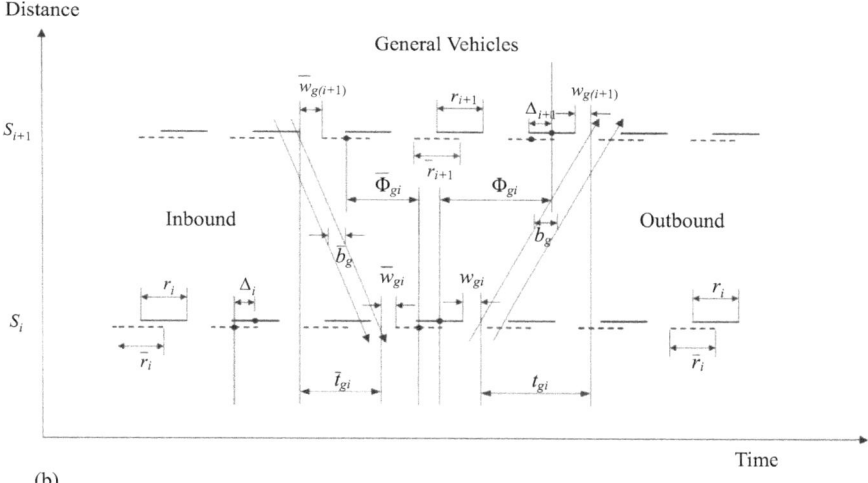

(b)

Figure 10.4 Time-space diagram for multimode model [74]

$$(1 - k_b)k_b b_b \geq (1 - k_b)\overline{b}_b \tag{10.24}$$

$$(1 - k_b)k_b b_g \geq (1 - k_b)\overline{b}_g \tag{10.25}$$

$$w_{bi} + b_b \leq 1 - r_i, i = 1, \ldots, n \tag{10.26}$$

$$\overline{w}_{bi} + \overline{b}_b \leq 1 - \overline{r}_i, i = 1, \ldots, n \qquad (10.27)$$

$$(w_{bi} + \overline{w}_{bi}) - (w_{b(i+1)} + \overline{w}_{b(i+1)}) + (t_{bi} + \overline{t}_{bi}) + \delta_i L_i - \overline{\delta}_i \overline{L}_i - m_{bi}$$
$$= (r_{i+1} + r_i) + \delta_{i+1} L_{i+1} - \overline{\delta}_{i+1} \overline{L}_{i+1}, i = 1, \ldots, n-1 \qquad (10.28)$$

$$w_{gi} + b_g \leq 1 - r_i, i = 1, \ldots, n \qquad (10.29)$$

$$\overline{w}_{gi} + \overline{b}_g \leq 1 - \overline{r}_i, i = 1, \ldots, n \qquad (10.30)$$

$$(w_{gi} + \overline{w}_{gi}) - (w_{g(i+1)} + \overline{w}_{g(i+1)}) + (t_{gi} + \overline{t}_{gi}) + \delta_i L_i - \overline{\delta}_i \overline{L}_i - m_{gi}$$
$$= -(r_{i+1} - r_i) + \delta_{i+1} L_{i+1} - \overline{\delta}_{i+1} \overline{L}_{i+1}, i = 1, \ldots, n-1 \qquad (10.31)$$

$$w_{bi} - w_{b(i+1)} + t_{bi} - (w_{gi} - w_{g(i+1)} + t_{gi}) = m_i, i = 1, \ldots, n-1 \qquad (10.32)$$

$$\overline{w}_{bi} - \overline{w}_{b(i+1)} + \overline{t}_{bi} - (\overline{w}_{gi} - \overline{w}_{g(i+1)} + \overline{t}_{gi}) = \overline{m}_i, i = 1, \ldots, n-1 \qquad (10.33)$$

$$t_{bi} = \begin{cases} t_{pi} + \sum_{1}^{N_i} t_{dij}, N_i > 0 \\ t_{pi}, N_i = 0 \end{cases}, i = 1, \ldots, n-1; j = 1, \ldots, N_i \qquad (10.34)$$

$$\overline{t}_{bi} = \begin{cases} \overline{t}_{pi} + \sum_{1}^{\overline{N}_i} \overline{t}_{dij}, \overline{N}_i > 0 \\ \overline{t}_{pi}, \overline{N}_i = 0 \end{cases}, i = 1, \ldots, n-1; j = 1, \ldots, \overline{N}_i \qquad (10.35)$$

$$b_{b\,min} \leq b_b \leq b_{b\,max} \qquad (10.36)$$

$$\overline{b}_{b\,min} \leq \overline{b}_b \leq \overline{b}_{b\,max} \qquad (10.37)$$

$$b_b \leq b_g \leq b_{g\,max} \qquad (10.38)$$

$$\overline{b}_b \leq \overline{b}_g \leq \overline{b}_{g\,max} \qquad (10.39)$$

$$T_{i\,min} Z + t_{dij\,min} \leq t_{bi} \leq T_{i\,max} Z + t_{dij\,max}, i = 1, \ldots, n-1; j = 1, \ldots, N_i \qquad (10.40)$$

$$\overline{T}_{i\,min} Z + \overline{t}_{dij\,min} \leq \overline{t}_{bi} \leq \overline{T}_{i\,max} Z + \overline{t}_{dij\,max}, i = 1, \ldots, n-1; j = 1, \ldots, \overline{N}_i \qquad (10.41)$$

$$w_{bi}, \overline{w}_{bi}, w_{gi}, \overline{w}_{gi} \geq 0, i = 1, \ldots, n \qquad (10.42)$$

$$m_{bi}, m_{gi}, m_i, \overline{m}_i, \text{integer}, i = 1, \ldots, n \qquad (10.43)$$

$$\delta_i, \overline{\delta}_i, \text{binary integer}, i = 1, \ldots, n \qquad (10.44)$$

where $N_i(\overline{N}_i)$ is the outbound (inbound) number of bus stops between the intersections S_i and S_{i+1}; $b_b(\overline{b}_b)$ is the outbound (inbound) bandwidth for a bus (cycles). $b_{b\,max}(\overline{b}_{b\,max})$ and $b_{b\,min}(\overline{b}_{b\,min})$ represent the lower limit and upper limit, respectively, for the outbound (inbound) bandwidth for a bus; $b_g(\overline{b}_g)$ is the outbound (inbound)

Algorithms and models for signal coordination 219

bandwidth for a general vehicle (cycles). $b_{g\ max}(\overline{b}_{g\ max})$ and $b_{g\ min}(\overline{b}_{g\ min})$ are the lower limit and upper limit, respectively, for the outbound (inbound) bandwidth for a general vehicle; $w_{bi}(\overline{w}_{bi})$ is the time from the right (left) side of $r_i(\overline{r}_i)$ at S_i to the left (right) edge of $b_b(\overline{b}_b)$ for a bus (cycles); $w_{gi}(\overline{w}_{gi})$ is the time from the right (left) side of $r_i(\overline{r}_i)$ at S_i to the left (right) edge of $b_g(\overline{b}_g)$ for a general vehicle (cycles); $t_{bi}(\overline{t}_{bi})$ is the travel time from S_i to S_{i+1} outbound (S_{i+1} to S_i inbound); $t_{pi}(\overline{t}_{pi})$ is the first part of the travel time for a bus from S_i to S_{i+1} outbound (S_{i+1} to S_i inbound), which consists of the travel time with an average bus speed, acceleration time when a bus leaves a bus stop, the deceleration time when a bus arrives at a bus stop (cycles); $T_i(\overline{T}_i)$ is the first part of the travel time for a bus from S_i to S_{i+1} outbound (S_{i+1} to S_i inbound)(s). $T_{imax}(\overline{T}_{imax})$ and $T_{imin}(\overline{T}_{imin})$ are the lower limit and upper limit, respectively; $\sum_1^{N_i} t_{dij} \left(\sum_1^{\overline{N}_i} \overline{t}_{dij}\right)$ is the second part of the travel time for a bus from S_i to S_{i+1} outbound (S_{i+1} to S_i inbound) = sum of dwell times at all outbound (inbound) bus stops between S_i and S_{i+1}(cycles), where t_{dij} is the dwell time at the outbound bus stop j and $\overline{t}_{dij}$ is the dwell time of at the inbound bus stop j; $t_{dij\ max}(\overline{t}_{dij\ max})$ and $t_{dij\ min}(\overline{t}_{dij\ min})$ are the lower limit and upper limit, respectively, where $j = 1, \ldots, N_i$; $t_{gi}(\overline{t}_{gi})$ is the travel time for a general vehicle from S_i to S_{i+1} outbound (S_{i+1} to S_i inbound). The travel time can be derived from the distance and the general vehicle speed between S_i and S_{i+1} (cycles); k_b is the ratio of the total inbound bus volumes to the total outbound bus volumes; k_g is the ratio of the total inbound general vehicle volumes to the total outbound general vehicle volumes.

Equations (10.2) and (10.22–10.44) compose the multimodal signal optimization model proposed by Dai and Wang. The authors have made many changes to suit the need of different systems. The feasibility and flexibility of bus management surpasses that of general vehicles. Plus, certain features of bus system, such as speed and dwell time, can be adjusted to enlarge the feasible region of the mixed-integer linear programming for the model. Therefore, to confine the outbound and inbound travel times for buses within a reasonable region and improve the performance of bus systems, the minimum total weighted travel times for buses in both directions is established as the new objective function. Through splitting the decision variables in to two sections for buses and general vehicles, a model is obtained that is able to administer optimization for a multimodal traffic scenario. The model can generate green bands for buses and general vehicles, which are mutually compatible and, thus, able to coexist in the same timing plan.

10.3.3 Path-based method

It is noted by researchers that congestion is often caused by the turning volumes on some links. Li and Wang [75] proposed an improved version of the multipath progression model, which presents solutions that optimize the bandwidths for trams and general vehicles simultaneously. As a matter of fact, the model is capable of dealing with traffic system where there are multiple demands from different traffic

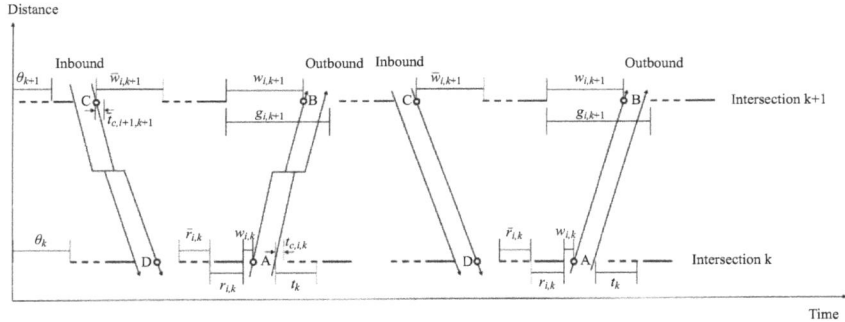

Figure 10.5 Time-space diagram for multipath model [75]

modes even though they do not always share the same route but head toward different directions (Figure 10.5).

$$Max\left(\sum_i \varphi_i b_i + \sum_i \overline{\varphi}_i \overline{b}_i\right), \forall i \in (Car + \overline{Car}) \quad (10.45)$$

s.t.

$$\theta_k + r_{i,k} + w_{i,k} + n_{i,k} + t_{i,k} = \\ \theta_{k+1} + r_{i,k+1} + w_{i,k+1} + n_{i,k+1} + \tau_{i,k+1}, \forall i \in \overline{I}; \forall k \in K_i \quad (10.46)$$

$$\overline{t}_{i,k} - \theta_k + \overline{r}_{i,k} + \overline{w}_{i,k} + \overline{n}_{i,k} + \overline{\tau}_{i,k} = \\ -\theta_{k+1} + \overline{r}_{i,k+1} + \overline{w}_{i,k+1} + \overline{n}_{i,k+1}, \forall i \in \overline{I}; \forall k \in K_i \quad (10.47)$$

$$x_{l,m,k} = \begin{cases} 1, \text{ if phase } l \text{ is before phase } m \text{ within} \\ \quad \text{the same cycle of intersection } k; \\ 0, \text{ otherwise} \end{cases} \quad (10.48)$$

$$x_{l,l,k} = 0, \forall l; \forall k \quad (10.49)$$

$$x_{l,m,k} + x_{m,l,k} = 1, \forall l \neq m; \forall k \quad (10.50)$$

$$x_{l,n,k} \geq x_{l,m,k} + x_{m,n,k} - 1, \forall l \neq m \neq n; \forall k \quad (10.51)$$

$$\beta_{i,l,k} = \begin{cases} 1, \text{ if path } i \text{ obtains green time in phase } l \text{ at intersection } k; \\ 0, \text{ otherwise} \end{cases}$$
$$\quad (10.52)$$

$$0 \leq w_{i,k} + b_i \leq \sum_l \beta_{i,l,k}\phi_{l,k}, \forall i \in Car; \forall k \in K_i \quad (10.53)$$

$$0 \leq \overline{w}_{i,k} + \overline{b}_i \leq \sum_l \beta_{i,l,k}\phi_{l,k}, \forall i \in \overline{Car}; \forall k \in K_i \quad (10.54)$$

$$0 \leq w_{i,k} + b_i \leq \sum_l \beta_{i,l,k}\phi_{l,k} - t_{c,i,k}, \forall i \in Tram; \forall k \in K_i \quad (10.55)$$

$$0 \leq \overline{w}_{i,k} + \overline{b}_i \leq \sum_l \beta_{i,l,k}\phi_{l,k}, \forall i \in \overline{Tram}; \forall k \in K_i \quad (10.56)$$

$$\overline{t}_{c,i,k} \leq \overline{w}_{i,k}, \forall i \in \overline{Tram}; \forall k \in K_i \quad (10.57)$$

$$r_{i,k} \leq \sum_l \beta_{i,m,k}\phi_{l,k} x_{l,m,k} + M(1 - \beta_{i,m,k}), \forall i \in I + \overline{I}; \forall k \in K_i; \forall m \quad (10.58)$$

$$\overline{r}_{i,k} \leq \sum_l \beta_{i,m,k}\phi_{l,k} x_{l,m,k} + M(1 - \beta_{i,m,k}), \forall i \in I + \overline{I}; \forall k \in K_i; \forall m \quad (10.59)$$

$$r_{i,k} + \overline{r}_{i,k} + \sum_l \beta_{i,l,k}\phi_{l,k} = 1, \forall i \in I + \overline{I}; \forall k \in K_i \quad (10.60)$$

$$\frac{L_k}{v_{i,\max}} Z \leq t_{i,k} \leq \frac{L_k}{v_{i,\min}} Z, \forall i \in Car \quad (10.61)$$

$$\frac{L_k}{v_{i,\max}} Z \leq \overline{t}_{i,k} \leq \frac{L_k}{v_{i,\min}} Z, \forall i \in Car \quad (10.62)$$

$$v_{i,\min} \leq v_{i,k} \leq v_{i,\max}, \forall i \in Tram \quad (10.63)$$

$$v_{i,\min} \leq \overline{v}_{i,k} \leq v_{i,\max}, \forall i \in \overline{Tram} \quad (10.64)$$

$$t_{p,i,k} = \left(\frac{L_k}{v_{i,k}} + \frac{N_k v_{i,k}}{2a} + \frac{N_k v_{i,k}}{2b}\right) Z, \forall i \in Tram \quad (10.65)$$

$$\overline{t}_{p,i,k} = \left(\frac{\frac{L_k}{\overline{v}_{i,k}} + \overline{N}_k \overline{v}_{i,k}}{2a} + \frac{N_k v_{i,k}}{2b}\right) Z, \forall i \in \overline{Tram} \quad (10.66)$$

$$t_{p,i,\min} + \sum_1^{N_k} t_{t,i,k,j} \leq t_{i,k} \leq t_{p,i,\max} + \sum_1^{N_k} t_{t,i,k,j}, \forall i \in Tram \quad (10.67)$$

$$\overline{t}_{p,i,\min} + \sum_1^{\overline{N}_k} \overline{t}_{t,i,k,j} \leq \overline{t}_{i,k} \leq \overline{t}_{p,i,\max} + \sum_1^{\overline{N}_k} \overline{t}_{t,i,k,j}, \forall i \in \overline{Tram} \quad (10.68)$$

$$\sum_{k=1}^{K_i} t_{i,k} = \sum_{k=1}^{K_i} \overline{t}_{i,k}, \forall i \in (Tram + \overline{Tram}) \quad (10.69)$$

$$C_{\min} \leq C \leq C_{\max} \quad (10.70)$$

$$b_i \geq b_{i,\min}, \forall i \in Tram \quad (10.71)$$

$$b_i = \overline{b}_i, \forall i \in (Tram + \overline{Tram}) \quad (10.72)$$

where C is the common cycle length (s); $Z = 1/C$; k refers to intersection k; L_k is the distance between intersection k and intersection $k+1$; θ_k is the offset of intersection k; $g_{i,k}(\overline{g}_{i,k})$ is the maximum green duration that the outbound (inbound) path i can obtain at intersection k; $r_{i,k}$ is the total red duration at the left side of the green band for path I; $\overline{r}_{i,k}$ is the total red duration at the right side of the green band for path i; $\emptyset_{l,k}$ is the length of phase l at intersection k; $w_{i,k}(\overline{w}_{i,k})$ is the part of a green duration before (after) the green band for outbound (inbound) path i at intersection k; $b_i(\overline{b}_i)$ is the green bandwidth for outbound (inbound) path I; $\varphi_i(\overline{\varphi}_i)$ is the weighting factor for outbound (inbound) path I; $n_{i,k}(\overline{n}_{i,k})$ is an integer variable to represent the number of signal cycles; $\tau_{i,k}(\overline{\tau}_{i,k})$ is the clearing time for outbound (inbound) path I at intersection k; $I(\overline{I})$ is the set of outbound (inbound) paths; $Car(\overline{Car})$ is the set of outbound (inbound) paths of general vehicles; $Tram(\overline{Tram})$ is the set of outbound (inbound) paths of trams; K_i is the set intersections path i passes; $t_{i,k}(\overline{t}_{i,k})$ is the travel time of outbound (inbound) path i between intersection k and $k+1$; $t_{c,i}(\overline{t}_{c,i})$ is the clearing time for outbound (inbound) path i of trams at intersection k; M is a large positive number; $N_k(\overline{N}_k)$ is the number of stations the outbound (inbound) tram stops by between intersection k and $k+1$; $v_{i,k}(\overline{v}_{i,k})$ is the average speed of outbound (inbound) path i between intersection k and $k+1$; $t_{p,i,k}(\overline{t}_{p,i,k})$ is the first part of travel time for outbound (inbound) tram between intersection k and $k+1$, including time of constant speed, acceleration time, and deceleration time; $t_{t,i,k,j}(\overline{t}_{t,i,k,j})$ is the second part of travel time the outbound (inbound) tram path i stops for at the station j between intersection k and $k+1$; a is the average acceleration of trams (m/s^2); b is the average deceleration of trams (m/s^2); $x_{l,m,k}$ is a binary decision variable which represents the phase sequences at intersection k; $\beta_{i,l,k}$ is a binary variable which decide whether path i obtains green signal in phase l at intersection k;

Previous models or tools for arterial signal control are mostly devoted to either the maximization of the progression for two-way through traffic flows or the minimization of their total delay. Consequently, they cannot adequately account for some heavy-path flows that need to take multiple turning movements along the arterial. By considering not only arterial phases but also secondary roads, this model can more properly simulate the reality and make allowances for turning flows. Actually, the multipath progression model is to some extent an expansion of MAXBAND especially when it comes to the optimization of phase sequence. Because it can arrange phase sequence appropriately by taking a set of binary variables and is enabled to change the sequence in both arteries and secondary roads. And it is also capable of trimming trivial parts of the solution where bandwidth might be too small to take effect. Therefore, the model is relatively flexible and pragmatic.

10.4 MAXBAND for network system

Coordination of traffic signals within a network is often seen as the heart of the smooth and efficient flow of traffic. Topologically different, a traffic-grid-optimization problem cannot be solved by models for arteries but needs more revision. It is quite intuitive to extend the models for arterial signal coordination and obtain applicable methods for a grid.

In 1965, John Little [18] applied his work to network problems. He gave an example at first and then the formulation of the problem. The network is shown in Figure 10.6, and all the constraints are quoted below. Figure 10.6 shows a network of seven signals, in which five arteries: 13, 35, 56, 16, and 47 have been selected for consideration by the author.

$$MaxB = \sum_{ij}(a^{ij}b^{ij}) \tag{10.73}$$

s.t.

$$b^i \geq k^i b^0, i = 1, \ldots, q \tag{10.74}$$

$$w_k^{ij} + b^{ij} \leq 1 - r_k^{ij} \tag{10.75}$$

$$\overline{w_k^{ij}} + \overline{b^{ij}} \leq 1 - r_k^{ij} \tag{10.76}$$

$$(w_k^{ij} + \overline{w_k^{ij}}) - (w_l^{ij} + \overline{w_l^{ij}}) + d(k,l)(u^{ij} + \overline{u^{ij}}) = m(k,l) - (r_k - r_l) \tag{10.77}$$

$$(1/f^{ij})z \leq u^{ij} \leq (1/e^{ij})z \tag{10.78}$$

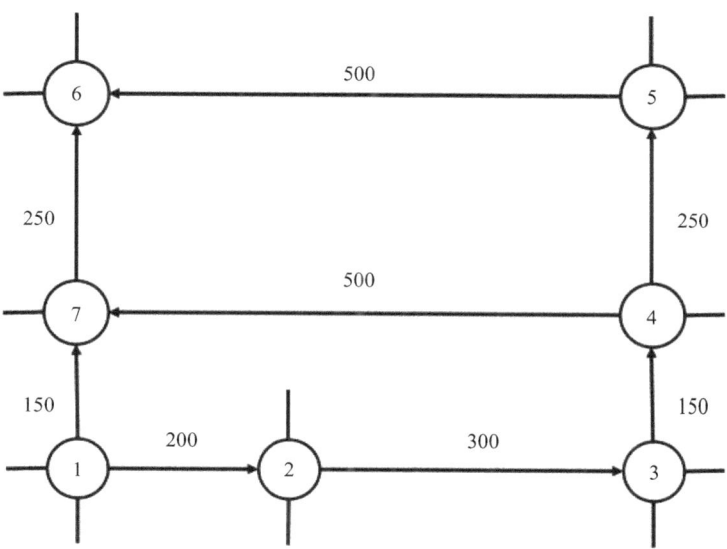

Figure 10.6 A network of seven signals

$$(1/\overline{f^{ij}})z \leq \overline{u^{ij}} \leq (1/\overline{e^{ij}})z \tag{10.79}$$

$$\begin{aligned}c(k_1,\ldots,k_p) = \\ \phi(k_1,k_2) + \phi(k_2,k_3) + \ldots + \phi(k_{p-1},k_p) + \phi(k_p,k_1) + p/2\end{aligned} \tag{10.80}$$

$$E_k^{ij} \leq r_k^{ij} \leq F_k^{ij} \tag{10.81}$$

$$G_k^{ij}z \leq r_k^{ij} \leq H_k^{ij}z \tag{10.82}$$

$$m(k,l), c(k_1,\ldots,k_p), \text{integers} \tag{10.83}$$

where an outbound artery ij is defined by one of its extreme ends, S_i, to the other, S_j. And b^{ij}, r_k^{ij}, $f^{ij}\left(\overline{f}^{ij}\right)$, $e^{ij}(\overline{e}^{ij})$, $u^{ij}(\overline{u}^{ij})$, and w_k^{ij} ($\overline{w}_k^{ij}$) are defined alike. Subscripts k and l are indicators of signals at adjacent intersections; $u^{ij}(\overline{u}^{ij}) = 1/v^{ij}T$ is the reciprocal speed outbound (inbound) on artery ij; k_1 represents the signal at the intersection of arteries i_p, j_p and i_1, j_1, k_2 at the intersection of i_1, j_1 and i_2, j_2, etc.; the cycle will then be denoted $C(k_1,\ldots,k_p)$, for which there is an integer variable $c(k_1,\ldots,k_p)$; E_k^{ij}, F_k^{ij} are the lower and upper limits on r_k^{ij} (cycles); G_k^{ij}, H_k^{ij} are the lower and upper limits $r_k^{ij}T$ (seconds). Note that ij is reduced to i for convenience in constraint (10.75).

Equations (10.3) and (10.72–10.83) constitute the MAXBAND model for network problems. Equation (10.80) functions as a key part in describing the topological structure of the traffic grid and the rest of the constraints are derivatives of the original model. The model is a primitive attempt to deal with optimization concerning traffic grids. And as the author pointed out, that the full n-signal objective function must be used even for an r-signal problem. Otherwise, the upper bound obtained is inappropriate. Limited by the capability of calculation back then, it is then unknown how large a program could reasonably be solved by the methods because solving the model entails the branch-and-bound methods. And, the normal way was to set specific values to some of the integer variables to acquire a subset of the solutions. It was not until 1981 did John Little, with the help of computer, develop a software [19] that eventually made the model useful enough for real problems.

To make allowance for the problems that might occur when it comes to area coordination with outer loops, a relaxation-enabled band model is designed by Yao and Wang [76] to optimize the green-waves of a grid. Figure 10.7 shows a case featuring the problems to which the model is dedicated. The detailed model is given in Figure 10.7.

$$max = \sum_{j \in D} k^j \left(\sum_{i=1}^{n^j-1} (b_i^j + k_d^j \bar{b}_i^j) \right) \tag{10.84}$$

s.t.

$$a_1 x_1 + a_2 x_2 + \ldots + a_n x_n = b \tag{10.85}$$

$$a_1 x_1 + a_2 x_2 + \ldots + a_n x_n \leq b \tag{10.86}$$

Algorithms and models for signal coordination 225

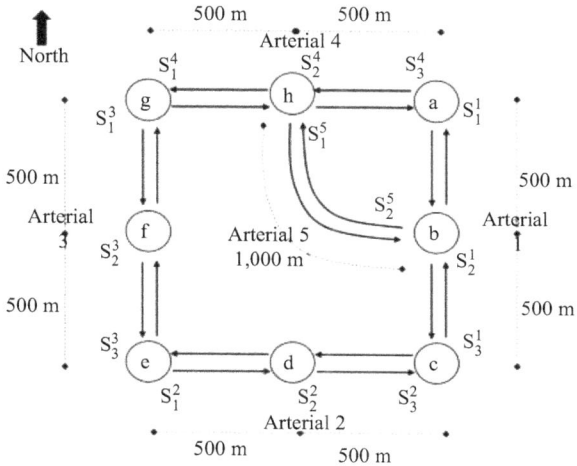

Figure 10.7 *Typical example of a network with loop constraints [76]*

$$-a_1 x_1 - a_2 x_2 - \ldots - a_n x_n \leq -b \tag{10.87}$$

$$a_1 x_1 + a_2 x_2 + \ldots + a_n x_n - Mu \leq b \tag{10.88}$$

$$Z \leq 1/C_{\min}; w_i^j(\bar{w}_i^j), b_i^j(\bar{b}_i^j), \delta_i^j(\bar{\delta}_i^j), u_i^j \leq 1 \tag{10.89}$$

$$-Z \leq -1/C_{\max}; -w_i^j(-\bar{w}_i^j), -b_i^j(-\bar{b}_i^j), -\delta_i^j(-\bar{\delta}_i^j), -u_i^j \leq 0 \tag{10.90}$$

$$l_i^j \delta_i^j - \bar{l}_i^j \bar{\delta}_i^j - \Delta_i^j = (l_i^j - \bar{l}_i^j)/2 \tag{10.91}$$

$$\bar{l}_{i_\alpha}^\alpha \bar{\delta}_{i_\alpha}^\alpha - \bar{l}_{i_\beta}^\beta \bar{\delta}_{i_\beta}^\beta - \Delta_{i_\alpha i_\beta}^{\alpha,\beta} = (\bar{l}_{i_\alpha}^\alpha - \bar{l}_{i_\alpha}^\alpha + 1)/2 \tag{10.92}$$

$$\Phi_i^j + \bar{\Phi}_i^j + \Delta_i^j - \Delta_{i+1}^j - m_i^j - Mu_i^j \leq 0 \tag{10.93}$$

$$-(\Phi_i^j + \bar{\Phi}_i^j + \Delta_i^j - \Delta_{i+1}^j - m_i^j) - Mu_i^j \leq 0 \tag{10.94}$$

$$w_i^j - w_{i+1}^j + t_i^j - \Phi_i^j - Mu_i^j \leq r_{i+1}^j/2 - r_i^j/2 \tag{10.95}$$

$$-(w_i^j - w_{i+1}^j + t_i^j - \Phi_i^j) - Mu_i^j \leq r_{i+1}^j/2 - r_i^j/2 \tag{10.96}$$

$$\bar{w}_i^j - \bar{w}_{i+1}^j + \bar{t}_i^j - \bar{\Phi}_i^j - Mu_i^j \leq \bar{r}_{i+1}^j/2 - \bar{r}_i^j/2 \tag{10.97}$$

$$-(\bar{w}_i^j - \bar{w}_{i+1}^j + \bar{t}_i^j - \bar{\Phi}_i^j) - Mu_i^j \leq \bar{r}_{i+1}^j/2 - \bar{r}_i^j/2 \tag{10.98}$$

$$w_i^j + 0.5 b_i^j - Mu_i^j \leq 1 - r_i^j \tag{10.99}$$

$$\bar{w}_i^j + 0.5 \bar{b}_i^j - Mu_i^j \leq 1 - \bar{r}_i^j \tag{10.100}$$

$$w_{i+1}^j + 0.5b_i^j - Mu_i^j \leq 1 - r_{i+1}^j \tag{10.101}$$

$$\bar{w}_{i+1}^j + 0.5\bar{b}_i^j - Mu_i^j \leq 1 - \bar{r}_{i+1}^j \tag{10.102}$$

$$-w_i^j + 0.5b_i^j - Mu_i^j \leq 0 \tag{10.103}$$

$$-\bar{w}_i^j + 0.5\bar{b}_i^j - Mu_i^j \leq 0 \tag{10.104}$$

$$-w_{i+1}^j + 0.5b_i^j - Mu_i^j \leq 0 \tag{10.105}$$

$$-\bar{w}_{i+1}^j + 0.5\bar{b}_i^j - Mu_i^j \leq 0 \tag{10.106}$$

$$(1-k_d^j)k_d^j b_i^j - (1-k_d^j)\bar{b}_i^j - Mu_i^j \leq 0 \tag{10.107}$$

$$-b_i^j - Mu_i^j \leq -b_{min} \tag{10.108}$$

$$-\bar{b}_i^j - Mu_i^j \leq -b_{min} \tag{10.109}$$

$$b_i^j + u_i^j \leq 1 \tag{10.110}$$

$$\bar{b}_i^j + u_i^j \leq 1 \tag{10.111}$$

$$t_i^j - (L_i^j/v_{min})Z - Mu_i^j \leq 0 \tag{10.112}$$

$$(L_i^j/v_{max})Z - t_i^j - Mu_i^j \leq 0 \tag{10.113}$$

$$\bar{t}_i^j - (L_i^j/v_{min})Z - Mu_i^j \leq 0 \tag{10.114}$$

$$(L_i^j/v_{max})Z - \bar{t}_i^j - Mu_i^j \leq 0 \tag{10.115}$$

where C is the length of the traffic signal cycle; $Z = 1/C$; j is the index of green-wave arterials; D is the collection of arterials; S_i^j is the i_{th} intersection of the arterial j; n^j is the total number of the intersections in the arterial j; L_i^j is the distance between S_i^j and S_{i+1}^j; u_i^j is a binary variable indicating whether the green-wave of section between S_i^j and S_{i+1}^j is broken (1-broken, 0-unbroken); M is a very big positive constant number; $r_i^j(\bar{r}_i^j)$ is the red time of S_i^j in outbound (inbound) direction; $l_i^j(\bar{l}_i^j)$ is the green time of left-turn phase of S_i^j in outbound (inbound) direction; R_i^j is the total green time of the minor street at S_i^j; $\delta_i^j(\bar{\delta}_i^j)$ is the binary variable of S_i^j for outbound (inbound) direction representing the scheme of the signal phases; Δ_i^j is the time difference between the center of inbound red and the center of the nearest outbound red of S_i^j; $w_i^j(\bar{w}_i^j)$ is the time difference between right hand (left) edge of the outbound (inbound) red of S_i^j and the central line of the nearest green-wave in outbound (inbound) direction; $t_i^j(\bar{t}_i^j)$ is the time of general traffic moving from S_i^j (S_{i+1}^j) to S_{i+1}^j (S_i^j) in outbound (inbound) direction; $\Phi_i^j(\bar{\Phi}_i^j)$ is the time difference between the outbound (inbound) centers of red in S_i^j and S_{i+1}^j, where the reds of S_i^j and S_{i+1}^j are the nearest and located left (right) to the same outbound (inbound) green-wave; $b_i^j(\bar{b}_i^j)$ is the bandwidth of the green-wave between S_i^j and S_{i+1}^j in the outbound (inbound) direction, unit: cycles. b_{min} denotes the lower bound of the bandwidth. $b_i^j(\bar{b}_i^j) \in [0, 1]$; $v_{min}(v_{max})$ is the lower (upper)

Algorithms and models for signal coordination 227

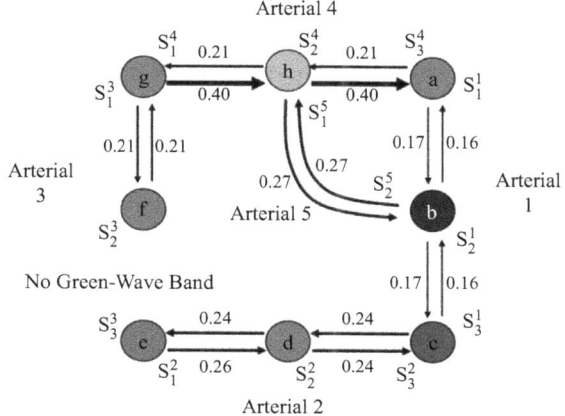

Figure 10.8 Green-wave bands of all sections in the network [76]

bound of moving speed of the general traffic between S_i^j and S_{i+1}^j in outbound (inbound) direction; k_{dir}^j is the weight of arterial j regarding to the band direction, which is equal to the ratio of the inbound flow to the outbound flow; $\Delta_{i_\alpha i_\beta}^{\alpha\beta}$, a variable in which $S_{i_\alpha}^\alpha$ and $S_{i_\beta}^\beta$ denote the same control node in different directions, which is the intersection of arterial α and β. $\Delta_{i_\alpha i_\beta}^{\alpha\beta}$ is time difference between the center of outbound red of $S_{i_\alpha}^\alpha$ and the nearest outbound red of $S_{i_\beta}^\beta$. If the center of outbound red of $S_{i_\alpha}^\alpha$ is on the left of that of $S_{i_\beta}^\beta$, $\Delta_{i_\alpha i_\beta}^{\alpha\beta} > 0$. Otherwise, $\Delta_{i_\alpha i_\beta}^{\alpha\beta} \leq 0$; m_i^j is an integer variable for constraint between S_i^j and S_{i+1}^j on arterial j; $m^{\alpha...\beta}$ is an integer variable for outer loop containing arterials $\alpha \ldots \beta$; k^j the weight showing importance of the green-wave of arterial j.

Unlike traditional models, the model offers an analytical way to break the loop constraints appropriately by introducing relaxation-enabled constraints consisting of binary variables for each road segment. Thereafter, the model can find optimal signal plan at the price of breaking the green-waves between some intersections, as shown in Figure 10.8.

10.5 Discussion and open issues

The transition of traffic control from single point control to network coordination is also a reflection of how concepts and methodologies evolve to suit the needs, which have changed greatly over the course of decades. Traditional models such as MAXBAND are invented in an analytic progression where constraints are always intuitively formulated because the number of factors is relatively small and the topology involved is rather primitive. As for now, arterial and network coordination are among the most heatedly discussed issue in the domain. Analytic models for arteries still are being enhanced with people adding more parameters and variables to make allowance for extra components. To put it another way, researchers are refining models by paying more attention to elements either poorly denoted or nonexistent to their predecessors. With sensors and monitors that are more advanced, the acquisition of more elaborate formulations is quite certain.

Nonetheless, it may not be the same case with network coordination. Researches on network decomposition are already partial to measuring the resemblance between different intersections without much regard to the topological structure as mentioned earlier. In essence, dynamical decomposition represents how traffic need is altering. Presently, indicators are mostly based in distance, volume, and cycle. And there is no agreed hypothesis of how to make the best of these data. Anyhow, one opportunity for breakthrough may lie in the development of information technology. People are already pouring intelligent algorithms into traffic optimization but that is far from enough. The simple graft of new algorithms hardly suffices for radical improvement. Traffic coordination is characterized with intrinsic properties. The most effective way to explore is to deepen the study on the basic constitution of a traffic network, nodes, roads, and topology, by way of example, by exerting mature tools borrowed from data science such as machine learning. With the aid of such tools, it is highly likely we can deduce the inherent mechanism from data and compare it with traditional analysis. Thus, we can further our comprehension of how a traffic network works. After that, we may proceed and create a prediction method for how subnetworks will alter over time. This is under no circumstances imaginable to people living prior to the information age since they neither lack the capacity of the collection of massive data nor the tools to bring out and understand the pattern that lies beneath.

As for coordination model, researcher nowadays may find it increasingly arduous to repeat the renewal as they do with MAXBAND and MULTIBAND. One reason consists in the structure. The fact that road grids are getting significantly more complicated than they ever are is rendering the analytic modeling of a network as a whole, more difficult. Another important element is that the targets of network optimization are of more uncertainty. For arteries, it is green-wave bandwidth or total delay of an artery. In terms of traffic control, an area is not merely a simple sum of arteries. To optimize an area, one needs to first consider how the intersections interact mutually, which can change dynamically in accordance with the variation of traffic scenario; then to decide how the network is to be optimized or what targets are most concerned. Specifically, popular optimization methods for network can be classified into two parts: one is to build a model with two or more layers respectively devoted to traffic distribution and signal control; the other seeks the modification and optimization of network parameters with the help of intelligent algorithm. Work dedicated in either the former or the latter involves to some degree how the parameters are evaluated and modified, whether wholly or separately to accord with their location or weight within the network. Both of them will find the decomposition prediction mentioned above greatly conducive to the decision-making. However, a standard of how to make the decision is yet to be established.

With all that being said above, a combination of data-driven prediction, topological analysis, and traditional analytic formulation may be a key to the possible

10.6 Conclusion

This chapter has given a description of single point signal control, arterial, and network coordination. The three ideas are first briefly introduced at a sequence conforming to their relationship, which is one of point, line, and area. Then some prominent achievements, such as MAXBAND and its various derivatives, are illustrated and compared, and differences and specialties among them are discussed. Moreover, probable directions of this field are mentioned at the end as a discussion on how this discipline may embrace the need and challenge of a new age.

References

[1] Clayton, A. J. H. "Road traffic calculations." Journal of the Institution of Civil Engineers 16.7(1941):247–264.
[2] Miller, and J. Alan. "Settings for fixed-cycle traffic signals." Journal of the Operational Research Society 14.4(1963):373–386.
[3] Webster, F. V. "Traffic signal settings." Rord Research Laboratory Technical Paper No.39 (1958).
[4] Alan, J. M. "A computer control system for traffic networks." (1963).
[5] Vincent, R. A., and C. P. Young. "Self-optimizing traffic signal control using microprocessors: The TRRL 'mova' strategy for isolated intersections." International Conference on Road Traffic Control 1986.
[6] Akcelik. "Traffic signals: Capacity and timing analysis: Rahmi Akcelik, Australian Road Research Board, 500 Burwood Highway, Vermont South, Victoria 3133, Australia, 1981, p. 108, $3.50." Transportation Research Part A: General 15.6(1981):505.
[7] Gartner, N. H. "OPAC: A demand-responsive strategy for traffic signal control." Transportation Research Record Journal of the Transportation Research Board No. 906.906 (1983):75–81.
[8] Sen, S., and K. L. Head. "Controlled optimization of phases at an intersection." INFORMS, 1997.
[9] Robertson, D. I., and R. D. Brethenon. "Optimum control of all intersection for any known sequence of vehicle arrival." Proceedings of the 2nd IFAC/IFIP/IFORS Symposium on Traffic Control and Transportation Systems, 1974.
[10] Hunt P. B., Robertson D. I., Bretherton R. D., *et al.* "The SCOOT on-line traffic signal optimization technique." Traffic Engineering & Control 1982, 23(4):190–192.
[11] Shelby, S. G. "Design and evaluation of real-time adaptive traffic signal control algorithms." The University of Arizona, 2001.
[12] Pappis, C. P., and E. H. Mamdani. "A fuzzy logic controller for a traffic junction." Systems Man & Cybernetics IEEE Transactions 7.10(1977):707–717.

[13] Favilla, J., A. Machion, and F. Gomide. "Fuzzy traffic control: Adaptive strategies." IEEE International Conference on Fuzzy Systems IEEE, 1993.
[14] Kaedi, M., N. Movahhedinia, and K. Jamshidi. Traffic signal timing using two-dimensional correlation, neuro-fuzzy and queuing based neural networks. Springer-Verlag, 2008.
[15] Teodorovic, D., P. Lucic, J. Popovic, S. Kikuchi, and B. Stanic. "Intelligent isolated intersection." IEEE International Conference on Fuzzy Systems IEEE, 2001.
[16] Pham, C. V., W. L. Xu, J. Potgieter, F. Alam, and F. C. Fang. "A probabilistic fuzzy logic traffic signal control for an isolated intersection." Mechatronics & Machine Vision in Practice IEEE (2012).
[17] Morgan, J. T., and J. D. C. Little. "Synchronizing traffic signals for maximal bandwidth." Operations Research 12.6(1964):896–912.
[18] Little, J. D. C. "The synchronization of traffic signals by mixed-integer linear programming." Operations Research 14.4(1966):568–594.
[19] Little, J. D. C. "MAXBAND: A versatile program for setting signals on arteries and triangular networks." Massachusetts Institute of Technology, Cambridge, MA, 1981.
[20] Gartner, N. H., S. F. Assmann, F. Lasaga, and D. L. Hou. "MULTIBAND—A variable-bandwidth arterial progression scheme." Transportation Research Record Journal of the Transportation Research Board 1287(1990):212–222.
[21] Gartner, N. H., S. F. Assmann, F. Lasaga, and D. L. Hou. "A multi-band approach to arterial traffic signal optimization." Transportation Research Part B Methodological 25.1(1991):55–74.
[22] Stamatiadis C., and N. H. Gartner. "MULTIBAND-96: A program for variable-bandwidth progression optimization of multiarterial traffic networks." Transportation Research Record Journal of the Transportation Research Board 1554.1(1996):9–17.
[23] Stamatiadis, C., and N. H. Gartner. "Progression optimization in large scale urban traffic networks: A heuristic decomposition approach." International Symposium on Transportation & Traffic Theory, 1999.
[24] Chang, E. C., and C. J. Messer. "Arterial signal timing optimization using PASSER II-90 - program user's manual." Expert Systems (1991).
[25] Chaudhary, N. A., and C. J. Messer. "PASSER IV-96, version 2.1, user/reference manual." Computer Program Documentation (1996).
[26] Chaudhary, N. A., and C. J. Messer. "PASSER IV-96, version 2.1, user/reference manual." Computer Program Documentation (1996).
[27] Sripathi, H. K., N. H Gartner, and C. Stamatiadis. "Uniform and variable bandwidth arterial progression schemes." Transportation Research Record 1494 (1995):135–145.
[28] Papola, N., and G. Fusco. "Maximal bandwidth problems: A new algorithm based on the properties of periodicity of the system." Transportation Research, Part B (Methodological) 32.4(1998):0–288.

[29] Pillai, R. S., A. K. Rathi, and S. L. Cohen. "A restricted branch-and-bound approach for generating maximum bandwidth signal timing plans for traffic networks." Transportation Research, Part B (Methodological) 32.8(1998):0–529.
[30] Hillier, J. A. "The synchronization of traffic signals for minimum delay." Transportation Science 1.2(1967):81–94.
[31] Lieberman, E. B., and J. L. Woo. "SIGOP II: A new computer program for calculating optimal signal timing patterns." Transportation Research Record (1976).
[32] Wallace, C. E., K. G. Courage, D. P. Reaves, G. W. Schoene, and G. W. Euler. "TRANSYT-7F user's manual." Fuel Consumption (1984).
[33] Husch, D., and J. Albeck. "Trafficware Synchro 6 user guide." Traffic Ware, Albany, CA, 2004.
[34] Liu, Y., and G. L. Chang. "An arterial signal optimization model for intersections experiencing queue spillback and lane blockage." Transportation Research Part C Emerging Technologies 19.1(2011):130–144.
[35] Yang, X. K., "Comparison among computer packages in providing timing plans for Iowa arterial in Lawrence, Kansas." Journal of Transportation Engineering 2001.127(4):311–318.
[36] Chang, J., B. Bertoli, and W. Xin. "New signal control optimization policy for oversaturated arterial systems." Transportation Research Board Meeting 2010.
[37] Wu, X., H. X. Liu, and D. Gettman. "Identification of oversaturated intersections using high-resolution traffic signal data." Transportation Research Part C Emerging Technologies 18.4(2010):626–638.
[38] Hu, H., X. Wu, and H. X. Liu. "Managing oversaturated signalized arterials: A maximum flow based approach." Transportation Research Part C: Emerging Technologies 36(2013):196–211.
[39] Hu, H., and H. X. Liu. "Arterial offset optimization using archived high-resolution traffic signal data." Transportation Research Part C: Emerging Technologies 37(2013):131–144.
[40] Wu, E. N., and Y. Xiaoguang. "Parameters co-optimization for artery coordinated control based on genetic algorithm." Journal of TONGJI University (Natural Science) (2008).
[41] Juan, X., and X. Lihong. "Hierarchy control algorithm and its application in urban arterial control problem." Journal of System Simulation (2008).
[42] Li, W., Y. Li, and D. He. "Arterial fuzzy compensated control based on linear programming." Journal of Highway and Transportation Research and Development 24.8(2007):110–114.
[43] Penic, M. A., and J. Upchurch. "TRANSYT-7F: Enhancement for fuel consumption, pollution emissions and user costs." Transportation Research Record Journal of the Transportation Research Board 1360(1992):104–111.
[44] Park, B., C. J. Messer, and T. Urbanik Ii. "Initial evaluations of new TRANSYT-7F version 8.1 program." Transportation Research Record Journal of the Transportation Research Board 1683.1(1999):127–132.
[45] Brianpark, B., N. M. Rouphail, and J. P. Hochanadel "Evaluating reliability of TRANSYT-7F optimization schemes." Journal of Transportation Engineering 127.4(2001):319–326.

[46] Ceylan, H. "Developing combined genetic algorithm—Hill-climbing optimization method for area traffic control." Journal of Transportation Engineering 132.8(2006):663–671.
[47] Chang E. C.-P., S. L. Cohen, C. Liu, N. A. Chaudhary, and C. Messer. MAXBAND-86: A program for optimizing left-turn phase sequence in multiarterial closed networks. In Transportation Research Record 1181, TRB, National Research Council, Washington, D.C., 1988, pp. 61–67.
[48] Cohen, S. L. Concurrent use of MAXBAND and TRANSYT signal timing programs for arterial signal optimization. In Transportation Research Record 906, TRB, National Research Council, Washington, D.C., 1983, pp. 81–84.
[49] Bretherton, R. D. "Optimizing networks of traffic signals in real time-the SCOOT method." Vehicular Technology IEEE Transactions on Vehicular Technology 40.1(1991):11–15.
[50] Wolshon, B., and W. C. Taylor. "Analysis of intersection delay under real-time adaptive signal control." Transportation Research Part C (Emerging Technologies) 7.1(1999):53–72.
[51] Lowrie, P. R. "Scats: A traffic responsive method of controlling urban traffic control." Roads and Traffic Authority (1992).
[52] Zhong, Z., and Deyong, G. "Haixin Hicon traffic signal control system." Technology and Production (2004).
[53] Gazis D. C., and R. B. Potts "The oversaturated intersection." International Symposium on Traffic Theory (1963).
[54] Diakaki, C., M. Papageorgiou, and K. Aboudolas. "A multivariable regulator approach to traffic-responsive network-wide signal control." Control Engineering Practice 10.2(2002):183–195.
[55] Kouvelas, A., K. Aboudolas, M. Papageorgiou, and K. Elias. "A hybrid strategy for real-time traffic signal control of urban road networks." IEEE Transactions on Intelligent Transportation Systems 12.3(2011):884–894.
[56] Aboudolas, K., M. Papageorgiou, and E. B. Kosmatopoulos. "Store-and-forward based methods for the signal control problem in large-scale congested urban road networks." Transportation Research Part C Emerging Technologies 17.2(2009):163–174.
[57] Nakamiti, G., R. Freitas, and F. Gomide. "Intelligent real-time traffic control." International Journal of Smart Engineering System Design 4.1(2002):49–62.
[58] Yu, X. H., and W. W. Recker. "Stochastic adaptive control model for traffic signal systems." Transportation Research Part C Emerging Technologies 14.4(2006):263–282.
[59] Dotoli, M., M. P. Fanti, and C. Meloni. "A signal timing plan formulation for urban traffic control." Control Engineering Practice 14.11(2006):1297–1311.
[60] McKenney, D., and T. White. "Distributed and adaptive traffic signal control within a realistic traffic simulation." Engineering Applications of Artificial Intelligence 26.1(2013):574–583.
[61] Dahal, K., K. Almejalli, and M. A. Hossain. "Decision support for coordinated road traffic control actions." Decision Support Systems 54.2(2013):962–975.

[62] Eltantawy, S., B. Abdulhai, and H. Abdelgawad. "Design of reinforcement learning parameters for seamless application of adaptive traffic signal control." Journal of Intelligent Transportation Systems 18.3(2014):227–245.

[63] Mengmeng, Z. "Study of urban traffic signal control system based on intelligent computation." Shangdong University, China, 2011.

[64] Walinchus, R. J. "Real-time network decomposition and subnetwork interfacing." Highway Research Record (1971).

[65] Yagoda, H. N., E. H. Principe, C. E. Vick, and B. Leonard. "Subdivision of signal systems into control areas." Traffic Engineering (1973)43(12):42–45.

[66] Pinnell, C., and M. R. L. Wilshire. "Areawide multilevel traffic control systems." IFAC Proceedings Volumes 9.4(1976):339–348.

[67] Moore, J. E., and P. P. Jovanis. "Statistical designation of traffic control subareas." Journal of Transportation Engineering 111.3(1985):208–223.

[68] Wangjing, M., L. Xiaodan, and Y.. Xiaoguang. "Incidence degree model of signalized intersection group based on routes." Journal of TONGJI University (Natural Science) 32.3(2008):1462–1466.

[69] Bie, Y., D. Wang, Q. Wei, and D. Ma. "Development of correlation degree model between adjacent signal intersections for subarea partition." 11th International Conference of Chinese Transportation Professionals (ICCTP), 2011, pp. 1170–1180.

[70] Nagase, H. "Modelling and optimal control of oversaturated transportation networks." Applied Mathematical Modelling 4.2(1980):101–108.

[71] Lo, H. K., Y. C. Chan, and A. H. F. Chow. "A new dynamic traffic control system: performance of adaptive control strategies for over-saturated traffic." Intelligent Transportation Systems, IEEE, 2001.

[72] Putha, R., L. Quadrifoglio, and E. Zechman. "Comparing ant colony optimization and genetic algorithm approaches for solving traffic signal coordination under oversaturation conditions." Computer-Aided Civil and Infrastructure Engineering 27.1(2012):14–28.

[73] Girianna, M., and R. F. Benekohal. "Using genetic algorithms to design signal coordination for oversaturated networks." Intelligent Transportation Systems Journal 8.2(2004):117–129.

[74] Dai, G., H. Wang, and W. Wang. "A bandwidth approach to arterial signal optimisation with bus priority." Transportmetrica A: Transport Science 11.7 (2015):579–602.

[75] Li, C., H. Wang, and D. Li. "Design of arterial green-wave considering right-turning trams." 19th COTA International Conference of Transportation Professionals, 2019.

[76] Yao, D., H. Wang, and D. Li. "A relaxation-enabled band model for network traffic signal optimization." 19th COTA International Conference of Transportation Professionals, 2019.

Chapter 11
Emerging technologies to enhance traffic signal coordination practices

Zong Tian[1] and Aobo Wang[1]

Traffic signal operations play a critical role in urban transportation management, which directly influence the overall system's travel delay, fuel consumption, emissions as well as motorists' perception to the quality of transportation service. In the United States, delays at signalized intersections contribute to an estimated 5 to 10% of overall travel delay which is equivalent to 295 million vehicle-hours annually on major roadways alone [1]. As traffic signal coordination allows vehicle platoons to travel along the main road with fewer stops, it has been recognized as one of the most effective and popular strategies for improving urban arterial operations.

Despite considerable efforts on promoting signal coordination through various initiatives over the past decades, the actual coordination operations remain far from ideal. A nationwide survey evaluated the signal coordination practices in the United States and only obtained a "C" grade, indicating that the nationwide signal coordination practices were performing at a barely adequate level [2].

Many agencies superficially suppose that traffic signal coordination would be easily realized because coordination development and maintenance are mostly a parametric calculation process without much infrastructure investment. However, achieving and maintaining the optimal coordination for a traffic signal system can be very time consuming and labor intensive, as reflected by the following aspects:

- Coordinating signals needs detailed traffic data such as 24-h volume counts and peak-period turning movement counts, which usually cost 500–1,000 dollars per intersection [3], while the total budget of signal timing is only about 3,000–5,000 dollars per intersection. Furthermore, the following field works such as timing plan implementation, fine tuning, and performance monitoring also involve largely manual processes that consume substantial project resources.
- Although many studies have been devoted toward signal coordination research, very few established tools or software applications can be directly

[1]Department of Civil and Environmental Engineering, College of Engineering, University of Nevada, Reno, USA

used for daily practices. Practitioners still heavily rely on their engineering judgment and spend a significant amount of time on each step of the procedure. Additionally, developing and maintaining high-quality signal coordination plans becomes a demanding job that only a few experienced practitioners are truly competent at.

- Measuring the performance of signal coordination is challenging through manual observations. Conducting such field observations may require a group of people to stand at intersections watching traffic flow or driving along arterials repeatedly before finalizing the timing plans. Few agencies properly assess the quality of signal coordination due to the lack of proper performance measurements, which may result in misleading information for elected officials and the general public.

The complex and costly procedure of the current practice presents major challenges in developing and regularly updating signal coordination timing plans. Transportation agencies have been seeking emerging technologies to address such challenges. New data sources, methodologies, and tools are expected to significantly improve the signal timing practice, particularly in the era of fast-growing traffic demands on urban streets.

This chapter specifically addresses how emerging technologies can be applied to enhance the current signal coordination practices. The discussions particularly focus on four critical aspects of signal timing, following the MODE acronym: (1) Managing a comprehensive signal timing database, (2) Optimizing without detailed traffic volumes, (3) Diagnosing erroneous signal timing, and (4) Evaluating the quality of coordination. It is noted that the signal timing practice referred to in this chapter is primarily based on conventional signal control and coordination systems, which will continue to be dominant in the foreseeable future. Adaptive signal control, although considered a major signal control technology, is not the focus of this chapter, as the market penetration of adaptive signal systems is currently limited [4], and their performances have not been widely acknowledged. In addition, this chapter mostly presents the practices in the United States where coordinated-actuated signal control is widely implemented. For readers who are not familiar with the terms used in this chapter, please refer to Chapter 7 of the Signal Timing Manual [5].

11.1 Coordination timing development and optimization

Developing a traffic signal coordination timing plan is a process of designing several key parameters such as cycle length, phase splits, phasing sequence, and offset so that optimal progression bands can be achieved to provide the best traffic progression along an arterial. Optimization of signal coordination is a classical traffic-engineering subject established almost 70 years ago. Based on the control objectives of reducing delays and progressing traffic, many methodologies [6,7] and tools [8–11] have been developed in the past decades. However, using these methods or tools to achieve the optimal signal coordination performance is still considered an arduous and difficult undertaking as it typically requires detailed traffic flow data, knowledge of optimization algorithms, familiarity with signal controller hardware, and experience of field operations [3].

In this section, two emerging techniques are introduced for improving the current timing development process by leveraging new data sources.

11.1.1 Developing cycle length and splits using controller event data

Determination of a common cycle length and phase splits for signal coordination is fundamental but involves a workload-intensive process. Manuals such as *Signal Timing Manual, Edition 2* [5] usually note that the practitioner should explicitly consider desired platoon travel time between intersections, traffic volumes, and turning movement counts when defining cycle length and splits, which necessitate laborious data-collection efforts.

Event data logged in modern traffic signal controllers are an emerging data source that can support a variety of signal coordination practices. Rather than storing the average values of data, individual time-stamped traffic events can be recorded, e.g., when a detector turns on or off or when a phase changes status (green, yellow, or red) are recorded by the controller or an external data collector. And based on the controller event data, it is feasible to obtain the historical traffic volume information, which can assist practitioners in finding adequate cycle length and phase splits for each signal. Using such controller-logged information overcomes one of the major shortcomings of the current practice which mainly relies on one-day peak hour turning traffic movement counts, since it is automatic and can cover a much longer time period.

Although some agencies in the United States have already been leveraging controller event data for developing cycle length and splits, this approach has not been widely seen in academic literature. Using a case in Reno, Nevada, this approach of using controller-logged information instead of conducting turning movements counts is presented below. The arterial is Sun Valley Boulevard, including a total of seven signals. These signals had been running free and new coordination plans needed to be developed.

- Step 1—Extract logged event data from field controllers or via the central management software. Care should be taken when selecting the time window in order to capture the typical traffic pattern. Data recorded on Tuesdays, Wednesdays, and Thursdays are usually recommended for developing weekday plans. Additionally, the practitioner should determine an appropriate amount of data to be collected. Enough data samples distributed among different weeks and months should be gathered. In this case, 220-h event data recorded between 7 a.m. and 6 p.m. on Tuesdays, Wednesdays, and Thursdays were obtained. Figure 11.1 exhibits a split history for a selected number of cycles extracted from Cubic ATMS, the central traffic signal management software used in Reno.
- Step 2—Process the extracted data. Theoretically, the cycle and split history can provide a good estimate of the proposed cycle length and splits for signal

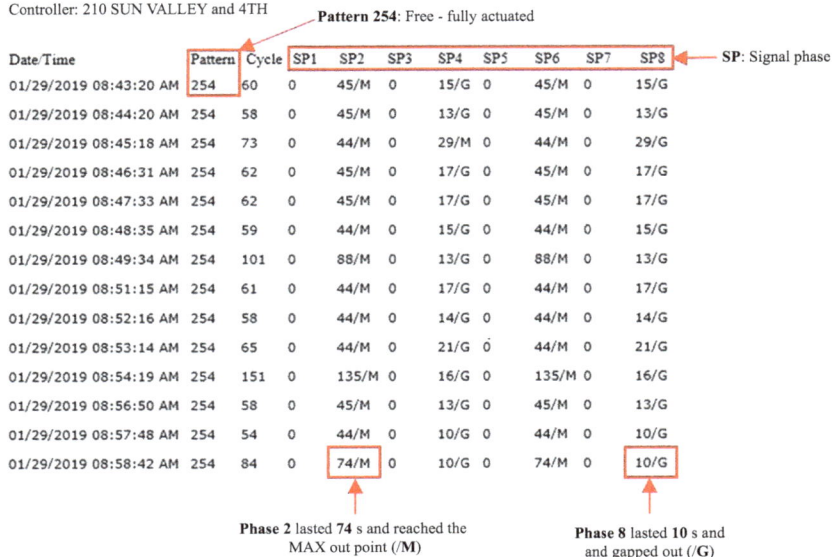

Figure 11.1 Split history extracted from Cubic ATMS

coordination. However, this calculation may be biased if the traffic demand is not fully served for some cycles, namely cycle failure or phase failure. Therefore, adjustments are necessary based on the frequency of cycle/phase failures. A simplified approach to approximating phase failures is to measure signal phase termination types. This involves comparing the proportion of gap-outs (which imply enough time to serve the demand) to max-outs or force-offs (which imply not enough time to serve the demand). Figure 11.2 shows an example of split termination logging.

The phases of each signal in the case study were all running in fully actuated control mode with Max Inhibit enabled to ensure enough green time was being provided to meet the demands.

- Step 3—Calculate cycle length and splits for coordination plans. A general approach is to set the arterial cycle length in proportion to flow levels at the most congested intersection on the arterial. Some practitioners also consider the geometric and operational traffic characteristics for attaining "Resonant Cycles" [12,13]. In the case study, the historical data of cycle lengths for each signal were obtained and the largest average cycle length among the total seven signals was 107 s. Hence, the final cycle length was determined to be 120 s considering the city requirements and the purpose of achieving robust progression over a range of traffic volumes. The splits for each intersection were proportionally assigned according to the extracted split history.

From the perspective of managing agencies, using controller event data can save a considerable amount of effort without having to collect turning

Emerging technologies to enhance traffic signal coordination practices 239

(a)

681	Record number		216				Sample period		15
682	Date		Wednesday, February 09, 2011				Mid-period time		17:07
683		Phase	1	2	3	4	5	6	7	8
684	Phase service		8	7	4	6	8	7	6	8
685	Ped service		0	1	0	1	0	1	0	2
686	Average green		18	44	20	39	20	42	8	32
687	Max-outs		0	0	0	0	1	0	0	0
688	Force-offs		4	6	0	4	2	5	2	8
689	Gap-outs		4	1	4	2	5	2	4	0

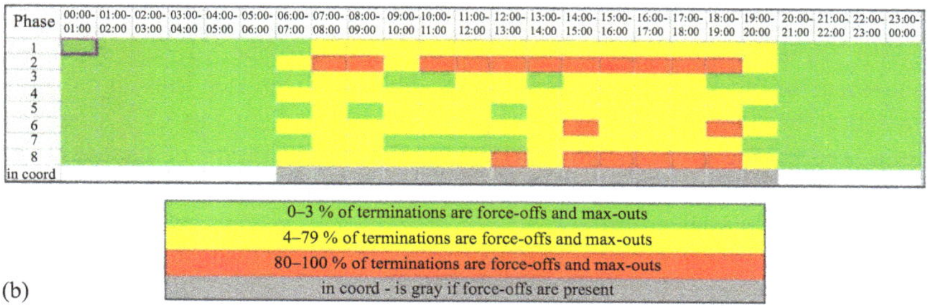

(b)

- 0–3 % of terminations are force-offs and max-outs
- 4–79 % of terminations are force-offs and max-outs
- 80–100 % of terminations are force-offs and max-outs
- in coord - is gray if force-offs are present

*Figure 11.2 Split termination logging (a) and visualization (b). (Source: Signal Timing Manual [5]). (a): The **Phase Service** row indicates the service times of each phase (1–8). The lower service times mean that the phase could have been skipped. The sum of **max-outs, force-offs**, and **gap-outs** equals to the phase service times. (b): Horizontal axis: 24-h time period; vertical axis: phases 1–8. The figure shows the hourly proportions of max-outs or force-offs for the eight phases. Red represents that 80–100% of phase terminations are max-outs or force-offs, yellow means the proportion ranges from 4 to 79%, and green indicates no more than 4% of terminations are max-outs or force-offs*

volumes. Moreover, phase splits estimated based on logged historical data better reflect a much wider traffic flow spectrum compared to manual traffic counts, which only represent traffic conditions of a short-period time window on specific days.

11.1.2 Optimizing offsets and phasing sequences based on travel-run trajectories

Theories regarding coordinated signal timing optimization primarily take two approaches: minimizing system-wide delay or maximizing main-street progression bandwidth. These two optimization approaches seem to function similarly; however, they are realized through different algorithms and require different data inputs.

Delay-based optimization is very sensitive to the accuracy of inputted traffic data. Delay-based optimization software requires very detailed traffic flow data and sophisticated modeling to generate feasible timing plans. On the other hand, bandwidth-based optimization can generate coordination timing plans with minimum required data inputs, mainly cycle and phase splits which could be obtained by the steps described in Section 8.1.1. Bandwidth-based optimization also best fits drivers' expectation by minimizing stops for the arterial through traffic; thus, it is a widely accepted approach as long as the phase splits for minor movements are sufficient for serving the demands. Additionally, the current practice for performance evaluation of signal coordination still primarily relies on arterial-based travel run studies.

With travel-run trajectory data readily available due to fast technology advancement in recent years, incorporating vehicle trajectories into signal timing optimization has been attempted [14]. Trajectories laid on top of a time-space diagram (see Figure 11.3) can not only provide a visualization of quality of progression, but also help engineers to identify locations where improvements can be achieved through signal retiming.

Currently, vehicle trajectory data can be automatically collected through various technologies. Paratransit and taxi vehicles equipped with GPS units generate a bulk of trajectory data. Transportation network companies such as Google, Uber, and vehicle manufactures also own a significant amount of trajectory data generated by navigation apps or built-in GPS. Advancement in connected and

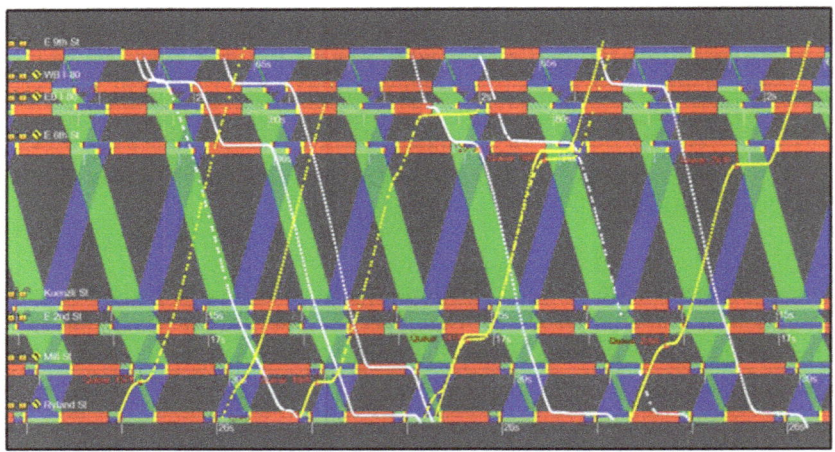

*Figure 11.3 Travel-run trajectories on a time space diagram. **Time Space Diagram:** Horizontal axis—time and cycle-length periods for reference; vertical axis—distance from a reference point. The trajectories of individual vehicles in motion are portrayed in this diagram by dashed lines and stationary vehicles are represented by horizontal lines. Two colored bands, one for each direction, show the time windows as the vehicles progress between intersections, shown as horizontal bars, without a stop*

autonomous vehicle technologies will further enhance the data source with additional information on traffic flow origin-destination. The application of trajectory-based bandwidth optimization is expected to be adopted by more agencies and practicing engineers in the future.

11.2 Field implementation and timing diagnosis

Field implementation is a process of transferring the designed timing parameters into the field controllers and ensuring the actual signal operations to achieve the anticipated arterial performance. During field implementation, a practitioner needs to continually diagnose signal timing issues and check whether the offsets or phasing sequences have been operating correctly and whether further adjustments are needed. Because many inherent controller settings could affect the outcome of a signal operation, diagnosing abnormal or erroneous signal timing is usually a very time-consuming task.

As a dominant signal control mode in the United States, semiactuated control during the period of signal coordination presents more challenges for quickly diagnosing signal timing problems. For example, if "phase early return" and "preemption" are not identified, the practitioner could be misled and draw erroneous conclusions in timing implementation. Phase early return is a phenomenon when minor phases gap out due to lack of demand, resulting in earlier return to the coordinated phase(s) than what has been programmed. Such a phase early return could be mistakenly interpreted as a wrong offset. Preemption events caused by emergency vehicles can disrupt signal coordination, but a practitioner may not immediately realize the event if the preemption vehicle is no longer at sight and the signal is still under transition. Another common case is the controller clock difference, due to either time drift or different set up when an arterial traverses different jurisdictions.

Although the abovementioned timing diagnosis can be accomplished by checking with local controllers or central management software, it becomes infeasible if the practitioner is doing observation while driving through the arterial or has no access to the central system. Hence, mobile-device-based tools could greatly help the timing diagnosis process. Figure 11.4 shows an example of an iOS mobile tool where the videos and real-time trajectories can be visualized in real time and recorded for later viewing during travel runs.

By comparing the differences between the signal indications shown in the video and on the time-space diagram, practitioners can clearly identify potential issues of signal timing. Figure 11.5 illustrates some cases where the actual signal operation did not match the coordination plan. The coordinated phases of three signals in the yellow box started early due to side street gapping out, which did not indicate any erroneous timing. This was due to normal semiactuated control. However, the signal in the red box clearly indicated an erroneous timing, a signal transition in this case caused by an emergency vehicle preemption, since the vehicle stopped at red when it was supposed to be green. Quickly identifying the

242 *Traffic information and control*

Figure 11.4   Synchronized time-space diagram, field-recorded videos, and vehicle trajectories. The left figure shows the video record during one field travel run and the right figure shows the trajectory on a time-space diagram recorded during the same travel run. The video and trajectory are synchronized automatically, which currently show the probe vehicle passing through an intersection

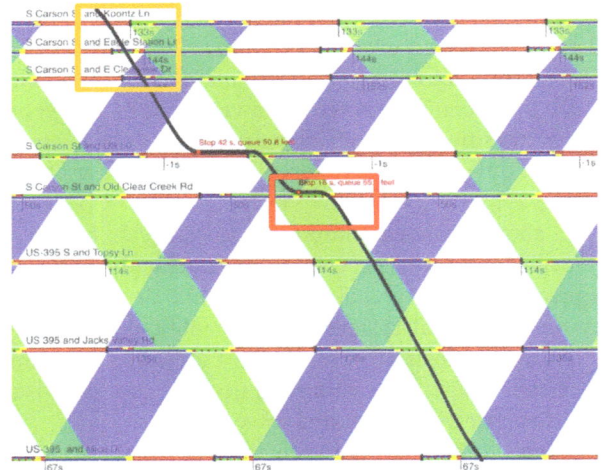

Figure 11.5   Impacts of actuations on signal operations

amount of phase early returns can greatly help practitioners fine-tune the plan settings, such as shifting offsets to accommodate early returns. Knowing whether a signal is in transition and how long the transition takes can help practitioners adjust transition parameters to reduce the potential transition influences.

Two abnormal operations are shown in Figure 11.5 (in colored boxes). In the upper yellow box, the trajectory indicates that the vehicle passed the intersection when the signal was supposed to be red. This could happen because the non-coordinated phases ended early and then the coordination phases started early, which can be verified by the video recorded concurrently. In the lower red box, the trajectory shows that the vehicle stopped when the signal was supposed to be green, which indicates that a signal transition occurred. This also can be verified through playing back the recorded videos.

11.3 Performance measures for assessing the quality of signal coordination

After field implementation of signal timing, two major tasks are generally involved: (1) conducting a before-and-after study (if desired or required) and (2) frequently monitoring and maintaining the timing. Establishing a set of performance measures is thus very important for performing these tasks. Performance measures are critical for signal coordination since they provide an opportunity to confirm whether the coordination objectives are met, and to best inform elected officials and the public about the success of a project. Performance measures can also be used to establish a benchmark for allocating future funds for locations with the highest benefit-cost ratio.

The conventional performance measures for signal coordination are based on analyzing travel time, stops, delay, and queuing. While some performance measures may be generated from software models and simulation, the most valuable information still comes from field data. Consequently, agencies need to assign specific manpower in order to collect sufficient and high-quality performance data. However, without a well-established performance measure methodology, this labor-intensive work usually proves to be in vain. Practitioners may ill-evaluate the quality of signal coordination by simply measuring travel time or number of stops while neglecting other important factors such as traffic volume, signal spacing, and pedestrian activities, which naturally affect travel time and stops regardless how well the timing has been done.

Many attempts have been made to develop performance measures for signal coordination in recent years. High-resolution controller data and GPS trajectories are two main data sources for establishing signal performance measures.

Implementation of Automated Traffic Signal Performance Measures (ATSPMs) is a major initiative led by the Federal Highway Administration (FHWA) using the high-resolution data-logging capabilities equipped with modern traffic signal controllers. The functions provide several data-analysis techniques for evaluating various aspects of a signal control system, including communication, detection, timing, and coordination. Agency professionals can use the information generated by ATSPMs to proactively identify and improve a signal system's deficiencies.

The FHWA has been promoting ATSPMs since the fourth round of the Every Day Counts (EDC-4) initiative [15]. According to the information published by the

FHWA, approximately 26 transportation agencies at both state and local levels have been involved in implementing ATSPMs. Recently, an open-source software package was developed as part of a Transportation Pooled Fund study, "Traffic Signal Systems Operations and Management" [16].

The Signal Performance Metrics (SPM) managed by Freeway and Arterial System of Transportation (FAST) in Las Vegas, NV, is one of the examples of ATSPMs. The SPM can show real-time and historical performance at signalized intersections [17]. Data visualization functions are also provided with several diagram layouts [18], which can be displayed on a website. Through such website interfaces, engineers and the general public can easily access the signal performance metrics.

Among other types of diagrams within FAST, there is one specifically addressing signal coordination called the "Purdue Coordination Diagram" (PCD). The PCD is a useful tool that enables practitioners to quickly visualize how well a coordinated signal is operating on a particular approach through quantified and graphical measures. As shown in Figure 11.7, the PCD combines the detection events (black dots) with the signal phase status, providing both visual and quantitative figures of proportion of vehicles arriving during red and green intervals. This chart plots time of day on the horizontal axis and time in cycle along the vertical axis. Each vertical strip in the data shows a single cycle. Begin of Green (BOG) time period is represented by the green line and End of Green (EOG) time period by the red line. The EOG line also shows the cycle length. The shaded region between BOG and EOG shows when the signal is green, and each dot represents a vehicle arrival as measured by a setback detector. By one chart, the PCD can illustrate the quality of progression for a coordinated approach over a 24-h period.

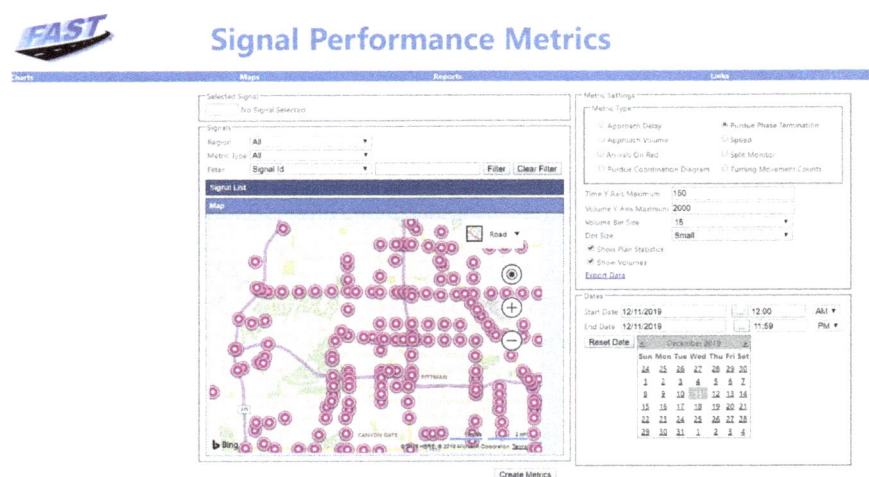

Figure 11.6   The web interface of FAST's signal performance metrics

Emerging technologies to enhance traffic signal coordination practices 245

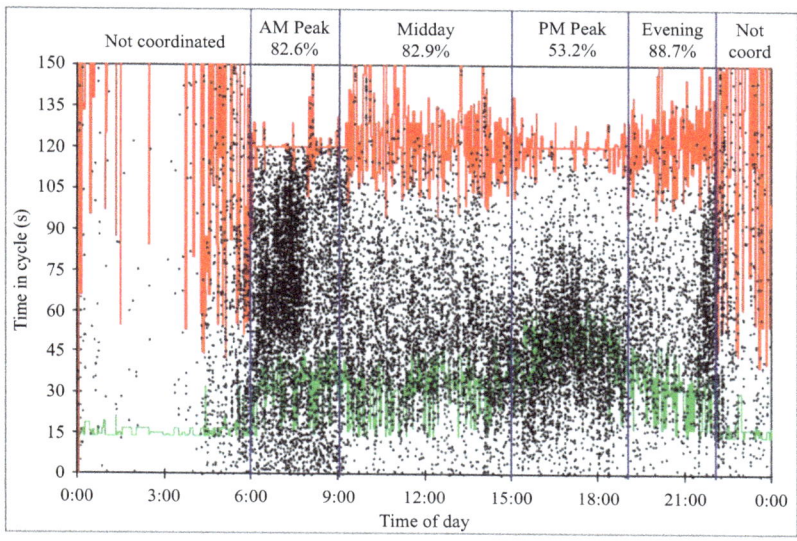

Figure 11.7   An example of the Purdue Coordination Diagram (Source: Signal Timing Manual)

Information available in a PCD includes the following:

- The green and red lines show the start and end of green, respectively, during each cycle. Vehicle arrivals that occur above the green line represent arrivals during a green indication, while those occurring below the green line are arrivals during a red indication. The practitioner can intuitively judge the quality of signal coordination based on a comparison between the dot densities above and below the green line.

- The PCD can also provide quantitative performance measures. For example, the percentage of vehicles arriving on green is shown for each coordination plan at the top of Figure 11.7. The total number of vehicles arriving on green is a useful metric for quantifying the quality of progression, especially when expressed as a percentage. A higher percentage of vehicles arriving on green typically represents a better quality of signal coordination. In addition, the percentage of vehicles arriving on green divided by the green-to-cycle ratio yields a new metric called "platoon ratio," which accounts for the fact that the greater the percentage of time that the signal is green, the more likely it is for vehicles to arrive on green. The metric therefore tends to reward a high percentage of vehicles arriving on green that is achieved with smaller greens and scales down the results of a high percentage of vehicles arriving on green that is achieved with extremely long arterial green times. A platoon ratio of 1.0 represents random arrivals; higher numbers represent good progression, while lower numbers represent poor progression.

As can be seen, ATSPMs is a powerful tool for generating performance measures for signal coordination. However, the massive data generated by ATSPMs may sometimes confuse policy makers and the public, as indicated by one of the comments from the current users of ATSPMs—"Purdue Coordination Diagram shows too many dots" [19]. Additionally, the PCD only contains information for one specific movement at an intersection, in which the measures such as Platoon Ratio mainly evaluate the progression between two intersections (or called link-based), but not for the whole arterial. ATSPMs also require advanced detection infrastructure and other add-in facilities, which makes the system less appealing to small agencies.

Besides high-resolution controller data used in ATSPMs, GPS trajectories are an emerging data source and are attracting attentions to their advantages for developing performance measures for signal coordination. Trajectory data are very valuable for measuring the quality of signal timing as they can reveal detailed characteristics of vehicle motion on arterials. And trajectory data can be conveniently collected through smart mobile devices, which requires little investment from the managing agency. In the future, connected vehicle technologies could make trajectory data widely obtainable at much reduced costs to collect.

The Orange County Transportation Authority (OCTA), in collaboration with the California Department of Transportation (Caltrans), and the local agencies within the county initiated a Signal Master Plan for the countywide synchronization endeavor in 2009. The Signal Master Plan has identified a new parameter to gauge the performance of signalized arterials, which is called Corridor Synchronization Performance Index (CSPI) [20,21].

The CSPI is a score-based method which evaluates the performance of signal coordination as per (1) average speed, with the highest possible score of 36, (2) the ratio of number of greens verses reds through signalized intersections, with the highest possible score of 40, and (3) the average number of stops per mile, with the highest possible score of 33 (see Figure 11.8, for the scoring scale). Based on the summation of the three scores, the general performance level can be obtained.

The limitations of CSPI should be noted. It does not consider the factors other than signal coordination, such as traffic volume level which can significantly influence travel speed and number of stops. Furthermore, the score does not take into account the aspect of drivers' perception, which is one of the most important goals in signal coordination improving driver experiences.

Compared with the CSPI methodology, the research team at the University of Nevada, Reno (UNR), developed an enhanced performance measure framework. This framework was based on a similar concept of the OCTA's approach but with several enhancements as described below.

- The impact of traffic volume level was considered by introducing a variable, Intersection Classification, into the calculation. Heavier traffic volume on the

CSPI Score	Signal synchronization description	Level
>=80	**Very good progression**—traveling through signalized intersections with minimal stops and favorable travel speeds.	Tier 1
70-80	**Good progression**—traveling through signalized intersections with few stops and good travel speeds.	Tier 2
60-70	**Fair progression**—traveling through signalized intersections with moderate stops and fair travel speeds.	Tier 3
50-60	**Limited progression***—traveling through signalized intersections with moderately high stops and slower travel speeds.	Tier 4
<50	**Very limited progression***—traveling through signalized intersections with frequent stops and slow travel speeds.	Tier 5

Speed (mph)	Score	Green/Red	Score	Stops per Mile	Score
34	36	5.0	40	0.7	33
32	33	4.5	36	0.9	31
30	30	4.0	32	1.1	29
28	27	3.5	28	1.3	27
26	24	3.0	24	1.5	25
24	21	2.5	20	1.7	23
22	18	2.0	16	1.9	21
20	15	1.5	12	2.1	19
15	8	1.0	8	2.3	17

Figure 11.8 CSPI scoring method and performance criteria [20,21]

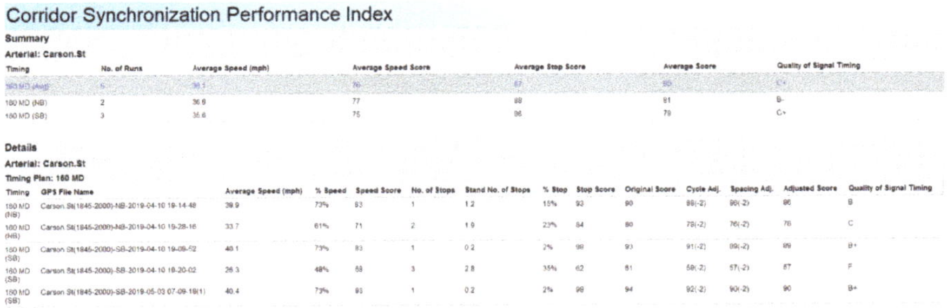

Corridor Synchronization Performance Index

Summary

Arterial: Carson.St

Timing	No. of Runs	Average Speed (mph)	Average Speed Score	Average Stop Score	Average Score	Quality of Signal Timing
160 MD (Avg)	5	36.1	76	[illegible]	80	C+
180 MD (NB)	2	36.9	77	88	81	B-
160 MD (SB)	3	35.6	75	96	79	C+

Details

Arterial: Carson.St

Timing Plan: 160 MD

Timing	GPS File Name	Average Speed (mph)	% Speed	Speed Score	No. of Stops	Stand No. of Stops	% Stop	Stop Score	Original Score	Cycle Adj.	Spacing Adj.	Adjusted Score	Quality of Signal Timing
160 MD (NB)	Carson St(1845-2000)-NB-2019-04-10 19-14-48	39.9	73%	93	1	1.2	15%	93	90	88(-2)	86(-2)	86	B
160 MD (NB)	Carson St(1845-2000)-NB-2019-04-10 19-28-16	33.7	61%	71	2	1.9	23%	84	80	78(-2)	76(-2)	76	C
160 MD (SB)	Carson St(1845-2000)-SB-2019-04-10 19-09-52	40.1	73%	93	1	0.2	2%	98	93	91(-2)	89(-2)	89	B+
160 MD (SB)	Carson St(1845-2000)-SB-2019-04-10 19-20-02	26.3	48%	68	3	2.8	35%	62	61	59(-2)	57(-2)	57	F
160 MD (SB)	Carson St(1845-2000)-SB-2019-05-03 07-09-19(1)	40.4	73%	93	1	0.2	2%	98	94	92(-2)	90(-2)	90	B+

Figure 11.9 An example of the UNR's performance measure calculation

main and minor streets would result in a higher Intersection Classification, which acted as an adjustment factor to rectify the evaluation results.

- Driver perception was considered by introducing a penalty for short-distance stops and normalization of the number of stops based on stop duration.
- Additional adjustments were made on the evaluation results in accordance with the specific geometric and operational conditions.

The performance level of signal coordination, or called Quality of Signal timing, was classified into Levels A, B, C, D, and F based on the proposed framework. Figure 11.9 illustrates an example where five trajectories (two in the northbound and three in the southbound) are shown along with the detailed performance measure calculations. In this example, the overall quality level of this timing plan is a C+.

11.4 Signal timing documentation

Well-documented signal coordination plans are vital to the long-term success of signal coordination practices. It is highly desirable to store the entire database of timing plans electronically and in a single data file so that practitioners can easily access, view, and modify the timing plans whenever necessary. Although a central management software system is generally available in most jurisdictions, such a system mainly functions as a timing data repository, monitoring, and uploading/downloading timing plans. Central management software systems lack optimization features, and it is generally difficult to select and view a coordination plan such as a time-space diagram. Furthermore, the central management system is generally not accessible by engineers outside of the jurisdiction.

Figure 11.10 shows an example of such a coordination timing data-management tool which permits an entire signal timing database for a city to be archived and edited through manipulating just a single file. The practitioner can easily access the plan information for any signals and time-space diagrams for user definable arterial segments. This data file can be stored on a mobile device or on a cloud server and is accessible by all of the potential users. As a result, the

Emerging technologies to enhance traffic signal coordination practices 249

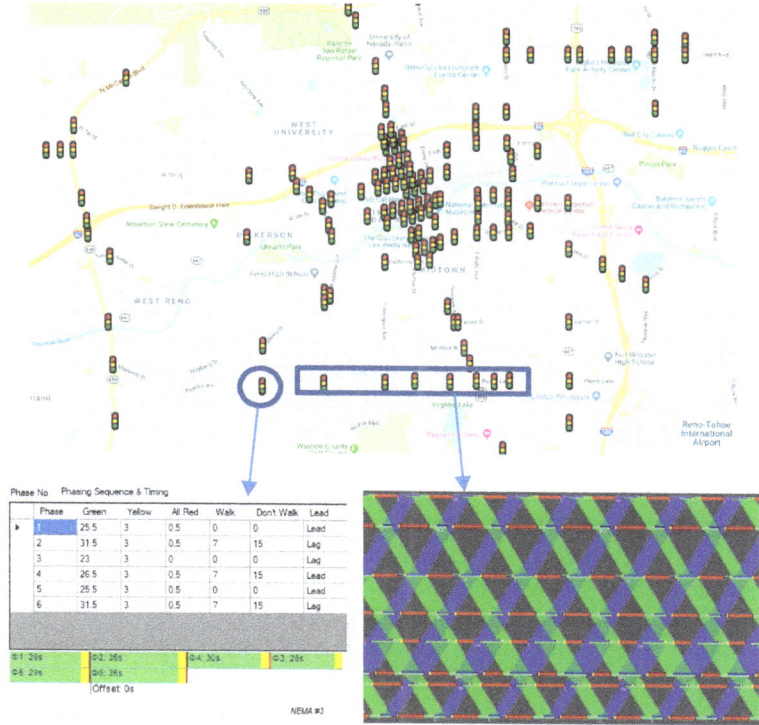

Figure 11.10   Signal coordination timing documentation for the city of Reno, NV

practitioner no longer needs to update, match, and verify the documented information among redundant forms and models.

Using such timing database, practitioners are able to view and edit through clicking the signal icons located at the real sites on a digital map. The software provides detailed timing information and time-space diagrams.

11.5   Summary

Emerging technologies and tools play a vital role for advancing the state of practice on signal timing and coordination. The discussions in this chapter primarily focused on addressing the four critical aspects and challenges facing management agencies and signal timing engineers: timing management and documentation, optimization without traffic volumes, trajectory-based timing diagnosis, and performance evaluation. Such emerging technologies and techniques help practitioners economize the laborious processes of the traditional signal timing process. It is anticipated that new technologies will continue to emerge and hold promises for significantly changing how traffic signal control issues are addressed. For example, with connected-vehicle technologies gaining increased market penetration, traffic flows would be predictable and guidable, which allow quicker and more accurate information exchange and signal control response.

Although emerging technologies are promising, practitioners should not simply deem that such technologies would automatically achieve better results. A careful investigation before applying any technologies and a comprehensive performance evaluation after the implementation are always necessary.

References

[1] Congestion Reduction Toolbox. U.S. Department of Transportation Federal Highway Administration. Accessible at www.fhwa.dot.gov/congestion/toolbox
[2] 2012 National Traffic Signal Report Card Technical Report. National Transportation Operations Coalition. Institute of Traffic Engineers. 2012.
[3] Henry, R. David. Signal timing on a shoestring. No. FHWA-HOP-07–006. 2005.
[4] About Adaptive Traffic Control and the map of ATCSs in the US. Accessible at http://latom.eng.fau.edu/research-reports/
[5] Urbanik, T., Tanaka, A., Lozner, B., *et al.* NCHRP Report 812: Traffic Signal Timing Manual, edition 2. Transportation Research Board of the National Academies, Washington D. C., 2015.
[6] Webster, F. V. Traffic Signal Settings. Road Research Technical Paper No. 39. H.M. Stationary Office, London, 1958
[7] Little, J. D. C., Kelson, M. D., and Gartner, N. H. (1981). MAXBAND: A program for setting signals on arterials and triangular networks. Transportation Research Record, 759, 40–46.
[8] User Manual for Synchro Studio, version 11. 2019.
[9] Stamatiadis, C., and Gartner, N. H. (1996). MULTIBAND-96: A program for variable-bandwidth progression optimization of multiarterial traffic networks. Transportation Research Record, 1554(1), 9–17.
[10] Wallace, C. E., Courage, K. G., Reaves, D. P., Schoene, G. W., and Euler, G. W. (1984). TRANSYT-7F user's manual (No. UF-TRC-U32 FP-06/07).
[11] Chaudhary, N. A., and Chi-Leung, C. New PASSER Program for Timing Signalized Arterials. Accessible at https://static.tti.tamu.edu/tti.tamu.edu/documents/4020-S.pdf
[12] Shelby, S. G., Bullock, D. M., and Gettman, D. (2005). Resonant cycles in traffic signal control. Transportation Research Record, 1925(1), 215–226.
[13] Ladrón de Guevara, F., Hickman, M., and Head, L. (2015). Resonant Cycles Under Various Intersection Spacing, Speeds, and Traffic Signal Operational Treatments (No. 15-4989).
[14] Yao, J., Tan, C., and Tang, K. (2019). An optimization model for arterial coordination control based on sampled vehicle trajectories: The STREAM model. Transportation Research Part C: Emerging Technologies, 109, 211–232.

[15] Lawrence Paulson, S. Managing Traffic Flow through Signal Timing. Federal Highway Administration Research and Technology. 2002.
[16] Traffic Signal Systems Operations and Management. Transportation Pooled Fund Program. 2017. Available at http://www.pooledfund.org/Details/Study/487
[17] Hand-out of Automated Traffic Signal Performance Measures. AASHTO Innovation Initiative. 2014. Available at http://aii.transportation.org/Documents/ATSPMs/atspms-handout-press.pdf
[18] Automated Traffic Signal Performance Measures Website. Utah Department of Transportation. Available at http://udottraffic.utah.gov/atspm/
[19] Patel, S. SPM in Las Vegas. Automated Traffic Signal Performance Measures Workshop. 2016. http://dx.doi.org/10.5703/1288284316028.
[20] Kulkarni, A. Corridor Operational Performance Report. Orange County Transportation Authority. 2015. Available at http://octec.net/downloads/OCTA.pdf
[21] 2013 Corridor Performance Report. Available at http://www.fehrandpeers.com/wp-content/uploads/2016/01/2013-Corridor-Performance-Report-Final_Main-Report.pdf

Chapter 12

Traffic signal control for short-distance intersections with dynamic reversible lanes

Haipeng Shao[1], Siyuan Song[2], Juan Yin[1] and Hui Jin[3]

Short-distance intersections are defined as intersection pairs with short link and strongly interact with each other, and they are often challenged with overflow due to the limited queuing length. In order to solve this kind of traffic bottleneck, a coordinated signal control method for short-distance intersections with dynamic reversible lanes on the link is proposed in this paper. The direction of the dynamic reversible lanes is switched in coordination to the signal phase in a signal cycle, which can increase the storage space of the link and enhance capacity in both directions to prevent overflow. Based on the signal sequence under counter-clockwise split phasing scheme, signal timing is modeled to avoid spillover on the link with clearance time prepared for the dynamic reversible lanes, which is calibrated and validated with VISSIM, a microscopic simulation modeling tool based on time interval and driving behavior, to significantly avoid spillover and reduce the number of vehicle stops. It is concluded that the coordinated control method for short-distance intersections with dynamic reversible lane performs better with a shorter link, fewer lanes, higher saturation or high left-turn ratio into the link. This research may provide insights on enhancing road capacity for short distance intersections.

12.1 Introduction

In densely developed urban area, intersections with short spacing are very common. These intersections interact with each other and result in significant reduction of road capacity as well as causing bottlenecks. Thus, it is critical to optimize the channelization and signal timing of short-distance intersections to better accommodate vehicles and to prevent spillover.

In recent years, scholars have carried out a series of researches to enhance the efficiency of short-distance intersections with the establishment of collaborative signal timing and lane assignment. But there were still shortcomings in the existing signal timing

[1]College of Transportation Engineering, Chang'an University, Xi'an, China
[2]Construction Bureau of Jiangsu Foho Hi-tech Zone, Suzhou, China
[3]School of Rail Transportation, Soochow University, Suzhou, China

and lane layout strategies, which were mainly for the intersections with balanced two-way traffic flow. For the signalized paired intersection system, some researchers began to study the traffic flow characteristics of short-link intersections, finding the spillover of downstream queues can greatly reduce the effective green time and passing cars in short-distance intersections [1,2]. To minimize delay, maximize the number of passing vehicles, and increase capacity between two close intersections, some researchers considered the damaging effects of downstream queues and established optimized models to alleviate delay. Yang *et al.* [3] introduced probability to the control of short-link intersections and proposed two different control strategies, i.e., double circle and responsive judgment, to enhance intersection efficiency. Other scholars indicated that the traffic flow properties, signal control parameters, and link geometry exerted appreciable influence to reduce delay [4], and chose signal offsets and green splits to alleviate queue length [5,6]. Because of changing traffic volumes, the offsets between intersections need to be dynamically adjusted based on each cycle [7]. Liu and Wang [8] included the additional bandwidth for a part of the connected intersections to maximize the integrated green wave bandwidth. With the development of artificial intelligence, some new technologies to solve signal coordination between two intersections were also proposed including fuzzy neural networks and agent technology [9–11]. However, these researches mainly aimed at designing the best signal timing plans, but not considering the traffic design lane layout plan on the link between two close intersections, so the optimized effects are not obvious.

Facing the contradiction between the increasing traffic demand and the limited road resources, more and more scholars emphasize the importance of traffic design to finely serve local traffic demand. Aiming at the directionally unbalanced traffic flows during peak periods, dynamic reversible lanes have been implemented in many cities [12–14], which is beneficial in reducing delay and increasing throughput [15]. Therefore, considering the signalized intersections with changing directional ratio of traffic demand, some scholars also proposed to establish a reversible approach lane [16,17]. And the dynamic reversible lanes have been deployed in coordination to signal timing at intersections with large left-turn traffic flow [18–20] and two intersections of diamond interchanges now [18]. Ensuring the safe operation of reversible lanes, there are some guidelines and standards for the establishment of the reversible lanes [21]. Moreover, the flow direction changing should be designed based on signal timing, coil detectors, and overhead changeable lane-use signals [22]. In general, these delicate traffic designs integrated with signal timing may better accommodate time-varying traffic demand and promote the enhancement of intersection capacity.

In this chapter, the dynamic reversible lane is adopted on the internal link of the short-distance intersections, which switches the lane direction with signal phase. An optimization method is proposed to coordinate dynamic reversible lane design and signal timing to improve traffic efficiency and to increase the storage space of the internal link.

12.2 Application of dynamic reversible lane

The dynamic reversible lane is set on the center lane of the in-between link of short-distance intersections, which changes direction twice in one signal cycle to

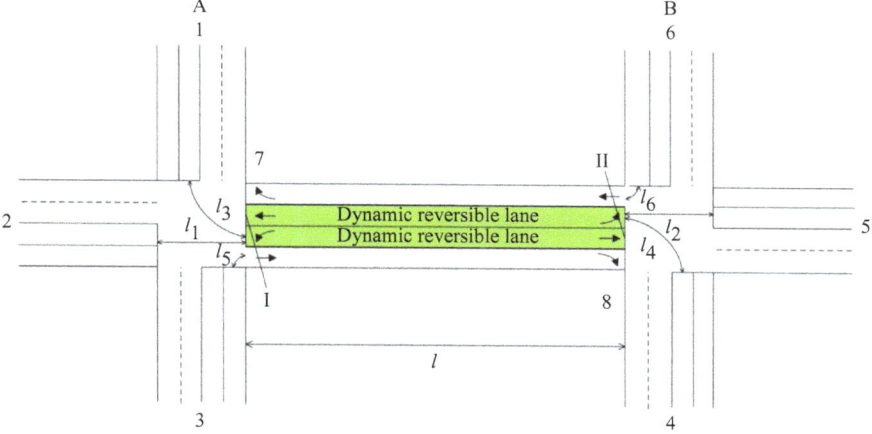

Figure 12.1 Two short-distance intersections with dynamic reversible lanes

increase the approach capacity in both directions. Because it takes time to clear the dynamic reversible lane on each direction switch, this method is only suitable for short-distance intersections with unbalanced directional traffic demand to serve the dominant side. Specifically, the proposed method is recommended to the situations as below:

1. The length of the link between the two intersections is less than 200 m.
2. Split phasing is adopted at the two signalized intersections.

Figure 12.1 shows two short-distance intersections on a four-lane two-way arterial road. The two center lanes are set as dynamic reversible to switch direction periodically with the signal phase with a variable sign, while the rest lanes still keep the original direction. Section I is the stop line of approach 7; section II is the stop line of approach 8; l is the length of the link; l_1 is the distance between the stop line of approach 2 and the section I; l_2 is the distance between the stop line of approach 5 and the section II; l_3 is the distance that the left-turn vehicles travel between the stop line of approach 1 to section I; l_4 is the distance that the left-turn vehicles travel between the stop line of approach 4 to section II; l_5 is the distance that the right-turn vehicles travel between the stop line of approach 3 to section I; and l_6 is the distance that the right-turn vehicles travel between the stop line of approach 6 to section II.

12.3 Model of signal timing

12.3.1 Signal phase and sequence

Vehicles entering the internal link between two intersections can move forward on the arterial or left/right-turning onto the minor-street. The green time of the three

256 *Traffic information and control*

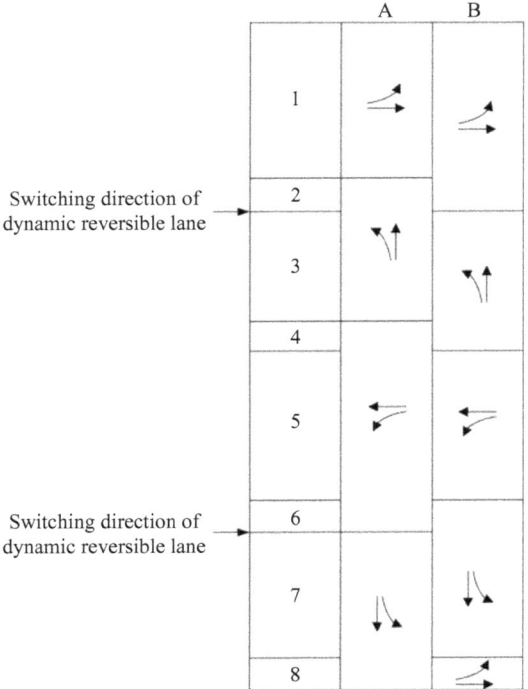

Figure 12.2   Signal phase sequence

movements should be scheduled within half a cycle. Compared with the clockwise split-phasing scheme, the counter-clockwise split-phasing scheme can reduce traffic conflicts during phase switching. Therefore, the four-timing scheme of counter-clockwise split phasing is adopted at the two intersections, with a total of 8 signal stages when combining the two intersections A and B. The signal control plan is shown in Figure 12.2. The dynamic reversible lanes are west to east in signal phases 7, 8, 1 and 2, and east to west in the signal phases 2, 3, 4, and 5. Table 12.1 shows the dissipate process of traffic movements at each signal phase.

Right-turning on each approach is controlled as: that on the approach 3 is allowed at the signal stages 7, 8, and 1, and that on the approach 6 at the signal stages 3, 4, and 5. Note that other right-turn movements are not controlled. Figure 12.3 shows the detailed signal control scheme at the short-distance intersections with dynamic reversible lane.

With respect to safety of pedestrians and non-motor vehicles, conflict with vehicles under the proposed signal scheme with dynamic lanes should be well addressed. Thus, when releasing the through and left-turning traffic on one approach, the pedestrians and non-motor vehicles cross the side street at the same time, scheme of which is shown in Figure 12.4.

Table 12.1 Traffic movements and signal stages

Stage	Traffic movements
1	The eastbound movement on the main road releases.
2	The green time of approach 2 ends, and the movement of approach 3 releases on intersection A; The green is on for Approach 8 at intersection B to clear vehicles on the link; The direction of the reversible lanes is switched on the end of this stage.
3	The approach 3 keeps green light on intersection A; The green time of approach 8 ends, and the movement of approach 4 releases at intersection B; The left-turning vehicles on approach 4 enter and queue on the link.
4	The green time of approach 3 ends, and the movement on approach 7 moves forward; The approach 4 keeps green on intersection B.
5	The approach 7 keeps green on intersection A; The green time of approach 4 ends, and the movement of approach 5 releases on intersection B.
6	The green time of approach 5 ends, and the movement of approach 6 releases on intersection B; The approach 7 keeps green on intersection A to empty vehicle on the link; The traffic direction of the reversible lanes is switched on the end of signal stage 6.
7	The green time of approach 7 ends, and the movement of approach 1 releases on intersection A; The left-turn vehicles of approach 1 enter and queue on the link; The approach 6 keeps green on intersection B.
8	The approach 1 keeps green on intersection A; The green time of approach 6 ends and the movement of approach 8 releases on intersection B.

12.3.2 Signal timing model

Figure 12.5 shows the signal timing process for two short-distance intersections with dynamic reversible lanes.

In the signal timing of two adjacent intersections, it is necessary to ensure that the queuing length on the link is well controlled to avoid overflow. Taking the signal stage 3 as an example, two cases are considered:

- Case 1: when there are only left-turn and straight lanes on the link, the following conditions must be met to avoid spillover:

$$m_{71} \times \frac{l}{d} \geq n_{71} \qquad (12.1)$$

$$m_{72} \times \frac{l}{d} \geq n_{72} \qquad (12.2)$$

for left-turning and straight lanes, respectively.

- Case 2: When there is a straight-left shared lane on the link, the above constraints are replaced with:

$$m_7 \times \frac{l}{d} \geq n_{71} + n_{72} \qquad (12.3)$$

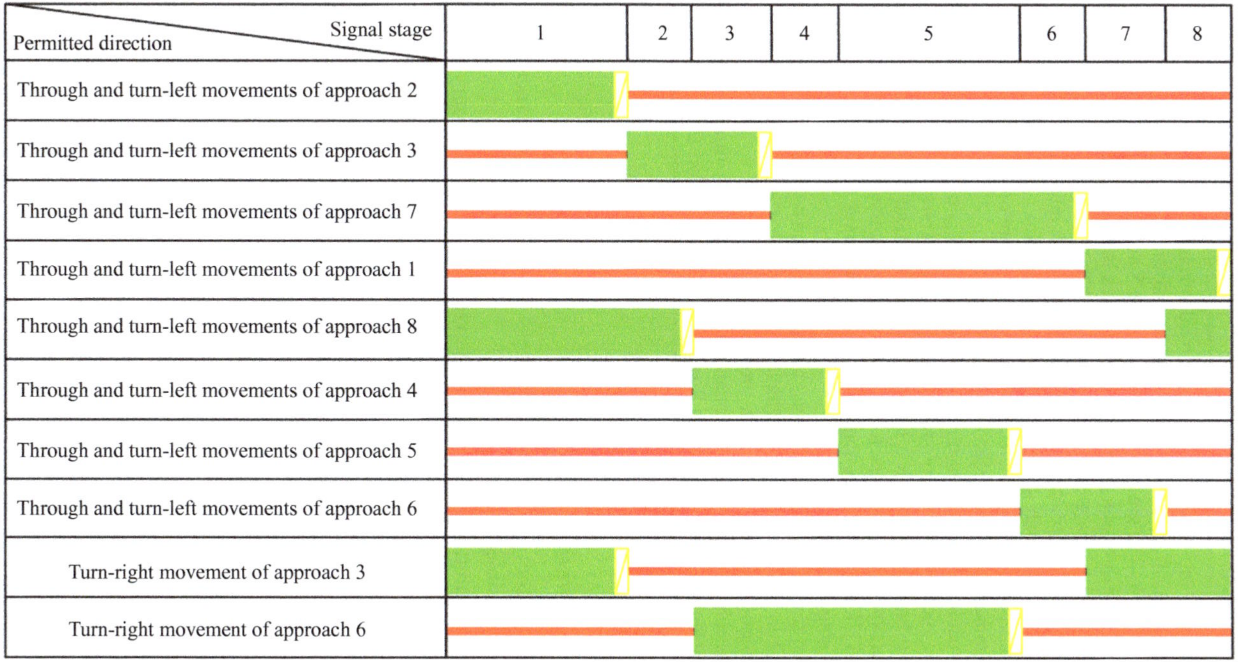

Signal stage Permitted direction	1	2	3	4	5	6	7	8
Through and turn-left movements of approach 2								
Through and turn-left movements of approach 3								
Through and turn-left movements of approach 7								
Through and turn-left movements of approach 1								
Through and turn-left movements of approach 8								
Through and turn-left movements of approach 4								
Through and turn-left movements of approach 5								
Through and turn-left movements of approach 6								
Turn-right movement of approach 3								
Turn-right movement of approach 6								

Figure 12.3 Signal control scheme

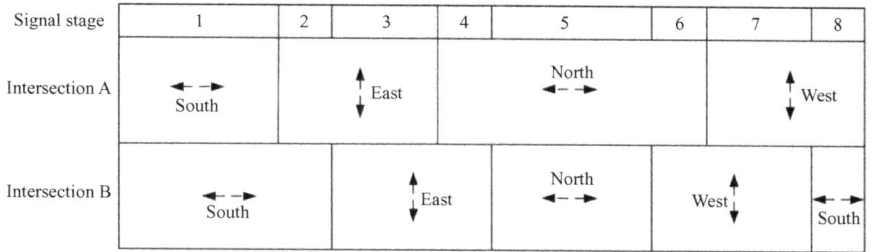

Figure 12.4 Phase scheme of pedestrian and non-motor vehicle crossing

as queues on the lanes are of the same length. Variable m_{iw} is the number of lanes serving w-movement on approach i, with $w = 1$ for left-turn movement, $w = 2$ for through movement, $w = 3$ for right-turn movement ; m_i is the total number of lanes on approach i; $\bar{d}$ is the average headway for queues; and n_{iw} is the maximum queue length of the w-movement vehicles on approach i on the link, calculation of which is given in the following four cases:

- Case 1: if $t_{41} \geq G_4 - G_3 - \frac{l_4}{\bar{v}_1}$ and $t_{63} \geq G_4 - G_3 - \frac{l_6}{\bar{v}_3}$, then:

$$n_{71} = S_{41}k_{411}\left(G_4 - G_3 - \frac{l_4}{\bar{v}_1}\right) + S_{63}k_{631}\left(G_4 - G_3 - \frac{l_6}{\bar{v}_3}\right) \quad (12.4)$$

$$n_{72} = S_{41}k_{412}\left(G_4 - G_3 - \frac{l_4}{\bar{v}_1}\right) + S_{63}k_{632}\left(G_4 - G_3 - \frac{l_6}{\bar{v}_3}\right) \quad (12.5)$$

- Case 2: if $t_{41} < G_4 - G_3 - \frac{l_4}{\bar{v}_1}$ and $t_{63} \geq G_4 - G_3 - \frac{l_6}{\bar{v}_3}$, then:

$$n_{71} = S_{41}k_{411}t_{41} + q_{41}k_{411}\left(G_4 - G_3 - \frac{l_4}{\bar{v}_1} - t_{41}\right)$$

$$+ S_{63}k_{631}\left(G_4 - G_3 - \frac{l_6}{\bar{v}_3}\right) \quad (12.6)$$

$$n_{72} = S_{41}k_{412}t_{41} + q_{41}k_{412}\left(G_4 - G_3 - \frac{l_4}{\bar{v}_1} - t_{41}\right)$$

$$+ S_{63}k_{632}\left(G_4 - G_3 - \frac{l_6}{\bar{v}_3}\right) \quad (12.7)$$

- Case 3: if $t_{41} \geq G_4 - G_3 - \frac{l_4}{\bar{v}_1}$ and $t_{63} < G_4 - G_3 - \frac{l_6}{\bar{v}_3}$, then:

$$n_{71} = S_{41}k_{411}\left(G_4 - G_3 - \frac{l_4}{\bar{v}_1}\right) + S_{63}k_{631}t_{63}$$

$$+ q_{63}k_{631}\left(G_4 - G_3 - \frac{l_6}{\bar{v}_3} - t_{63}\right) \quad (12.8)$$

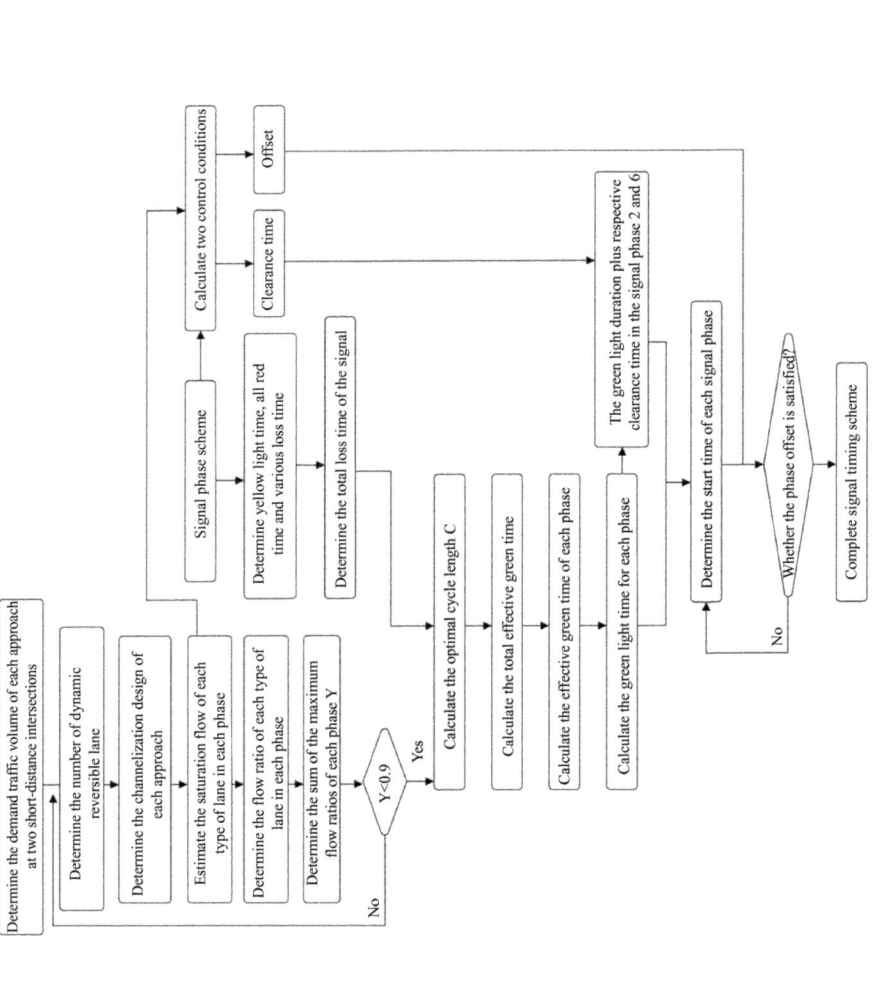

Figure 12.5 Signal timing of two short-distance intersections with dynamic reversible lanes

$$n_{72} = S_{41}k_{412}\left(G_4 - G_3 - \frac{l_4}{\bar{v}_1}\right) + S_{63}k_{632}t_{63}$$

$$+ q_{63}k_{632}\left(G_4 - G_3 - \frac{l_6}{\bar{v}_3} - t_{63}\right) \quad (12.9)$$

- Case 4: if $t_{41} \geq G_4 - G_3 - \frac{l_4}{\bar{v}_1}$ and $t_{63} < G_4 - G_3 - \frac{l_6}{\bar{v}_3}$, then:

$$n_{71} = S_{41}k_{411}t_{41} + q_{41}k_{411}\left(G_4 - G_3 - \frac{l_4}{\bar{v}_1} - t_{41}\right) + S_{63}k_{631}t_{63}$$

$$+ q_{63}k_{631}\left(G_4 - G_3 - \frac{l_6}{\bar{v}_3} - t_{63}\right) \quad (12.10)$$

$$n_{72} = S_{41}k_{412}t_{41} + q_{41}k_{412}\left(G_4 - G_3 - \frac{l_4}{\bar{v}_1} - t_{41}\right) + S_{63}k_{632}t_{63}$$

$$+ q_{63}k_{632}\left(G_4 - G_3 - \frac{l_6}{\bar{v}_3} - t_{63}\right) \quad (12.11)$$

where $\bar{v}_1$ is the average speed of left-turn vehicles at the intersection and $\frac{l_4}{\bar{v}_1}$ is the time for the first left-turn vehicle on approach 4 to arrive section II to join the internal link, and t_{iw} is the time for w-movement on approach i to release at saturated flow rate, given by:

$$t_{41} = \frac{q_{41}[C - (G_5 - G_3)]}{S_{41} - q_{41}} \quad (12.12)$$

$$t_{63} = \frac{q_{63}[C - (G_6 - G_3)]}{S_{63} - q_{63}} \quad (12.13)$$

where q_{iw} is the arrival rate of vehicles with w-movement on approach i; S_{iw} is the saturated flow rate of w-movement on approach i; k_{i1w} is the ratio of the w-movement vehicles at the next intersection in the left-turn vehicles on approach i; k_{i3w} is the ratio of the w-movement vehicles at the next intersection in the right-turn vehicles on approach i; G_n is the beginning of green light at signal stage n with $G_1 = 0$, and C is cycle length. As shown in Figure 12.6, the green time for approach 4 is between G_3 and G_5, where saturated left-turn flow persists until t_{41}, when queue on the approach dissipates. Between t_{41} and G_5, left-turn flow is released at arrival rate q_{41}. After G_5, vehicles begin to queue up after the stop line until $C + G_3$. Similarly, the right-turn movement of approach 6 is from G_3 to G_6. With respect to t_{63}, queuing right-turning vehicles move at saturated rate q_{63}, after which flow rate drops to the arrival rate.

Figures 12.7 and 12.8 show the arrival cumulative curve of the left-turn traffic on approach 4 to section II when $t_{41} \geq G_4 - G_3 - \frac{l_4}{\bar{v}_1}$ and $t_{41} < G_4 - G_3 - \frac{l_4}{\bar{v}_1}$, respectively; Figures 12.9 and 12.10 show the arrival cumulative curve of the right-

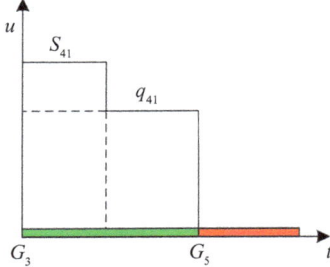

Figure 12.6   The flow rate of left-turn vehicle on approach 4

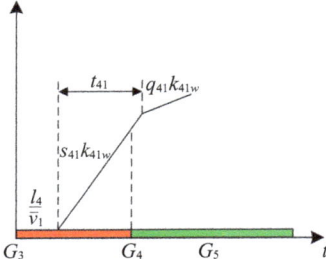

Figure 12.7   Cumulative arrival curve of left-turning to section II $\left(t_{41} \geq G_4 - G_3 - \frac{l_4}{v_1}\right)$

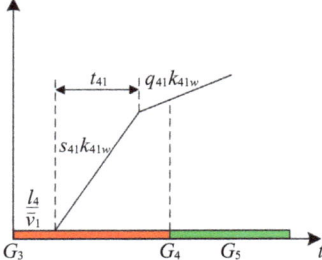

Figure 12.8   Cumulative arrival curve of left-turning to section II $\left(t_{41} < G_4 - G_3 - \frac{l_4}{v_1}\right)$

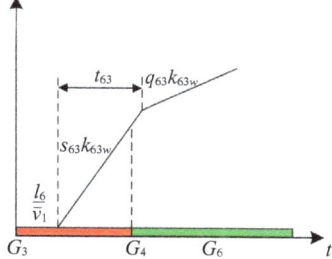

Figure 12.9 Cumulative arrival curve of right-turning to section II $\left(t_{63} \geq G_4 - G_3 - \frac{l_6}{v_3}\right)$

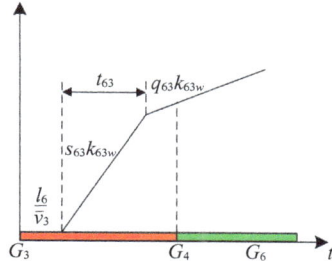

Figure 12.10 Cumulative arrival curve of right-turning to section II $\left(t_{63} < G_4 - G_3 - \frac{l_6}{v_3}\right)$

turn traffic on approach 6 to section II when $t_{63} \geq G_4 - G_3 - \frac{l_6}{v_3}$ and $t_{63} < G_4 - G_3 - \frac{l_6}{v_3}$, respectively.

Moreover, before switching the direction of the dynamic reversible lane, it is necessary to clear the vehicles on the dynamic reversible lanes, with the green light constrained with:

$$G_3 - G_2 \geq \frac{l + l_1}{v_{\min}} \quad (12.14)$$

$$G_7 - G_6 \geq \frac{l + l_2}{v_{\min}} \quad (12.15)$$

for signal stages 2 and 6, separately. Taking signal stage 2 as an example, it is secured that the last car that leaves at G_2 from approach 2 passes through the stop line before G_3 at approach 8 at a minimum speed $v_{\min}$.

12.4 Calibration and validation

12.4.1 Simulation scenarios

To validate the proposed method of coordinated control for short-distance intersections with dynamic reversible lanes proposed, the simulation package VISSIM is used to compare the performance of the proposed control scheme (Plan 1) with the bidirectional green wave scheme (Plan 2) and the uncoordinated one (Plan 3). Figures 12.11 and 12.12 show the simulation scenario with short-distance intersections connected to four-lane two-way roads. Channelization of the in-between lane without and with dynamic reversible lanes is also given. Note only cars are

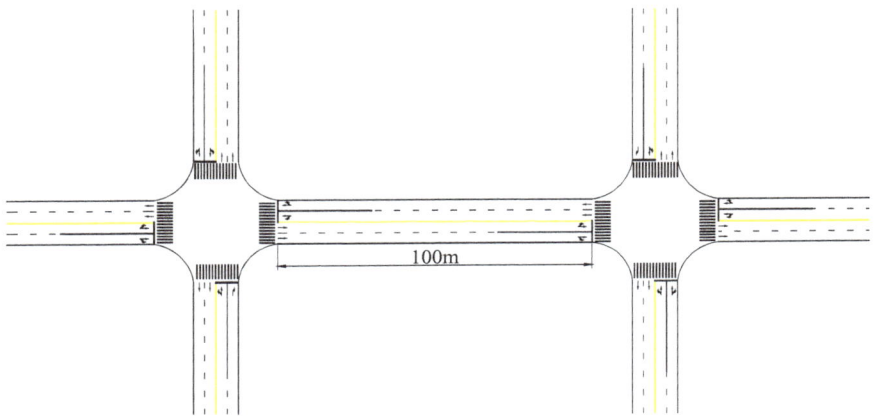

Figure 12.11 Simulation scenario without dynamic lanes

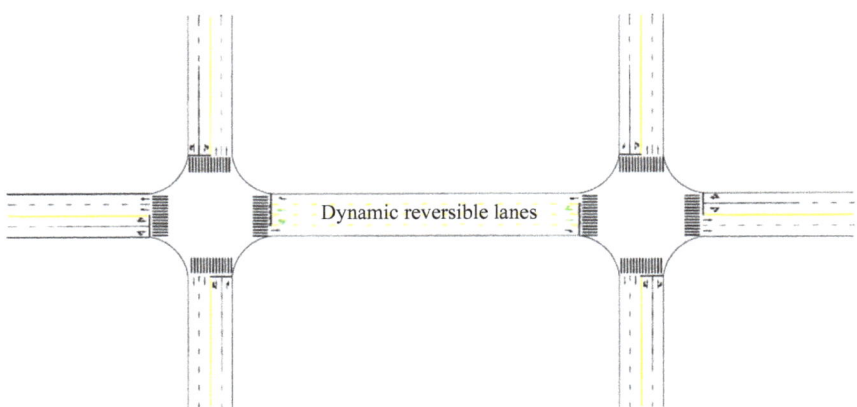

Figure 12.12 Simulation scenario with dynamic lanes

considered in the case with speed ranging from 40 to 45 km/h and clearance time is set as 15 s. Table 12.2 summarizes the three levels of traffic demand (low, medium, and high) for the simulation.

12.4.2 Validation of the proposed plan

Tables 12.3 to 12.5 give the signal timing (all red time being zero) for the simulation scenario with different traffic levels. VISSIM is then used to test the timing plans with results averaged from ten simulations to overcome randomness, duration of the simulation is set to 30 min and the first 15 min is not considered to eliminate the instable traffic flow at the beginning of the simulation. Figure 12.13 shows the performance of the three planes with the parameters of vehicle throughput, average delay, the number of stops, and the maximum queue length, which are compared in Table 12.6.

The results show that the bidirectional green wave control can perform better than the uncoordinated control at low flow rate, while the proposed scheme performs best at medium and high flow rate, especially with respect to the number of stops. It is explained that, with the same signal cycle, the proposed scheme allows vehicles to pass through the two intersections at one time and the number of stops can be significantly reduced regardless of traffic level. In comparison, the proposed scheme is less effective in reducing vehicle delay because the clearance time can reduce effective green time.

12.5 Adaptability analysis

The proposed model is tested with varying road and traffic conditions to provide guidance on the adaptability of the proposed control method with dynamic reversible lane.

12.5.1 Road conditions

Table 12.7 summarizes six scenarios of varying road conditions with respect to lane number and the link length. The signal cycle of each group is 180 s, and the clearing times are set 15 s, 20 s, and 23 s when the links are 100 m, 150 m, and 200 m, respectively. Table 12.8 shows the simulation results of each scenario.

Figure 12.14 shows the comparison of different control methods toward the number of vehicles and average delay. It can be seen that the performance of the proposed scheme decreases with the increase of the link length and the number of road lanes. For example, when the intersection spacing is 100 m, the number of vehicles by the proposed method on the four and six-lane road is higher than that by the other two methods; while the proposed method performs less well when the intersection spacing is longer. Moreover, compared with the uncoordinated control, the proposed method can increase the number of vehicles by 25.4 per cent, and reduce the average delay by 46.5 per cent for the four-lane road with 100 m links, while increase the number of vehicles by 5.7 per cent and reduce average delay by 1.6 per cent for the six-lane road. The reason for the

Table 12.2 Three traffic demand levels

Approach		1			2			3		
Movement		LT	TH	RT	LT	TH	RT	LT	TH	RT
Traffic volume (veh/h)	High	189 (63/63/63)	375	189	250	500 (125/250/125)	250	189	189	375 (125/125/125)
	Medium	150 (50/50/50)	300	150	200	400 (100/200/100)	200	150	150	300 (100/100/100)
	Low	114 (38/38/38)	225	114	150	300 (75/150/75)	150	114	114	225 (75/75/75)
Approach		4			5			6		
Movement		LT	TH	RT	LT	TH	RT	LT	TH	RT
Traffic volume (veh/h)	High	189 (63/63/63)	375	189	250	500 (125/250/125)	250	189	189	375 (125/125/125)
	Medium	150 (50/50/50)	300	150	200	400 (100/200/100)	200	150	150	300 (100/100/100)
	Low	114 (38/38/38)	225	114	150	300 (75/150/75)	150	114	114	225 (75/75/75)

Note: The traffic turning volume at the next intersection is shown in the brackets, in the order of left-turning, straight, and right-turning, represented with LT, TH, and RT, respectively.

Table 12.3 Signal timing of the proposed scheme (plan 1) at three traffic levels

Traffic demand levels	Cycle length (s)	Approach Permitted direction	2 LT/TH/RT	3 LT/TH	7 LT/TH	1 LT/TH/RT	8 LT/TH	4 LT/TH/RT	5 LT/TH/RT	6 LT/TH	3 RT	6 RT
High	180	Start of green (s)	0	42	75	147	165	57	90	132	147	57
		End for green (s)	39	72	144	177	54	87	129	162	39	129
		Duration of green (s)	39	30	69	30	69	30	39	30	72	72
Medium	150	Start of green (s)	0	34	60	124	135	49	75	109	124	49
		End for green (s)	31	57	121	147	46	72	106	132	31	106
		Duration of green (s)	31	23	61	23	61	23	31	23	57	57
Low	100	Start of green (s)	0	20	35	85	85	35	50	70	85	35
		End for green (s)	17	32	82	97	32	47	67	82	17	67
		Duration of green (s)	17	12	47	12	47	12	17	12	32	32

Table 12.4 Signal timing of bidirectional green wave scheme (plan 2) at three traffic levels

Traffic demand levels	Cycle length (s)	Approach Permitted direction	2 LT/TH/RT	3 LT/TH	7 LT/TH	1 LT/TH/RT	8 LT/TH	4 LT/TH/RT	5 LT/TH/RT	6 LT/TH
High	180	Start of green (s)	0	50	89	141	0	52	91	141
		End for green (s)	47	86	138	177	49	88	138	177
		Duration of green (s)	47	36	49	36	49	36	47	36
Medium	150	Start of green (s)	0	42	74	118	0	44	76	118
		End for green (s)	39	71	115	147	41	73	115	147
		Duration of green (s)	39	29	41	29	41	29	39	29
Low	100	Start of green (s)	0	27	49	78	0	29	51	78
		End for green (s)	24	46	75	97	26	48	75	97
		Duration of green (s)	24	19	26	19	26	19	24	19

Table 12.5 Signal timing of uncoordinated scheme (plan 3) at three traffic levels

Traffic demand levels	Cycle length (s)	Approach Permitted direction	2 LT/TH/RT	3 LT/TH	7 LT/TH	1 LT/TH/RT	8 LT/TH	4 LT/TH/RT	5 LT/TH/RT	6 LT/TH
High	180	Start of green (s)	0	50	89	141	89	141	0	50
		End for green (s)	47	86	138	177	138	177	47	86
		Duration of green (s)	47	36	49	36	49	36	47	36
Medium	150	Start of green (s)	0	42	74	118	74	118	0	42
		End for green (s)	39	71	115	147	115	147	39	71
		Duration of green (s)	39	29	41	29	41	29	39	29
Low	100	Start of green (s)	0	27	49	78	49	78	0	27
		End for green (s)	24	46	75	97	75	97	24	46
		Duration of green (s)	24	19	26	19	26	19	24	19

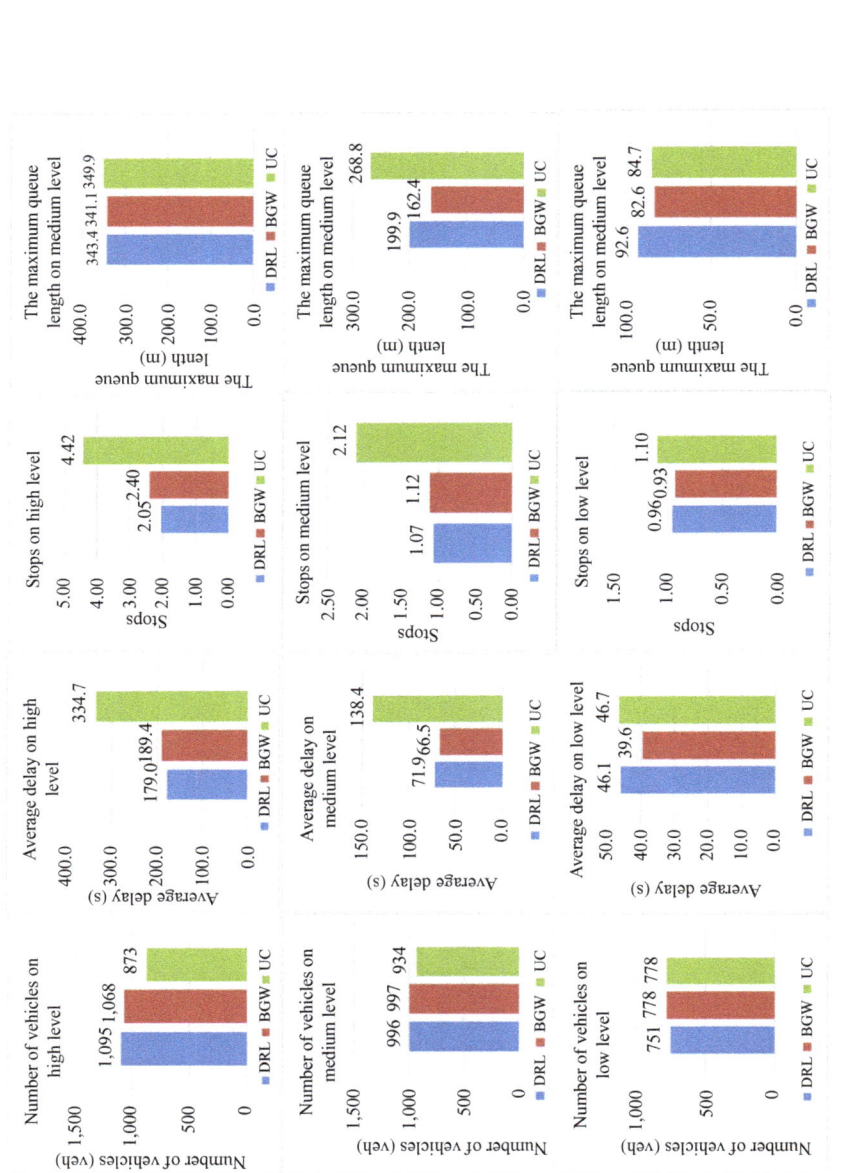

Figure 12.13 Comparison of four indexes of each control plan under three traffic levels

Table 12.6 Comparison the control effect of three timing plans under three traffic levels

Volume levels	Control method	Vehicle throughput (veh)	Optimization effect	Average delay (s)	Enhancement	Stops	Enhancement	The maximum queue length (m)	Enhancement
High	Dynamic reversible lane control	1095	25.5%[1]	179.0	46.5%[3]	2.05	53.6%	343.4	1.9%
	Bidirectional green wave control	1068	22.4%[2]	189.4	43.4%[4]	2.40	45.7%	341.1	2.5%
	Uncoordinated control	873	—	334.7	—	4.42	—	349.9	—
Medium	Dynamic reversible lane control	996	6.6%	71.9	48.1%	1.07	49.4%	199.9	25.6%
	Bidirectional green wave control	997	6.7%	66.5	51.9%	1.12	47.3%	162.4	39.6%
	Uncoordinated control	934	—	138.4	—	2.12	—	268.8	—
Low	Dynamic reversible lane control	751	−3.5%	46.1	1.2%	0.96	12.9%	92.6	−9.3%
	Bidirectional green wave control	778	−0.1%	39.6	15.3%	0.93	14.7%	82.6	2.4%
	Uncoordinated control	778	—	46.7	—	1.10	—	84.7	—

Note: Enhancement of vehicle throughput is calculated by (1) (1095−873)/873 × 100%; (2) (1068−873)/873 × 100%; enhancement of average delay is calculated by (3) (334.7−179)/334.7 × 100% and (4) (334.7−189.4)/334.7 × 100%, which applies to that of the maximum queue length.

Table 12.7 Scenarios with varying road conditions

Scenario	Number of lanes	Length of internal link (m)
1	4	100
2	4	150
3	4	200
4	6	100
5	6	150
6	6	200

Table 12.8 Simulation results under different road conditions

Group	Road condition	Control method	Number of vehicles (veh)	Average delay (s)	Stops	The maximum queue length (m)
1	4-lane 100-m	Dynamic reversible lane control	1095	179.0	2.05	343.4
		Bidirectional green wave control	1068	189.4	2.40	341.1
		Uncoordinated control	873	334.7	4.42	349.9
2	4-lane 150-m	Dynamic reversible lane control	1017	228.3	2.56	345.1
		Bidirectional green wave control	1137	145.6	1.87	335.1
		Uncoordinated control	1125	176.6	2.24	337.8
3	4-lane 200-m	Dynamic reversible lane control	970	256.2	2.93	346.1
		Bidirectional green wave control	1173	130.0	1.67	329.9
		Uncoordinated control	1198	140.4	1.77	323.5
4	6-lane 100-m	Dynamic reversible lane control	1766	117.7	1.20	287.7
		Bidirectional green wave control	1700	108.6	1.41	428.8
		Uncoordinated control	1671	119.6	1.54	430.2
5	6-lane 150-m	Dynamic reversible lane control	1703	146.6	1.38	288.7
		Bidirectional green wave control	1798	87.2	1.12	448.4
		Uncoordinated control	1746	108.4	1.44	478.7
6	6-lane 200-m	Dynamic reversible lane control	1637	172.0	1.53	289.1
		Bidirectional green wave control	1822	80.1	1.02	380.6
		Uncoordinated control	1796	100.7	1.32	428.9

above simulation results is that the long intersection spacing requires more clearance time to further reduce the effective green light time, and that more lanes alleviate the effect of dynamic reversible lanes on road capacity, which reduces the performance of the proposed method.

Traffic signal control for short-distance intersections 273

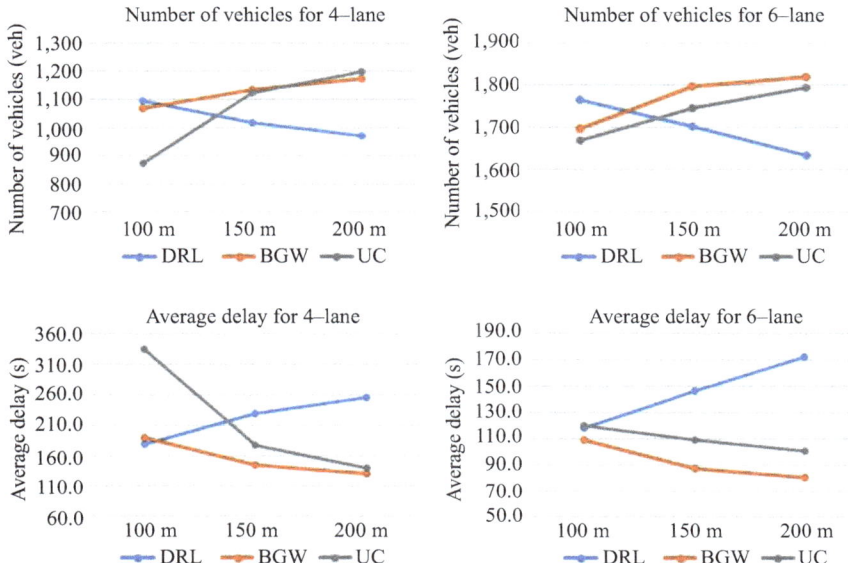

Figure 12.14 Comparison of different control plans under varying road conditions

12.5.2 Left-turning traffic proportion

This section tests the proposed method under varying ratio of left-turning vehicles onto the in-between link, i.e., vehicles from approach 1 to section I and from approach 6 to section II, as summarized in Table 12.9, for the two-way four-lane road with internal being 100 m at high traffic level. Figure 12.15 shows the variation of vehicle number and average delay with varying ratio of left-turning under the three control methods as in previous sections. It is observed that the proposed method is more suitable for short-distance intersections with a larger proportion of left-turning vehicles on the link, where vehicle throughput by the bidirectional green wave control decreases while that under the proposed scheme does not change significantly.

12.6 Conclusion

In this chapter, a coordinated control method is proposed for short-distance intersections with dynamic reversible lanes. Adopting counter-clockwise split phasing scheme, signal sequence is preset with a total of eight signal stages under the combination of intersections A and B. Then signal timing is modeled to avoid spillover on the short internal link with clearance time prepared for dynamic reversible lanes, which is calibrated and validated with VISSIM simulation to significantly avoid spillover and reduce the number of vehicle stops. It is concluded

Table 12.9 Simulation results under varying proportion of left-turning vehicles

Left-turn Proportion	Control method	Number of vehicles (veh)	Average delay (s)	Stops	Maximum queue length (m)
25%	Dynamic reversible lane control	1,095	179.0	2.05	343.4
	Bidirectional green wave control	1,068	189.4	2.40	341.1
	Uncoordinated control	873	334.7	4.42	349.9
30%	Dynamic reversible lane control	1,096	168.9	1.81	343.4
	Bidirectional green wave control	1,061	195.1	2.49	341.1
	Uncoordinated control	911	321.3	4.40	349.9
35%	Dynamic reversible lane control	1,095	170.4	1.85	343.4
	Bidirectional green wave control	1,013	212.7	2.71	343.4
	Uncoordinated control	837	355.3	4.86	349.9
40%	Dynamic reversible lane control	1,081	179.6	1.90	344.8
	Bidirectional green wave control	986	230.5	2.95	344.2
	Uncoordinated control	843	365.4	4.99	349.9
45%	Dynamic reversible lane control	1,075	180.4	2.00	344.9
	Bidirectional green wave control	970	238.7	3.06	346.4
	Uncoordinated control	765	398.8	5.38	349.9
50%	Dynamic reversible lane control	1,061	187.3	2.03	344.9
	Bidirectional green wave control	970	243.8	3.14	346.4
	Uncoordinated control	763	412.8	5.55	349.9

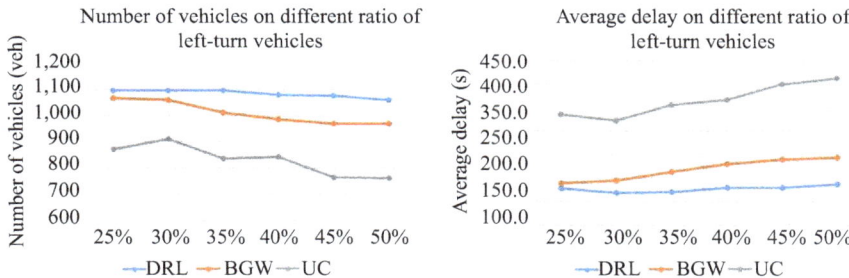

Figure 12.15 Comparison of different control schemes under varying ratios of left-turning traffic

that the coordinated control method for short-distance intersections with dynamic reversible lane proposed performs better with shorter internal link, fewer lanes, and more saturated traffic. Compared with the uncoordinated control, vehicle throughput under the proposed method increases by 22.5 per cent, and the average delay reduces by 46.5 per cent under the given condition; while compared with the bidirectional green wave control, the proposed method is more suitable for short-distance intersections with larger ratios of left-turning vehicles onto the internal link. Further research is directed at speed analysis or speed guidance of vehicles to reduce the clearance time required for the dynamic reversible lanes and to further enhance traffic efficiency.

References

[1] Prosser, N, and Dunne, M. A procedure for estimating movement capacities at signalised pair intersections. *International Symposium on Highway Capacity*, 1994.

[2] Messer, C. 'Extension and application of Prosser-Dunne model to traffic operation analysis of oversaturated, closely spaced signalized intersections'. *Transportation Research Record Journal of the Transportation Research Board*, 1998,1646(1):106–114.

[3] Yang, X., Yuan, C., and Wu, Z. Methodology and strategy of coordinated control for adjacent short-link intersections. *Intelligent Transportation Systems*. Proceedings, 2003.

[4] Ahmed, K., and Abu-Lebdeh, G. 'Modeling of delay induced by downstream traffic disturbances at signalized intersections'. *Transportation Research Record*, 2005,1920(1):106–117.

[5] Rouphail, N. M., and Akcelik, R. 'A preliminary model of queue interaction at signalised paired intersections'. *In Australian Road Research Board Ltd (ARRB) Conference*, 16th, 1992, Perth, Western Australia (Vol. 16, No. 5).

[6] Gu, Y. L., and Shao, C. F. 'Study on coordinated control of traffic signal for adjacent intersections'. *Advanced Materials Research*, 2012,433–440:4765–4770.

[7] Liqun, Q. I. 'Signal offset between short-spaced intersections under saturated traffic condition'. *Journal of Transport Information and Safety*, 2013.

[8] Liu, X. M., and Wang, L. 'Optimization method of intersection signal coordinated control based on integrated green wave bandwidth maximization'. *Journal of Jilin University (Engineering and Technology Edition)*, 2013,43(1):62–67.

[9] Ling-Xi, L. I. 'Control signal coordination of two adjacent traffic intersections'. *Acta Automatica Sinica*, 2003,6(6):6.

[10] Sun, X. L., Urbanik, T., and Han, L. D. 'Neurofuzzy control to actuated-coordinated system at closely-spaced intersections'. *Applied Mechanics & Materials*, 2013,321–324:1249–1258.

[11] Shou Feng, M. A., Ying, L. I., and Liu, B. 'Agent-based traffic coordination control method for two adjacent intersections'. *Journal of Systems Engineering*, 2003,18(3):272–278.

[12] Laurence, L., and Brian, W. 'Characterization and comparison of traffic flow on reversible roadways'. *Journal of Advanced Transportation*, 2010,44(2):113–122.

[13] Wu, J. J., Sun, H. J., Gao, Z. Y., *et al.* 'Reversible lane-based traffic network optimization with an advanced traveller information system'. *Engineering Optimization*, 2009,41(1):87–97.

[14] Zhao, J., Ma, W., Liu, Y., and Yang, X. 'Integrated design and operation of urban arterials with reversible lanes'. *Transportmetrica B: Transport Dynamics*, 2014,2(2):130–150.

[15] Zhao, J., Liu, Y., and Yang, X. 'Operation of signalized diamond interchanges with frontage roads using dynamic reversible lane control'. *Transportation Research Part C: Emerging Technologies*, 2015,51:196–209.

[16] Jing, Z., Jingyan, F. U., and Yang, X. 'Optimization model of dynamic lane assignment for isolated signalized intersections'. *Journal of Tongji University*, 2013,41(7):996–1001.

[17] Li, L. L., Yao, R. H., and Zhou, H. M., *et al.* 'Optimization model for pre-timed signals at an intersection with reversible approach lanes'. *Journal of Jilin University (Engineering and Technology Edition)*, 2015,45(1):75–81.

[18] Zhao, J., Yun, M., Zhang, H. M., and Yang, X. 'Driving simulator evaluation of drivers' response to intersections with dynamic use of exit-lanes for left-turn'. *Accident Analysis & Prevention*, 2015,81,107–119.

[19] Zhao, J., and Liu, Y. 'Safety evaluation of intersections with dynamic use of exit-lanes for left-turn using field data'. *Accident Analysis & Prevention*, 2017,102:31–40.

[20] Zhao, J., Wan-Jing, M. A., and Han, Y. 'Integrated optimization model of layouts and signal timings of exit-lanes for left-turn intersections'. *China Journal of Highway & Transport*, 2017.

[21] Wolshon, B., and Lambert, L. 'Reversible lane systems: Synthesis of practice'. *Journal of Transportation Engineering*, 2006,132(12):933–944.
[22] Li, X., Chen, J., and Wang, H. 'Study on flow direction changing method of reversible lanes on urban arterial roadways in China'. *Procedia-Social and Behavioral Sciences*, 2013,96:807–816.

Chapter 13

Multiday evaluation of adaptive traffic signal system based on license plate recognition detector data

Ruimin Li[1], Fanhang Yang[1] and Shichao Lin[1]

This study evaluates the performance of an adaptive traffic signal control system (ATSCS) based on license plate recognition (LPR) detector data. The LPR detector data can provide abundant individual vehicle-based information for comprehensive, detailed evaluation of traffic signal control. Several measurements including travel time delay, cumulative travel time frequency diagram, Purdue coordination diagram, 95th percentile travel time, and buffer index are applied to reveal the various aspects of traffic signal control performance. A before and after comparison was conducted in the eastbound direction of a road segment, in which the ATSCS was deployed in October 2016 while the time-of-day traffic signal planning was used previously. Results show the improvement in traffic condition in the morning and evening peaks after the deployment of the ATSCS. However, the traffic condition at midnight worsened on certain days after the deployment of the ATSCS.

13.1 Introduction

With the rapid development of economy and urbanization, traffic demand has been continuously increasing in many cities. A well-designed traffic signal control system, which can allocate the capacity reasonably [1] and improve the efficiency of traffic flow, plays a vital role in the road network of a city. Most cities apply time-of-day (TOD) traffic signal plans; however, these plans lack robustness when the volume of traffic fluctuates significantly. The adaptive traffic signal control system (ATSCS), such as Sydney Coordinated Adaptive Traffic System (SCATS) and Split Cycle Offset Optimization Technique (SCOOT), can adjust the signal timing plans according to real-time traffic conditions, which may potentially be beneficial for traffic efficiency [2–4], traffic environment [5], and traffic safety [6,7]. Measuring how the ATSCS improves the traffic condition is important in traffic-management system. The

[1]Institute of Transportation Engineering, Department of Civil Engineering, Tsinghua University, Beijing, China

assessment of a traffic signal control system helps engineers to understand the efficiency of the traffic signal control system and find an optimal traffic signal control plan for road networks.

In general, two main approaches, namely, simulation-based method and field study, are applied to evaluate the performance of a traffic signal control system. The simulation-based method is usually applied to build a model [1,8,9] for evaluation or evaluate new proposed or optimized traffic signal control models, algorithms, and plans [5,10–14]. In these simulation-evaluation systems, typical measurements, such as delay and numbers of stopping, are often used. Balasha and Toledo [1] built a mesoscopic traffic-simulation model to evaluate a traffic signal control system with delay. Abdelgawad et al. [13] applied a hardware in-the-loop simulation to assess a self-learning ATSCS Multi-Agent Reinforcement Learning (MARLIN), which indicated up to 20% reduction of the average travel time delay compared to the optimized and coordinated actuated signal timing plans. Kurihara et al. [5] evaluated the benefits of SCATS in a microsimulation environment, which presented a 2% saving of emission and fuel consumption due to the reduction of delay and stops compared to the TOD.

The field study method is used for specific evaluation measurements [15,16] or evaluation of traffic signal control systems [2,4,17,18]. For specific measurements, Day et al. [15] evaluated the traffic signal control system based on Bluetooth data using a visualization tool called Purdue coordination diagram (PCD). Quantitative measures, such as the percentage of vehicles that arrives on green, can also be extracted from the PCD. Freije et al. [16] considered detector occupancy, green occupancy ratio, and red occupancy ratio as traffic signal evaluation measurements to evaluate whether sufficient green times are being provided to avoid split failure. For a systematic evaluation, Peters et al. [2], Hu et al. [4], Tian et al. [17], and Hunter et al. [18] compared the ATSCS with the TOD; their studies involved evaluation measurements, such as delay, number of stops, and travel time reliability.

For the field study method of traffic signal control system evaluation, probe data [2,4], video detection data [17], radar data [19], and Bluetooth data [15] have been mainly applied in previous studies. The qualities and features of data are important to the evaluation system. Shladover and Li [20] verified the difficulties in estimating the measures of effectiveness related to traffic signal control at low market penetrations of probe vehicles. Tian et al. [17] claimed that the use of video detection data to evaluate SCATS might have significantly limited the results. With the development of intelligent transportation system (ITS), emerging detection technologies have provided opportunities in exploring new evaluation measurements. At present, license plate recognition (LPR) detector data can be collected in many cities in China and focus on the data of individual vehicles detected at intersections. Table 13.1 exhibits the features of several typical data sources. Loop data and radar cannot provide discrete travel time information, and indexes, such as 95th percentile travel time, cannot be acquired. GPS data [18] are widely applied as sampling data and the accuracy of this system depends on the sampling rate. That is, GPS data are inapplicable for cities where the popularity of GPS vehicle does not meet the demand. Radio frequency identification (RFID) detector data are promising data source with abundant information; however, these data sources are

Table 13.1 Features of different data sources

Data source	Traffic flow volume	Average travel time-related indexes	Discrete travel time-related indexes	Arrival time-related indexes	Detector occupancy	Limitation
Loop detector	√	×	×	×	√	No re-identification function
Radar detector	√	×	×	×	√	No re-identification function
GPS	×	√	√ (When sample rate is high)	√ (When sample rate is high)	×	Low sample rate
LPR detector	√	√	√	√	√	Fixed location
Bluetooth detector	√	√	√	√	×	Not widely applied in China
RFID dedicated short range communications (DSRC)	√	√	√	√	√	Not widely applied at present

newly developing technologies and have not been widely applied in the transportation field in China. Among these data sources, the LPR detector data have high sample rate and high accuracy, can reflect the characteristics of traffic flow accurately using the detailed information of each vehicle, and offer the data for traffic signal control performance evaluation. Therefore, abundant indexes can be used to evaluate the various performance of traffic signal control systems effectively based on the LPR detector data to understand the advantages and disadvantages of traffic signal control systems comprehensively.

In this study, the performance of a traffic signal control system before and after the deployment of the ATSCS is evaluated based on the LPR detector data using several indexes. This study distinguishes itself from previous studies as follows. (1) The travel time of each detected individual vehicle can be obtained; thus, the measurements that require discrete information, such as the 95th percentile travel time, the PCD, and cumulative frequency diagram of travel time, can be derived. (2) The LPR detector data can support the real-time measurements in this study. For example, to evaluate the reliability of travel time, the GPS data may only contain less than 100 records during 1 h, whereas the LPR detector data can collect the travel time of over 80% vehicles, which can reach hundreds of records in an hour. That is, the LPR detector data can support real-time evaluation of travel time reliability when high sample rate and high-quality information are provided. (3) Multiday evaluation of the ATSCS is conducted in this study and the features of different days of a week can be exhibited.

The remainder of this chapter is organized as follows. Section 13.2 describes the methodology, which is mainly estimation methods for several evaluation measurements. Section 13.3 introduces the evaluation site and the dataset. Section 13.4 shows the analysis results. Finally, Section 13.5 concludes the study.

13.2 Methodology

This study focuses on conducting a generalized process to use the LPR detector data for evaluating the performance of adaptive traffic control systems. Each record of the LPR detector data includes vehicle license plate number, passing time, intersection or road segment ID, approach ID, and lane number. Data preprocessing is applied in cleaning the original data and obtaining basic traffic information to calculate the performance measurements of the traffic signal control systems. Travel time delay, cumulative frequency diagram of travel time, the PCD and related indexes, 95th percentile travel time, and buffer index are considered in the performance evaluation of traffic signal control.

Travel time delay measures the improvement of the ATSCS on average traffic condition quantitatively. Cumulative frequency diagram of travel time indicates the distribution of travel time, which is fulfilled by abundant individual travel time information. The PCD measures the improvements on the coordination of the ATSCS. The traffic signal control system is considered good when high percentage of vehicles approach an intersection during the green period. The 95th percentile travel time, which indicates the absolute range of travel time, and buffer index, which reveals the difference between the 95th percentile and the average travel time (relative range of travel time), mainly represent the reliability of travel time.

13.2.1 Travel time delay

Delay is the travel time loss caused by traffic signals and other factors. Travel time delay is the time difference between the free-flow and real travel times of a road section, which reflects the performance of the traffic signals for the signalized road. Generally, free-flow travel time may be defined as the travel duration of a vehicle in the road segment at the free-flow speed. Free-flow speed is the average speed of traffic on a segment when volume and density approach zero [21], which is usually the declared or expected speed limit. Therefore, (13.1) computes the travel time delay of a vehicle:

$$D_i = T_i - \frac{l}{v_{\lim}} \qquad (13.1)$$

where D_i is the travel time delay of a road section, T_i is the real travel time of a road section, l is the length of a road section, and $v_{\lim}$ is the declared speed limit of a road section.

Through the license plate matching process, the passing times can be obtained through the upstream and downstream detectors of an individual vehicle. The real travel time can be calculated by the difference between the downstream and upstream detector passing times. Travel time delay measures the average traffic condition quantitatively and reflects the coordination of traffic signal control systems. For a well-coordinated arterial, most of the vehicles pass the intersection without stopping, thereby reducing the total travel time delay on the arterial.

13.2.2 Travel time-based measurements

The travel times for 80–90% of vehicles can be obtained based on the LPR detector data. The high sample rate of discrete travel time data can not only acquire certain aggregated indexes, such as average travel time and speed, but also certain discrete measurements, such as scatter and cumulative frequency diagrams of travel time.

13.2.2.1 Scatter diagram of travel time (delay)

The index of travel time is the duration of a vehicle to pass through a road segment. Travel time can estimate whether an individual vehicle stops at a red light or passes through the intersection directly at a green light. The LPR detector data support the formation of the scatter diagram of travel time with abundant individual information, whereas the traditional aggregated data cannot provide necessary information for such measurements. The horizontal axis for a scatter diagram of travel time is the time of day and the vertical axis is the travel time of an individual vehicle. Scatter diagrams of travel time can reflect the distributions of individual travel time, thereby representing the general regularity of the traffic flow.

13.2.2.2 Cumulative frequency diagram of travel time

In addition, the cumulative frequency diagram of travel time of vehicles for a road segment during a time interval can be drawn based on individual travel time dataset, which reflects the frequency distributions and reliability of travel time. The horizontal and vertical axes of a cumulative frequency diagram of travel time are the travel speed and cumulative frequency of travel time, respectively.

Several features related to travel time and travel time reliability can be derived. First, the cumulative frequency curves reflect the general distribution of travel time, in which the curves on the left side have smaller travel time than those on the right side. The typical travel time can be easily obtained from the curves. The median travel time (where the cumulative frequency curve crosses the 0.5 line) is a typical measure of central tendency for many distributions and several typical percentile travel times (95th percentile) can reflect the reliability of travel time. Moreover, the slope of cumulative frequency curve indicates the degree of variability and reliability of travel time. A steep slope reveals that the travel times concentrate in small ranges with minimal variability (high reliability), whereas a shallow slope shows that the travel times are distributed in wide ranges with further variability (low reliability).

13.2.3 PCD and related indexes

The PCD, proposed by Purdue University, is a powerful visualization tool for visualization and the quantitative evaluation of signal performance [15]. The PCD indicates the state of traffic signal when an individual vehicle arrives at a certain road section near the intersection (usually approximately 120 m away from the intersection). If the vehicles arrive in the road section during green period, then the traffic signals have good coordination.

The PCD for one single signal cycle is presented in Figure 13.1. The vertical axis is the time in cycle and the horizontal axis is the event time. The green line is the beginning time of the green period in a cycle, whereas the red line is the beginning time of the red period. The black points represent the vehicles and different colors of

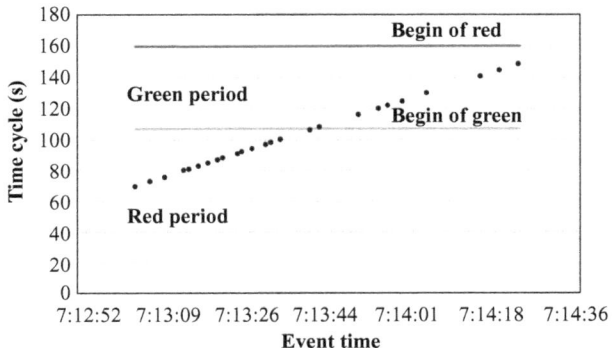

Figure 13.1 The PCD for a single cycle

the points can also be applied to present traffic features, such as travel time. If the points are between the red and green lines, then the vehicles arrive during the green period. The number of vehicles that arrives during the green period can be easily calculated through the PCD, thereby providing basic information to the quantitative analysis. One typical index is the percentage on green (POG), calculated by (13.2), which expresses the ratio of the vehicles that arrive during the green period:

$$POG_i = \frac{N_{g,i}}{N_i} \tag{13.2}$$

where POG_i is the ratio of vehicles that arrives at the section during the green period in time interval i, $N_{g,i}$ is the traffic flow of the vehicles that approaches the intersection during the green period in time interval i, and N_i is the total traffic flow of the vehicles that approaches the intersection in time interval i.

In this study, the LPR detector data can only provide the vehicle passing times of the upstream and downstream intersections. The event time when a vehicle arrives at the road section 120 m away from the intersection, which is calculated by the sum of upstream passing time and travel time, should be estimated. The travel time between the upstream intersection and the road section can be estimated by using the Robertson model [22], which considers the traffic flow dispersion. In the Robertson model presented in (13.3) and (13.4), the travel time follows a geometric distribution, and the parameters are related to the expected travel time and road condition:

$$g(t) = F(1-F)^t \tag{13.3}$$
$$F = 1/(1 + \alpha t_{\exp}) \tag{13.4}$$

where F is the dispersion parameter, which is decided by the expected travel time $t_{\exp}$ (28.8 s in this study) and related to road condition constant α (0.35 in this study); and $g(t)$ is the probability distribution function for travel time t.

13.2.4 Travel time reliability indexes

Before starting a trip, travelers will commonly estimate an expected travel time. An extra travel time will usually be added to the estimated travel time due to the uncertainty of a trip. The added travel time will decrease with the increase in the reliability of travel time. Real-time travel time reliability, which is instructive for the real-time travel time estimation, can be estimated based on the abundant LPR detector data. Several indexes, such as 90th or 95th percentile travel time, buffer index, and planning time index, can measure the reliability of travel time. The 95th percentile travel time and buffer index are applied in this study.

13.2.4.1 The 95th percentile travel time

The 95th percentile travel time is a simple method used to measure the absolute reliability of travel time because it is unrelated to the average and median travel times. The 95th percentile indicates that the travel time will not surpass a certain value with the probability of 95%. That is, the 95th percentile travel time can be regarded as the worst expectation of a trip.

13.2.4.2 Buffer index

Buffer index is related to the average traffic condition, which is computed by (13.5):

$$BI = \frac{TT_{95}}{TT_{ave}} - 1 \tag{13.5}$$

where BI is the buffer index, TT_{95} is the 95th percentile travel time, and TT_{ave} is the average travel time.

Buffer index shows the difference between the 95th percentile and the average travel time, which considers the average traffic condition. Buffer index usually represents the additional travel time the travelers must add when planning a trip. Particularly, the travel time is reliable when the buffer index is small.

13.3 Case description and dataset

13.3.1 Case description

This study evaluates the efficiency of the ATSCS deployed on Hefei Road/Jinsong 1st Road intersection and Hefei Road/Jinsong 3rd Road intersection in Qingdao, Shandong Province, based on the abundant LPR detector data. The geographic information of the two intersections is illustrated in Figure 13.2. Hefei Road is an east-west signalized arterial in Qingdao City with high traffic flow volume during morning and evening peaks. Heavy congestion occurs in the eastbound direction during the morning and evening peak periods. Therefore, the coordination and optimization of traffic signal control are essential. The westbound direction has lower traffic volume and better traffic signal coordination compared to the eastbound direction. Thus, the before and after comparison of the TOD and the ATSCS in the eastbound direction is considered in this study. The length and the speed limit on the road section are 520 m and 50 km/h, respectively. Camera detectors are set after the stop line in the two intersections to record the passing time and the license plate information of the vehicles that pass through the stop lines.

13.3.2 Dataset

The two intersections are under the fixed time control based on the TOD traffic control before October 1, 2016. After October 1, 2016, the two intersections are under the ATSCS developed by Hisense. The LPR detector data of the TOD are collected during July 1–28, 2016, whereas those of the ATSCS are collected during October 10–14 and October 21–28, 2016. The sample rate of the LPR detector data applied in this study can reach up to 80–90%.

Each original LPR detector data record can provide the vehicle license plate number, passing time, direction, and lane number. A total of 138,001 and 56,835 pair records of passing time are presented in July and October, respectively, based on the original LPR detector data records in the eastbound direction and license matching process for the two intersections. Each pair record includes the passing

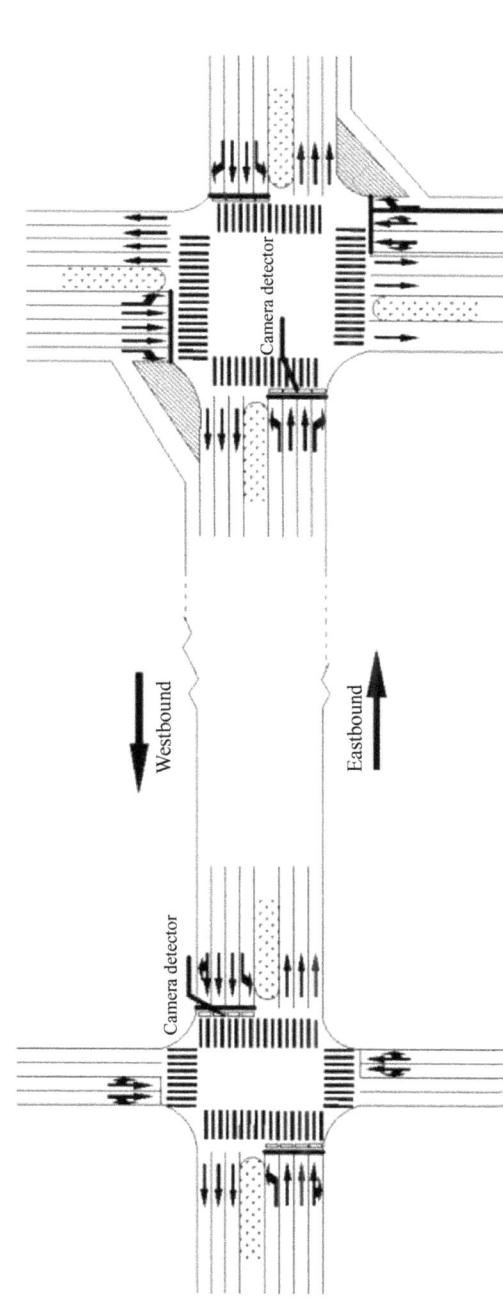

Figure 13.2 Geographic information of the two intersections

times of upstream and downstream intersections for a certain vehicle. Travel time can be calculated by the difference between the downstream and upstream passing times. The division of the day of the week (Monday to Sunday) and time of day (morning peak, midday, evening peak, and midnight) are considered because some measurements are analyzed by time interval. Morning peak, midday, evening peak, and midnight are defined as 7 AM to 10 AM, 10 AM to 4 PM, 4 PM to 7 PM, and 7 PM to 7 AM, respectively.

13.4 Results

13.4.1 Evaluation of travel time delay

13.4.1.1 Scatter diagram analysis of travel time delay

The improvement in travel time delay is discussed in this subsection. Scatter diagrams are plotted to describe the general difference in the average travel time delay of a 5-min interval in the eastbound direction between October and July. The horizontal axis of the scatter diagram represents the ordinal time interval of a day (50 represents the 50th 5 min of a day, which is 4:10–4:15, and 288 time intervals in a day), whereas the vertical axis represents the average travel time delay of vehicles, which can be calculated based on the LPR pair records.

Generally, the regularity of the changes in travel time delay varies with the day of the week and time of day. In this study, two typical scatter diagrams are presented in Figure 13.3 (Figure 13.3(a) for workday and (b) for weekend). The data for July are the average of four Wednesdays and Sundays in July, and the data for October are the average of two Wednesdays and one Sunday in October. On Wednesday, improvements of travel time delay are observed in time zones (1) and (2) in October, which are the morning and evening peaks of the day. For the other time periods, the travel time delays are similar before and after the deployment of the ATSCS. On Sunday, a significant deterioration in October is observed in time zone (3), which can represent the midnight period. In time zones (4) and (5), improvements can be observed in October.

13.4.1.2 Quantitative analysis of travel time delay improvement

The differences of the average travel time delay include two general aspects. One aspect is the trend of the change, positive or negative, and the other aspect is the degree of the change. From the scatter diagrams, only the trend of the changes can be depicted, whereas the percentage of changes cannot be exhibited. Therefore, the quantitative analysis of the travel time delay in different intervals is performed. The average travel time delays are calculated based on the data of 4 weeks in July and nearly 2 weeks (only one weekend) in October. The changing percentages of the average travel time delay are calculated, in which negative and positive values represent the decrease and increase of travel time delay in October, respectively. The changes in different time intervals and days of a week are summarized in Table 13.2, and a Kolmogorov-Smirnov test is utilized with $\alpha = 0.05$.

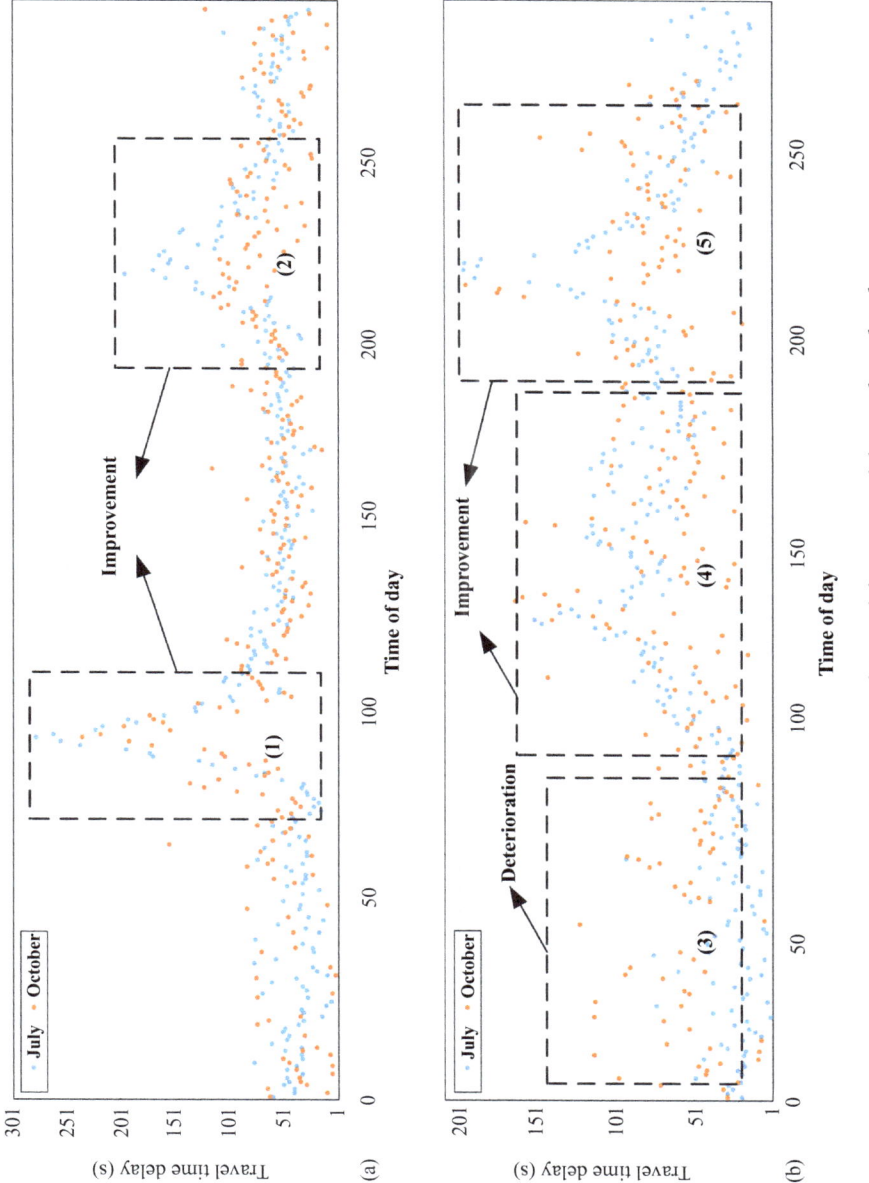

Figure 13.3 Travel time delay on workday and weekend

Table 13.2 Average travel time delay change after the deployment of the ATSCS

Time interval	Monday (%)	Tuesday (%)	Wednesday (%)	Thursday (%)	Friday (%)	Saturday (%)	Sunday (%)
Morning peak	−30.62	−40.67	−19.54	−39.00	−1.47	−45.90	−19.39
Midday	**7.10**	−6.86	**0.53**	**−4.62**	−12.07	−45.82	−12.39
Evening peak	−3.82	−30.08	−20.36	−19.39	−33.82	−38.82	−22.45
Midnight	16.09	**−2.42**	**5.20**	**9.00**	−13.05	24.51	25.16
Whole day	−11.18	−26.87	−12.32	−22.06	−14.87	−36.29	**−8.01**

Note: Non-statistically significant differences are represented in bold font.

Travel time delays decrease significantly during all the morning and evening peaks in October. All the improvements are over 19%, in which most of these improvements reach the extent of over 30%, except for Monday evening and Friday morning peaks. At middays, the improvements range from 4.62 to 45.82%, whereas a minimal increase in travel time delays is observed on Monday and Wednesday. Most situations at midnights show an increase in travel time delays. The whole days of all the week improve from 8.01 to 36.29%, which denotes that the ATSCS generally exhibits a good performance.

13.4.2 Cumulative frequency diagram of travel time

The cumulative frequency diagrams of travel time show the distribution and reliability of the travel time. The cumulative frequency diagrams of travel time in the eastbound direction for 7 days of a week and different time intervals are illustrated in Figure 13.4. In a cumulative frequency diagram of travel time, the columns and curves represent the frequency and the cumulative frequency of a certain travel time, respectively. Each curve can present the travel time distribution of the corresponding day, including the travel times of all the individual vehicles detected during the day.

Four travel time cumulative frequency curves are presented on each day of the week in July, as shown by the blue curves. Two curves for the workday and one curve for the weekend are presented in October, as shown by the orange curves. For the whole day, most of the cumulative frequency curves of travel time in October are on the left side of those in July, which indicate that the travel times after the deployment of the ATSCS are generally smaller than before. For the morning and evening peaks, significant improvements can also be observed on most of the days after the deployment of the ATSCS. However, for the morning peak on Friday, one curve in October is on the right side of the curves in July. For the midday, the curves in October basically overlap the curves in July on workdays, which indicates the small changes in travel time distribution after the deployment of the ATSCS. However, the curve in the midday on Saturday in October is on the left side of the curves in July. The curve in the midday on Sunday in October is partly on the left side and partly on the right side of the curves in July. For the midnight, most of the curves in October are on the right side of the curves in July, which indicates a generally large travel time after the deployment of the ATSCS.

To summarize, improvements can be observed on all the days of a week for the whole day. However, different time intervals have different features. Improvements in travel time distribution can be observed in the morning and evening peaks on most of the day after the deployment of the ATSCS. The changes of travel time in the midday are small on workdays, whereas a better performance in travel time distribution is presented in the midday on Saturday after the deployment of the ATSCS. However, performances of travel time distribution are worse in the midnight in October than those in July.

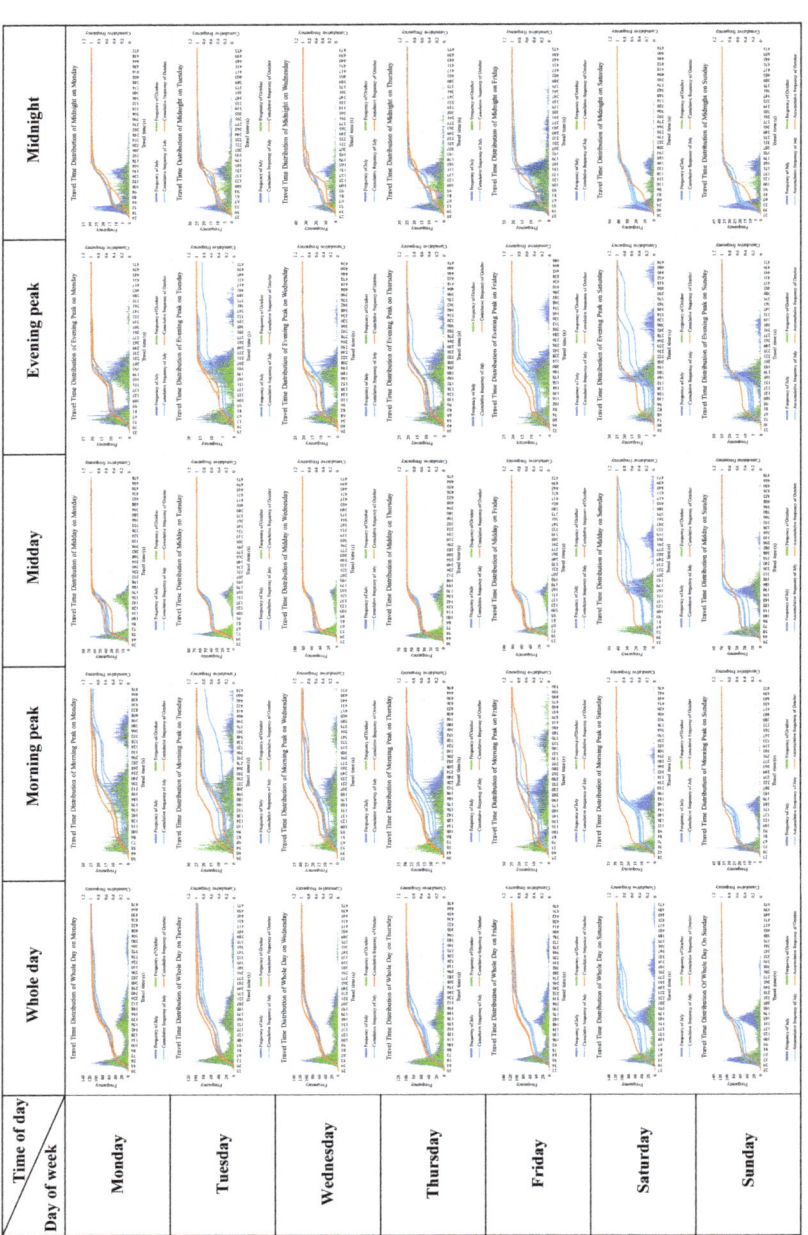

Figure 13.4 Cumulative frequency diagrams of travel time in the eastbound direction

13.5 Evaluation of the PCD

The PCD is an event data-based tool used to evaluate the coordination of the traffic signals. In this study, considering the PCDs for 7 days for a week and 24 h for a day is unnecessary; thus, the PCDs for the morning peak and 1–3 PM during the midday on a typical workday (Wednesday, July 20 and October 20) before and after the deployment of the ATSCS are presented in Figure 13.5. Figure 13.5(a) and (b) represents the morning peaks before and after the deployment of the ATSCS, respectively, whereas Figure 13.5(c) and (d) represents the middays before and after the deployment of the ATSCS, respectively.

The POG is calculated using the PCDs, as shown in Table 13.3. In the morning peak on Wednesday, an improvement in POG is observed after the deployment of the ATSCS. However, in the midday, that is, from 1 PM to 3 PM, the POG decreases after the deployment of the ATSCS. For further information, the dots with different colors represent the vehicles with different travel times. The blue, black, and red dots are the vehicles with travel time under 100 s, between 100 s and 200 s, and over 200 s, respectively. The decrease of the red dots in the morning peak, especially from 7:20 AM to 8:30 AM, is significant after the deployment of the ATSCS. During the midday, the changes in travel do not change considerably.

The changes in the coordination of traffic signals vary with traffic condition after the deployment of the ATSCS. In the morning peak, the traffic flow volume is high, and congestion occurs under the TOD traffic signal control. After the deployment of the ATSCS, the coordination of traffic signal control improves, with more vehicles arriving during the green period and less vehicles having a travel time of over 200 s. From 1 PM to 3 PM in the midday, the coordination of TOD traffic signal control is good with low traffic flow volume. However, after the deployment of the ATSCS, the performance of the traffic signal control worsens, with less vehicles arriving during the green period.

13.5.1 Travel time reliability evaluation
13.5.1.1 The 95th percentile travel time

In Table 13.4, the 95th percentile travel time change after the deployment of the ATSCS is presented quantitatively. A positive value in Table 13.4 indicates an increase in the 95th percentile travel time and the deterioration in reliability after the deployment of the ATSCS, whereas a negative value indicates the decrease in the 95th percentile travel time and the improvement in reliability.

The changes in the 95th percentile travel time vary with day of week and time interval. Generally, for whole days, the 95th percentile travel times represent a decreasing trend from Monday to Saturday after the deployment of the ATSCS, whereas the 95th percentile travel time increases by 6.15% on Sunday. For the morning and evening peaks, most of the 95th percentile travel times dramatically reduce after the deployment of the ATSCS, and several 95th percentile travel times, such as the travel time during the morning peak on Wednesday, fluctuate in a normal range. The Friday morning peak has a different trend, with 25.05% increase in

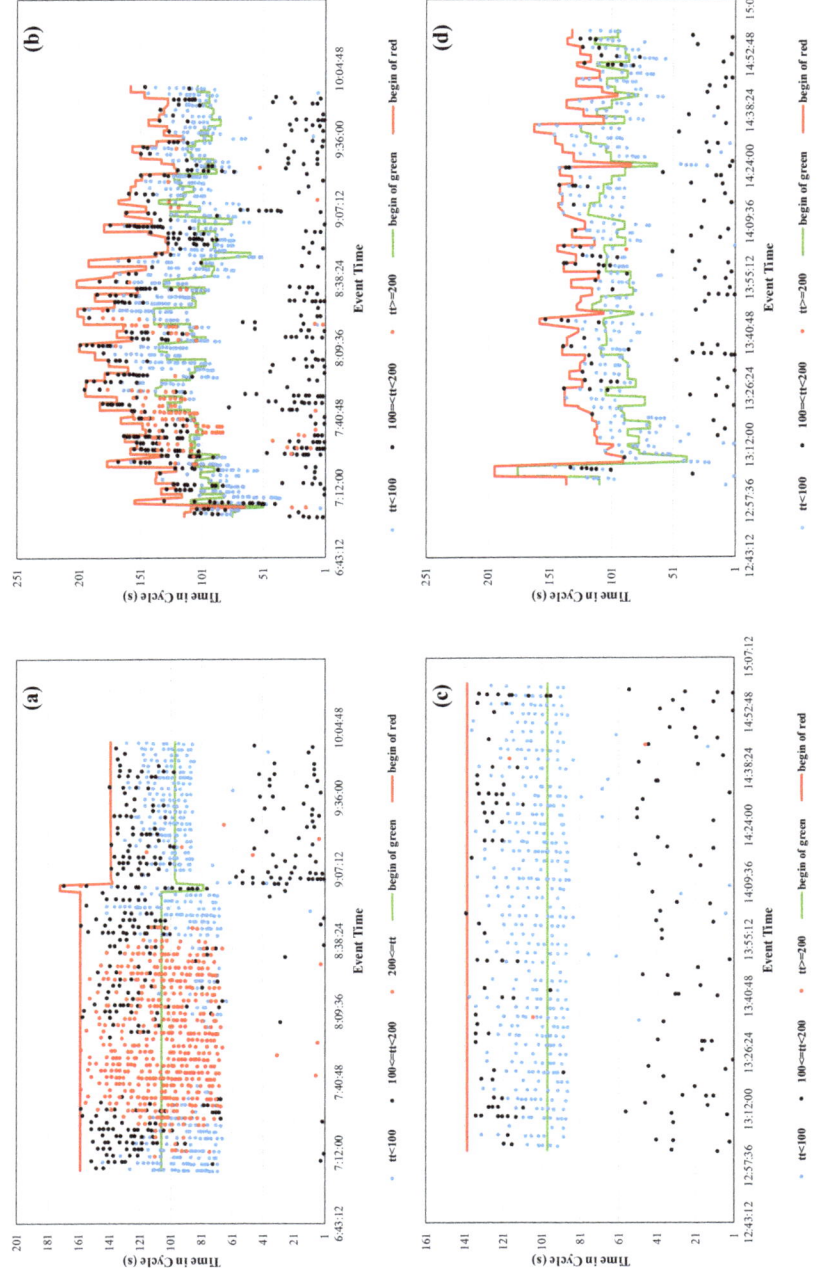

Figure 13.5 PCDs in the morning peak and during the midday before and after the deployment of ATSCS

Table 13.3 The POG in the morning peak and during the midday before and after the deployment of ATSCS

Period	July	October
Morning peak	0.45	0.53
Midday (1–3 PM)	0.57	0.37

Table 13.4 95th percentile travel time change after the deployment of ATSCS

Time interval	Monday (%)	Tuesday (%)	Wednesday (%)	Thursday (%)	Friday (%)	Saturday (%)	Sunday (%)
Morning peak	−10.89	−28.75	0.68	−40.40	25.05	−35.95	−0.57
Midday	−2.84	−2.33	1.95	0.58	−0.58	−41.37	8.72
Evening peak	3.21	−46.71	−17.37	1.62	−39.77	−50.35	−29.21
Midnight	10.18	1.10	18.39	39.36	−22.61	18.60	8.14
Whole day	−31.09	−40.32	−5.02	−12.97	−4.13	−40.57	6.15

the 95th percentile travel time after the deployment of the ATSCS. The changes in the 95th percentile travel times for the midday are in a normal range after the deployment of the ATSCS, whereas a significant decrease in the 95th percentile travel time on Saturday is observed. For the midnight, most of the 95th percentile travel times have increased considerably after the deployment of the ATSCS, with the decrease in travel time on Friday by 22.61% after the deployment of the ATSCS.

Generally, from the perspective of the 95th percentile travel time, an overall improvement in reliabilities after the deployment of the ATSCS is observed. The improvements in travel time reliability are mainly observed during the morning and evening peaks, whereas deteriorations in travel time reliability are mainly observed during the midnight. The reliability of travel time does not normally change considerably during the midday.

13.5.1.2 Buffer index

Buffer index reflects the difference between the 95th percentile and the average travel times, and a small buffer index indicates a reliable travel time. The buffer index changes after the deployment of the ATSCS are exhibited quantitatively in Table 13.5. A positive value denotes the increase in buffer index and deterioration in travel time reliability after the deployment of the ATSCS, whereas a negative value denotes the decrease in the buffer index and the improvement in the travel time reliability.

The changes in buffer index after the deployment of the ATSCS are fluctuant. For the whole day, the buffer indexes decrease by over 30% on Monday, Tuesday, and Saturday after the deployment of the ATSCS, whereas dramatic increases in buffer indexes appear on Friday and Sunday. The buffer indexes on Wednesday and Thursday do not change considerably after the deployment of the ATSCS for the whole day. For the morning peak, the buffer indexes show a significantly increasing trend on 5 days of a week after the deployment of the ATSCS. The buffer index during evening peak shows a reverse trend compared to those during the morning peak, with 5 days of a week having a decreasing buffer index after the deployment of the ATSCS. The changes in buffer index during the midday and midnight are irregular.

From the perspective of buffer index, the travel time reliabilities after the deployment of the ATSCS greatly fluctuate for the whole day, with obvious improvements on 3 days of a week, dramatic deteriorations on 2 days of a week, and normal changes on the remaining 2 days of a week. Similar fluctuations appear during the midday and midnight. The reliability of travel time has a general deterioration during the morning peak and a general improvement during the evening peak after the deployment of the ATSCS.

13.6 Conclusion

The rapid development of technologies provides opportunities for traffic signal control system and the ATSCS has been applied in many cities. However, studies

Table 13.5 Buffer index changes after the deployment of ATSCS

Time interval	Monday (%)	Tuesday (%)	Wednesday (%)	Thursday (%)	Friday (%)	Saturday (%)	Sunday (%)
Morning peak	42.51	13.66	29.69	−33.33	36.15	−8.79	28.48
Midday	−15.43	10.25	4.40	8.46	15.87	−17.53	54.77
Evening peak	15.23	−52.18	−10.51	21.46	−36.37	−50.61	−30.33
Midnight	0.75	−7.76	33.60	72.94	−28.15	6.99	−12.47
Whole day	−39.53	−39.50	4.14	0.00	19.54	−32.18	31.16

on the evaluation of traffic signal control performance based on real data are few and limited by data quality. The emerging LPR detector data provide abundant information, thereby allowing the possibility of evaluating the traffic signal control system comprehensively and accurately. Certain measurements based on the LPR detector data are investigated in this study to evaluate the different aspects of the traffic signal control system. The measurements are applied in a coordinated road segment with heavy congestion in the peak hours and the measurements under the TOD and the ATSCS are compared. The results are as follows. (1) Travel time delay and cumulative frequency diagram of travel time are significantly improved for 7 days of a week in the morning and evening peaks after the deployment of the ATSCS, whereas the changes in travel time delay are not evident on most days in the midday. In the midnight, the deterioration of the measurements can be observed after the deployment of the ATSCS. (2) The PCD reflects the specific situation of each detected individual vehicle and improvements in the peak hours are observed after the deployment of the ATSCS. (3) The reliabilities of travel time based on the 95th percentile travel time indicate the improvement in peak hours and deteriorations in the midnight after the deployment of the ATSCS, whereas the reliabilities of travel time based on buffer index indicate the fluctuant changes of travel time reliabilities. (4) For the weekend, the improvements of travel time delay in the midday are also significant after the deployment of the ATSCS. The results indicate that the ATSCS can alleviate the congestion in the peak hours by allowing more vehicles to arrive during the green period and reducing the average travel time.

Most of the measurements applied in this study require high data quality, and the LPR detector data are accurate, discrete, and have high sample rate. For example, the high sample rate (80–90%) of the LPR detector data is the fundamental for measurements, such as cumulative frequency diagram of travel time and the PCD, and the other data sources, such as GPS data, which have low sample rate that cannot support these types of measurements. For the evaluation of traffic signal control, many other measurements can also be derived from the LPR detector data. For example, the detector occupancy in the green time period reflects the efficiency of green time and the LPR detector data provide the basic information (i.e., the headway of each vehicle) for this measurement. The LPR detector data (also known as automated vehicle identification data) can be applied to other studies, such as real-time queue length estimation [23] and origin-destination (OD) estimation [24] because of their features. Furthermore, similar data sources, such as Bluetooth and RFID data, can be applied to evaluate the traffic signal control system based on the methodology in this study.

This study presents the methods that can be used to evaluate the traffic signal control performance based on the LPR detector data, which not only highlight the quality of the traffic signal control system but also indicate the part where certain optimization analyses can be dedicated. Thus, a thorough understanding can be obtained by evaluating the traffic signal control system and the traffic-management department might have an efficient system in cities, which is helpful for further optimizing the costs for trips, energy consumption, and vehicle emissions. Improved technologies that constantly acquire accurate data emerge due to the development of the ITS; thus, the deployment of infrastructure that aims to obtain high-quality data, such as the LPR detector data, can become easy for many cities in the future.

Acknowledgments

The authors are grateful to the following organizations for their sponsorship and support: the National Key Research and Development Program of China (Grant No. 2018YFB1601000).

References

[1] Balasha, T., and Toledo, T., 2015. A mesoscopic traffic simulation model to evaluate and optimize signal control plans. *Transportation Research Board Annual Meeting*.
[2] Peters, J.M., Monsere, C.M., Li, H., and Mahmud, M., 2008. 08–0774 - Field-based evaluation of corridor performance after deployment of an adaptive signal control system in Gresham, Oregon. *Transportation Research Board 87th Annual Meeting*.
[3] Ren, Y., Wang, Y., Yu, G., Liu, H., and Xiao, L., 2016. An adaptive signal control scheme to prevent intersection traffic blockage. *IEEE Transactions on Intelligent Transportation Systems*, 1–10.
[4] Hu, J., Fontaine, M.D., Park, B.B., and Ma, J., 2016. Field evaluations of an adaptive traffic signal—Using private-sector probe data. *Journal of Transportation Engineering* 142, 04015033.
[5] Kurihara, S., Aoyagi, S., and Onai, R., 2012. Environmental benefits of an adaptive traffic control system: Assessment of fuel consumption and vehicular emissions, *Trb Meeting, Transportation Research Board*, pp. 112–127.
[6] Stevanovic, A., Kergaye, C., and Haigwood, J., 2011. Assessment of safety benefits of an adaptive traffic control system, *International Conference on Road Safety and Simulation*, pp. 112–127.
[7] Fink, J., Kwigizile, V., and Oh, J.S., 2016. Quantifying the impact of adaptive traffic control systems on crash frequency and severity: Evidence from Oakland County, Michigan. *Journal of Safety Research* 57, 1–7.
[8] Katakura, M., Mukai, S., and Sakurada, Y., 1987. A simulation for the evaluation of signal control at traffic networks. *IFAC Proceedings Volumes* 20, 245–250.
[9] Burghout, W., and Wahlstedt, J., 2010. Hybrid traffic simulation with adaptive signal control. *Transportation Research Record Journal of the Transportation Research Board* 1999, 191–197.
[10] Abdelgawad, H., Abdulhai, B., El-Tantawy, S., Hadayeghi, A., and Zvaniga, B., 2015a. Assessment of self-learning adaptive traffic signal control on congested urban areas: Independent versus coordinated perspectives. *Canadian Journal of Civil Engineering* 42, 353–366.

[11] Lee, J., Abdulhai, B., Shalaby, A., and Chung, E., 2005. Real-time optimization for adaptive traffic signal control using genetic algorithms. *Journal of Intelligent Transportation Systems* 9, 111–122.
[12] Sawada, S., and Nogami, S., 2011. Proposal of an adaptive control using traffic signal aspect pattern and its evaluation. *Ieice Technical Report* 111, 7–12.
[13] Abdelgawad, H., Rezaee, K., El-Tantawy, S., Abdulhai, B., and Abdulazim, T., 2015b. Assessment of adaptive traffic signal control using hardware in the loop simulation, *IEEE International Conference on Intelligent Transportation Systems*, pp. 1189–1195.
[14] Skabardonis, A., and Gomes, G., 2010. Effectiveness of adaptive traffic control for arterial signal management: Modeling results. *Path Research Report*.
[15] Day, C.M., Haseman, R., Premachandra, et al., 2010. Evaluation of arterial signal coordination methodologies for visualizing high-resolution event data and measuring travel time. *Transportation Research Record Journal of the Transportation Research Board* 2192, 37–49.
[16] Freije, R.S., Hainen, A., Stevens, A.L., *et al.*, 2014. Graphical performance measures for practitioners to triage split failure trouble calls. *Transportation Research Record Journal of the Transportation Research Board* 2439, 27–40.
[17] Tian, Z., Ohene, F., and Hu, P., 2011. Arterial performance evaluation on an adaptive traffic signal control system. *Procedia - Social and Behavioral Sciences* 16, 230–239.
[18] Hunter, M.P., Wu, S.K., Kim, H.K., and Suh, W., 2012. A probe-vehicle-based evaluation of adaptive traffic signal control. *IEEE Transactions on Intelligent Transportation Systems* 13, 704–713.
[19] Santiago-Chaparro, K.R., Chitturi, M., Bill, A.R., and Noyce, D.A., 2012. Real time performance measures from radar-based vehicle detection systems, *19th ITS World Congress*.
[20] Shladover, S.E., and Li, J.Q., 2011. Evaluation of probe vehicle sampling strategies for traffic signal control, *International IEEE Conference on Intelligent Transportation Systems*, pp. 1753–1758.
[21] National Research Council, 2010. *HCM2010: Highway capacity manual*. 5th ed. Transportation Research Board.
[22] Robertson, D.I., 1969. TRANSYT method for area traffic control. *Traffic Engineering & Control* 10, 181–182.
[23] Zhan, X., Li, R., and Ukkusuri, S.V., 2015. Lane-based real-time queue length estimation using license plate recognition data. *Transportation Research Part C Emerging Technologies* 57, 85–102.
[24] Sun, J., and Feng, Y., 2011. A novel OD estimation method based on automatic vehicle identification data. *Communications in Computer & Information Science* 135, 461–470.

Chapter 14

Conclusion

Zhengbing He[1] and Ruimin Li[2]

In the age of the explosion of new technology and information, traditional human-dependent and feedback-based traffic engineering are rapidly stepping into the highly automated and data-rich stage with advanced technologies such as wireless communication and artificial intelligence. The book is organized in the following two parts to timely reflect the advancement of the modern traffic information technology and control.

Part I of the book is *Modern Traffic Information Technology*, which introduces recent advancement of traffic engineering facilitated by modern information and big-data technologies, consists of seven chapters.

- Chapter 2 focuses on the extraction of traffic information from wide-available social media and websites in the age of mobile internet. A traffic analytic system for traffic information extraction from online web data is proposed, which involves various advanced data analytic technologies such as word embedding, Bayesian networks, and deep learning. Practical application results are presented to demonstrate the capability and advantage of the proposed system.
- Chapter 3 presents a floating car data-based method of evaluating the overall extent of the traffic congestion for an entire city. An indicator named as the network-level trip speed is proposed, which is a simple but efficient indicator to reflect the network speed and with no need of map-related procedures such as map-matching.
- Deep learning, which is a rather hot topic over the past 10 years, provides us a variety of new tools of prediction and understanding traffic dynamics. In this book, to predict short-term traffic speed under both normal and abnormal traffic flow conditions, Chapter 4 proposes a Temporal Spatial Traffic Graph Attention network. The newly proposed deep learning method can learn both spatial and temporal propagation rules for traffic on a network; in addition, high prediction accuracy is demonstrated through practical applications and model comparisons.

[1]Beijing Key Laboratory of Traffic Engineering, Beijing University of Technology, Beijing, China
[2]Department of Civil Engineering, Tsinghua University, Beijing, China

- Moreover, Chapter 5 summarizes several influential deep learning methods, including Recurrent Neural Networks, Convolutional Neural Networks, and Generative Adversarial Networks. To make short-term traffic time prediction, a series of long short-term memory neural networks with deep neural layers is established using 16 settings of hyperparameters, and a comprehensive comparison is made.
- After obtaining real-time traffic information, diversion is one of the important ways of mitigating traffic congestion. To address the problem, Chapter 6 proposes a traffic diversion model based on dynamic origin-destination estimation that is updated to improve the efficiency and accuracy of the dynamic traffic diversion model.
- Lane change maneuvers are main causes of traffic turbulence at highway bottlenecks. To mitigate their impacts on traffic, Chapter 7 proposes a dynamic game framework for cooperative vehicles interacting at a merging bottleneck. Cooperative vehicles on the highway mainline seek for optimal strategies to minimize their cost, while taking into account potential future interaction.
- For the future of fully automation, Chapter 8 introduces a novel lane change-free road network. Through the negotiation of right-of-way when approaching an intersection/ramp, autonomous vehicles that move in such a road system are able to reach their destinations without making any on-road lane changes. It is expected that, by introducing such a system, the connected and automated vehicle technology could be greatly simplified and driving safety and efficiency could be improved.

Part II of the book is *Modern Traffic Signal Control*, mainly focusing on traffic signal coordination of multi intersections, includes five chapters.

- As two most important signal control systems, Split Cycle Offset Optimizing Technique (SCOOT) and Sydney Coordinated Adaptive Traffic System (SCATS), have been widely deployed worldwide. To better understand those two systems and look into the future, Chapter 9 first introduces them with basic architecture, principles, and main functions, and then discusses the challenge faced by the systems in the upcoming age of connected vehicles.
- Signal coordination is one of the most important topics in traffic signal control. Chapter 10 presents an introduction of the single point signal control, arterial, and network coordination. Particularly, the signal coordination algorithm called MAXBAND and its various derivatives are carefully revisited.
- To enhance the current signal coordination practices, Chapter 11 focus on four critical aspects of signal timing, following the MODE acronym: (1) managing a comprehensive signal timing database, (2) optimizing without detailed traffic volumes, (3) diagnosing erroneous signal timing, and (4) evaluating the quality of coordination.
- Realizing the challenging of managing short-distance intersections, Chapter 12 proposes a coordinated control method with dynamic reversible lanes on the in-between link. The direction of the dynamic reversible lanes is switched in coordination to the signal phase in a signal cycle, which can increase the

storage space of the internal link and enhance capacity in both directions to prevent overflow.
- Chapter 13 evaluates the performance of an adaptive traffic signal control system based on license plate recognition detector data. A before-and-after comparison is conducted and the results show that the system improves the traffic condition in the morning and evening peaks while worsens the traffic at midnight on certain days.

Limited by the space, many advanced technologies and methods are not included in the book, such as connected and automated vehicles-based trajectory optimization and Cooperative Vehicle Infrastructure System. Nevertheless, it is not difficult to imagine that, with the rapid development of various new technologies, not only traditional traffic information and control technologies will be improved, but also many novel schemes and applications will emerge. Those improvement and innovation will greatly promote life experience. In particularly, the emergence of connected and automated vehicles will fundamentally change the way of travel as well as human life. Right of way including the usage of roads and intersections would be more likely to be reserved before stepping out of doors, resulting in more scheduled traveling. Leveraging highly coordinated and automated vehicle and infrastructure system, traffic will be "organized" with more order and higher efficiency. The arrival will be more accurate and the waste on roads will be largely reduced.

Index

adaptive traffic signal control system
   (ATSCS) 279
   case description 286
   dataset 286
   Purdue coordination diagram (PCD),
      evaluation of 290
      buffer index 294–5
      95th percentile travel time 291–4
      and related indexes 284–5
   travel time, cumulative frequency
      diagram of 290
   travel time-based measurements 283
      cumulative frequency diagram of
         travel time 283–4
      scatter diagram of travel time
         (delay) 283
   travel time delay 282–3
      quantitative analysis of
         improvement of 288
      scatter diagram analysis of 288–9
   travel time reliability indexes 285
      buffer index 285–6
      95th percentile travel time 285
Advanced Traffic-Management
   System (ATMS) 189
Advanced Traveller Information
   Systems 79
Aimsun software 126
Akaike Information Criterion (AIC)
   33, 41, 44
Amap 6
Application Programming Interface
   (API) services 6
ARIMA (Autoregressive Integrated
   Moving Average) models 33,
   41–3, 72–3
   analysis based on 45–52
   hybrid ARIMA–SVM model,
      analysis based on 52–5
artificial intelligence technique 1
Attention Graph Convolutional
   Sequence-to-Sequence
   model 86
Australia, urban traffic control systems
   in 189–90
Autocorrelation Function (ACF) 40
Automated Traffic Signal Performance
   Measures (ATSPMs) 243, 246
automated vehicle 163
Automatic Cruise Control (ACC)
   163
Autoregressive (AR) model 42
Autoregressive Integrated Moving
   Average (ARIMA) time-series
   models 81
Autoregressive Moving Average
   (ARMA) model 42

bandwidth-based optimization 240
Bayesian Information Criterion (BIC)
   44
Bayesian network 13–14
Bayes network algorithm 83
Begin of Green (BOG) time period
   244
Bell model 126
Bert model 126
Bing Map 6
buffer index 285–6, 294–5

Caltrans Performance Measurement
   System (PeMS) 59

Central Management Computer (CMC) 199
Chang and Wu model 125
Conflict Avoidance-based Cooperative Driving (CACD) strategy 173
conflict avoidance-based cooperative driving strategy
  conflict region 176–7
  coordination 177–9
Connected and Automated Vehicle (CAV) technology 148, 152, 163–5
connected vehicles 163
Continuous Bag-of-Word (CBOW) model 10, 12–13
Controlled Optimization of Phases (COP) 207
controller event data 237–9
Convolutional LSTM (ConvLSTM) 64
convolutional neural network 16–17
Convolutional Neural Network (CNN) 17, 63–4, 86, 92
Cooperative Adaptive Cruise Control 147
Cooperative Automatic Cruise Control 163
Cooperative Diverging Area (CDA) 179
cooperative driving strategies
  at intersections 166
    reservation-based strategy 167–8
    safety driving pattern-based strategy 166–7
    trajectory optimization-based strategy 168–70
  at on-ramps 170
    slot-based strategy 171–2
    virtual vehicle mapping strategy 170–1
Cooperative Intersection with All-direction turn lanes (CI-A) 173
Cooperative Merging Area (CMA) 179

Cooperative Vehicle Infrastructure System 301
correlative relationships between variables 38–9
Corridor Synchronization Performance Index (CSPI) 246
crawler-based methods 8
Cremer–Keller model 124
crossing conflict 176

data 1
data collection 8–9
data preprocessing 9–10
Deep Belief Networks (DBN) method 62
deep learning 14
  convolutional neural network 16–17
  long short-term memory (LSTM) 14–16
    and convolutional neural network 17
  prediction of traffic using: *see* prediction of traffic using deep learning techniques
  short-term travel-time prediction by: *see* short-term travel-time prediction by deep learning
degree of saturation (DS) 196, 198
delay-based optimization 239
Dempster Shafer theory 118
Diffusion Convolutional Recurrent Neural Network (DCRNN) 87–8
directed graph, defined 89
discrete probability selection model 117–18
diversion strategy, optimization of 120
  iterative guidance strategy 122–3
  responsive guidance strategy 121–2
driver's choice behavior model 120
driver's diversion behavior, simulation of 129–30
duplicate text removal 9
dynamic lane change game formulation 153–5

Dynamic O–D Estimation (DODE) model 123–5
 dynamic traffic diversion model based on 131–3
 expressway model 124–5
 intersection model 124
 module of 128–9
 network model 125–6
dynamic reversible lane, application of 254–5
dynamic traffic diversion model 126
 based on Dynamic O–D Estimation (DODE) 131–3
 Dynamic O–D Estimation (DODE) model, module of 128–9
 experimental design (case study) 134–6
  and results of DODE 140–1
  and results of traffic diversion 136–40
 METANET model, module of 127–8
 model solution 133–4
 urban expressway, traffic diversion model of
  driver's diversion behavior, simulation of 129–30
  influence of diversion on traffic flow of exit ramp 130
  road network performance, evaluation index of 130–1
DYNASMART system 126, 189

Electronic Road Direction Adjustment System (TELCS) 189
End of Green (EOG) time period 244
En-route guidance 115
Europe, urban traffic control systems in 187–8
Expectation Maximization algorithm 13
expected utility theory 117
exponential smoothing method 41, 82

Facebook 5, 7
Federal Highway Administration (FHWA) 243–4

First-Come First-Served (FCFS) rule 168
Floating Car Data (FCD) 31
 macroscopic traffic performance indicator
  correlative relationships between variables 38–9
  empirical data for analyses 34–7
  influential factors, descriptive analyses of 37–8
  Network-Level Trip (NLT) speed, mathematical form of 33–4
 Network-Level Trip (NLT) speed time series
  ARIMA models, analysis based on 45–52
  decomposition of 44–5
  exponential smoothing methods, analysis based on 45
  hybrid ARIMA–SVM model, analysis based on 52–5
  modeling performance, evaluation criteria of 43–4
 time series analysis, methods of
  ARIMA (Autoregressive Integrated Moving Average) method 41–3
  concept and basic features 40
  exponential smoothing method 41
  support vector machine (SVM) method 43
F-measure 19–20
free-flow travel time 282
future direction 179–81
fuzzy logic model 119

game theoretic lane change strategy for cooperative vehicles 147
 dynamic lane change game formulation 153–5
 equilibrium, existence of 155–6
 highway traffic system dynamics 150
  closed-loop dynamics 152–3

lane change and dynamic
    communication topology 152
  lateral dynamics 152
  longitudinal dynamics 151
lane change dynamic game,
    properties of 156–7
  numerical examples
    courtesy lane change 159–61
    delayed merge 158–9
    experimental setting 158
  problem formulation 150
Gated Recurrent Unit (GRU) 61–2,
    87, 92
Generative Adversarial Network
    (GAN) 64–6
Gold Beach Road 26
Google Map 6
Gradient Boosting Decision Tree
    (GBDT) 102
Graph Attention Network 94
Graph Convolutional Networks
    (GCNs) 81, 93–4

hierarchical softmax 12
highway traffic system dynamics
    150
  closed-loop dynamics 152–3
  lane change and dynamic
    communication topology 152
  lateral dynamics 152
  longitudinal dynamics 151
Historical Average (HA) algorithms
    81, 102
Huffman tree 12
Hybrid Tree model 120

ICTCLAS 9
Inductive Loop Detectors (ILDs) 96
Information and Communications
    Technologies 1
Instagram 8
Intelligent Transportation System
    (ITS) 59, 79, 116–17
Internet 6
iterative guidance strategy 122–3

Kakutani's theorem 155
Karush-Khun-Tucker conditions 170
k-Nearest Neighbour (kNN) 84, 102
Kolmogorov-Smirnov test 288
KPSS test 47–8

lane change and dynamic
    communication topology 152
lane change dynamic game, properties
    of 156–7
lane change-free road transportation
    system 172–4
  conflict avoidance-based
    cooperative driving strategy
    conflict region 176–7
    coordination 177–9
  overall approaching process 174–6
  simulation test 179
lane change maneuvers 147
lane change strategy, defined 153
latent Dirichlet allocation (LDA)
    10–12
LeakyReLU 94
left-turning traffic proportion 274–6
LeNet-5 16
license plate recognition (LPR)
    detector data 279–81, 285, 296
Ljung–Box test 45, 50–2
Logit model 117–18
Long Short-Term Memory (LSTM)
    14–16, 61–2, 86, 92, 102
Long Short-Term Memory with
    Deep Neural Networks
    (LSTM-DNN) models 66–7, 75
  comparison with benchmarks 72–4
  hyperparameter settings for 68–72
LOTUS (Logistic Regression Trees
    with Unbiased Selection) 120

macroscopic traffic performance
    indicator
  correlative relationships between
    variables 38–9
  descriptive analyses of influential
    factors 37–8

empirical data for analyses 34–7
Network-Level Trip (NLT) speed,
  mathematical form of 33–4
MapReduce 13
MAXBAND 208
  basic approach 211–14
  extended approach 214
    multimode band method 216–19
    path-based method 219–22
    variable bandwidth method
      214–16
  for network system 222–7
Mean Absolute Error (MAE) 95
Mean Absolute Percentage Error
  (MAPE) 68, 73, 75, 136
Mean Square Error (MSE) 67, 95, 136
Merging conflict 176
METANET model, module of 127–8
MIDAS system 96
Minimum Practical Cycle Time
  (MPCY) 196
Mixed Integer Linear Programming
  (MILP) 149
Mixed Logical Dynamical (MLD)
  model 149
Moving Average (MA) model 42–3
MULTIBAND 208, 214, 216
Multilayer Perceptron (MLP) model
  24
multimode band method 216–19

Nash equilibrium 155–7
negative sampling techniques 12
neocognitron 16
Network-Level Trip (NLT) speed 32
  mathematical form of 33–4
Network-Level Trip (NLT) speed time
  series
  ARIMA models, analysis based on
    45–52
  decomposition of 44–5
  exponential smoothing methods,
    analysis based on 45
  hybrid ARIMA–SVM model,
    analysis based on 52–5

modeling performance, evaluation
  criteria of 43–4
network system, MAXBAND for
  222–7
Neural Networks (NNs) 83
news-LDA 24
95th percentile travel time 285, 291–4

Odd-Even-Plate-Number (OEPN)
  travel restriction 34–5
online web data, traffic analytics with
  5
  advantages 6
  applications 10
  Bayesian network 13–14
  data collection 8–9
  data preprocessing 9–10
  deep learning 14
    convolutional neural network
      16–17
    Convolutional Neural Network
      (CNN) 17
    Long Short-Term Memory
      (LSTM) 17
    long short-term memory neural
      network (LSTM) 14–16
  latent Dirichlet allocation (LDA)
    10–12
  modeling and mining 10
  semantic reasoning for traffic
    congestion 24–7
  traffic event detection 18–23
  traffic sentiment analysis and
    monitoring system 18
  traffic status prediction 24
  word embedding 12–13
Optimization Policies for Adaptive
  Control (OPAC) 207
Orange County Transportation
  Authority (OCTA) 246

path-based method 219–22
path choice behavior, model of 116
  discrete probability selection model
    117–18

driver's choice behavior model 120
    fuzzy logic model 119
    Hybrid Tree model 120
    prospect theory model 118–19
payoff function 156
percentage on green (POG) 284, 291
Performance Measurement System (PeMS) 59
plan-level optimization 190
platoon ratio 245
prediction accuracy, quantification of 95
prediction of traffic
    under disrupted conditions 82–3
    under normal conditions 81–2
    using deep learning techniques 86
        attention mechanism 93–5
        data representation 87–8
        future research 108–9
        loss function and parameter optimisation 95
        model structure 90–5
        problem formulation 90
        quantification of prediction accuracy 95
        spatial dependencies 92–3
        spatio-temporal features 88–9
        temporal dependencies 92
        traffic network representation on a graph 89–90
        see also short-term traffic data prediction
pre-trip guidance 115
Principle Component Analysis 83
Probit model 117–18
prospect theory model 118–19
Purdue coordination diagram (PCD) 244–5, 280, 290
    and related indexes 284–5
    travel time reliability evaluation buffer index 294–5
    95th percentile travel time 291–4
Python 67

Qingdao traffic sentiment analysis and monitoring system 18

random utility theory 117
real-time demand-based traffic diversion 115
    diversion strategy, optimization of 120
    iterative guidance strategy 122–3
    responsive guidance strategy 121–2
    Dynamic O–D Estimation (DODE), research on 123
        expressway model 124–5
        intersection model 124
        network model 125–6
    dynamic traffic diversion model 126
        based on DODE 131–3
        DODE model, module of 128–9
        experimental design (case study) 134–41
        METANET model, module of 127–8
        model solution 133–4
        urban expressway, traffic diversion model of 129–31
    path choice behavior, model of 116
        discrete probability selection model 117–18
        fuzzy logic model 119
        prospect theory model 118–19
real-time-level optimization 190
Real-Time Optimization (RTO) system 188
Rectified Linear Unit 17
Recurrent Neural Network (RNN) 60–3, 84, 92
regional computer 199–200
reservation-based strategy 167–8
resilient backpropagation (Rprop) 95
responsive guidance strategy 121–2
Revealed Preference (RP) 117
RHODES system 189
RNN (Recurrent Neural Network) 14–15
road network performance, evaluation index of 130–1

road traffic broadcasting 116
Robertson model 285
Root Mean Square Error (RMSE) 44, 68, 75
R software 45

safety driving pattern-based strategy 166–7
SCATS system 189, 280
Seasonal and Trend decision using Loess (STL) decomposition method 40
seasonal ARIMA (SARIMA) 82
semantic reasoning for traffic congestion 24–7
short-distance intersections, control for 253
  adaptability analysis 272
    left-turning traffic proportion 274–6
    road conditions 272–4
  dynamic reversible lane, application of 254–5
  signal timing model 257–64
    signal phase and sequence 255–7
  simulation scenarios 264
  validation of the proposed plan 264–72
short-term traffic data prediction 96
  prediction model, preparation for
    baseline methods for comparison 102
    graph representation 100–2
    traffic speed data preprocessing 96–100
  short-term traffic speed prediction under non-incident conditions
    model setups 102
    prediction results 102–5
    traffic speed data 96
  under incidents 105
    prediction results during disruptions 106–8
    traffic incident data 105

short-term traffic prediction under disruptions using deep learning 79
  attention mechanism 93–5
  data representation 87–8
  disrupted conditions, traffic prediction under
    traffic characteristics under disrupted conditions 82
    traffic prediction under disrupted conditions 82–3
  future research 108–9
  loss function and parameter optimisation 95
  normal conditions, traffic prediction under 81–2
  preparation for the prediction model
    baseline methods for comparison 102
    graph representation 100–2
    traffic speed data preprocessing 96–100
  problem formulation 90
  quantification of prediction accuracy 95
  spatial dependencies 92–3
  spatio-temporal features 88–9
  temporal dependencies 92
  traffic network representation on a graph 89–90
  traffic speed data 96
  under incidents 105
    prediction results during disruptions 106–8
    traffic incident data 105
  under non-incident conditions
    model setups 102
    prediction results under non-incident conditions 102–5
short-term travel-time prediction by deep learning 59
  datasets 68
  evaluation metrics 68
  future work 75

Long Short-Term Memory with
    Deep Neural Networks
    (LSTM-DNN) models 66–7
  comparison with benchmarks
    72–4
  hyperparameter settings for
    68–72
  traffic time series estimation with
    deep learning 60
  Convolutional Neural Network
    (CNN) 63–4
  Generative Adversarial Network
    (GAN) 64–6
  Recurrent Neural Network
    (RNN) 60–3
signal coordination 207
  basic MAXBAND approach
    211–14
  extended MAXBAND approach
    214
    multimode band method 216–19
    path-based method 219–22
    variable bandwidth method
      214–16
  MAXBAND for network system
    222–7
  open issues 227–9
signal-free intersection/on-ramp
  management 165
Signal Performance Metrics (SPM)
  244
signal timing model 257–64
  signal phase and sequence
    255–7
simple exponential smoothing (SES)
  method 41
Singular Spectrum Analysis (SSA) 84
Skip-Gram models 12
slot-based strategy 171–2
social transportation data 5
Spatio-Temporal Graph Convolutional
  Network 88
Split Cycle Offset Optimization
  Technology (SCOOT) 188,
  208–9

basic principles 192–3
bus priority 197
gating 197
optimization process
  congestion prediction 195
  cycle optimizer 196
  demand detection 194
  offset optimizer 197
  performance prediction 195
  queue prediction 194–5
  signal optimization 196
  split optimizer 196
system architecture 193–4
Stacked Autoencoder model 62
State Preference (SP) 117
stationary, concept of 42
stochastic probability distribution
  model 117
support vector machine (SVM)
  method 33, 43, 83
  hybrid ARIMA–SVM model,
    analysis based on 52–5
Sydney Coordinated Adaptive
  Traffic System (SCATS) 198,
  209–10
  basic principles 198–9
  optimization process 200
    cycle determination 201
    demand detection 200–1
    offset determination 201–2
    split determination 201
  system architecture 199
    Central Management Computer
      (CMC) 199
    regional computer 199–200
    traffic signal controller 200
Synchro 209

Tavana model 126
Temporal Spatial Traffic Graph
  Attention network
  (TS-TGAT) 102–3, 106–8
TensorFlow 67
time-of-day (TOD) traffic signal
  plans 279

Index   313

time series analysis, methods of
  ARIMA (Autoregressive Integrated Moving Average) method 41–3
  concept and basic features 40
  exponential smoothing method 41
  support vector machine (SVM) method 43
Tong An Road 26
Total Travel Time (TTT) 136
Total Waiting Time (TWT) 136
traditional traffic sensing methods 7
traffic accident 32
traffic diversion guidance 115
traffic event detection 18–23
traffic movements and signal stages 257
traffic network representation on a graph 89–90
Traffic Network Study Tool, version 7F (TRANSYT 7-F) 209
traffic sentiment analysis and monitoring system 18
traffic signal control 1
traffic signal controller 200
traffic signal coordination practices 235
  coordination timing development and optimization 236
    developing cycle length and splits using controller event data 237–9
    optimizing offsets and phasing sequences 239–41
  field implementation and timing diagnosis 241–3
  performance measures 243–8
  signal timing documentation 248–9
traffic speed data preprocessing 96–100
traffic status prediction 24
trajectory optimization-based strategy 168–70
TRANSYT 188
travel time 283

cumulative frequency diagram of 283–4, 290
scatter diagram of 283
travel time delay 282–3
  quantitative analysis of improvement of 288
  scatter diagram analysis of 288–9
travel time reliability indexes 285
  buffer index 285–6
  95th percentile travel time 285
TRRL in British released Traffic Network Study Tool (TRANSYT) 209
Twitter 5–7

United States, urban traffic control systems in 189
urban traffic 31
urban traffic control systems 79, 187
  in Australia 189–90
  classification 190–1
  in Europe 187–8
  future analysis of 202
    computational power, demand of 204
    standardization 204–5
    system environments, changes in 203–4
    traffic control objects and variables 204
    traffic data 203
  level of traffic control system 191
  Split Cycle Offset Optimization Technology (SCOOT) 188
    basic principles 192–3
    bus priority 197
    gating 197
    limitation analysis 202
    optimization process 194–7
    system architecture 193–4
  Sydney Coordinated Adaptive Traffic System (SCATS) 198
    basic principles 198–9
    limitation analysis 202
    optimization process 200–2

system architecture 199–200
  in United States 189
user-generated traffic-related data 5
user optimum (UE) guidance strategy
  123

variable bandwidth method 214–16
variable message board (VMS)
  116–17
virtual vehicle mapping strategy
  170–1

VISSIM simulation 253, 264, 276
VMS (variable message boards) 119

Webster timing model 210
Webtris platform 96
Weibo 5–6, 10
Weibo-LDA model 11, 24
word embedding 12–13
word segmentation 9

Yan An San Road 26–7

www.ingramcontent.com/pod-product-compliance
Ingram Content Group UK Ltd.
Pitfield, Milton Keynes, MK11 3LW, UK
UKHW020435200426
11946UKWH00035B/52